The Essence of
Crystallography

Essential Textbooks in Chemistry

ISSN: 2059-7738

The *Essential Textbooks in Chemistry* explores the most important topics in Chemistry that all Physical Sciences students need to know to pass their undergraduate exams (years 1, 2 and 3 of the BSc).

Written by senior academics as well lecturers recognised for their teaching skills, they offer in around 200 to 250 pages a theoretical overview of fundamental concepts backed by problems and worked solutions at the end of each chapter.

Their lively style, focused scope and pedagogical material make them ideal learning tools at a very affordable price.

Published

Problems of Instrumental Analytical Chemistry: A Hands-On Guide
    by JM Andrade-Garda, A Carlosena-Zubieta, MP Gómez-Carracedo, MA Maestro-Saavedra, MC Prieto-Blanco and RM Soto-Ferreiro

Astrochemistry: From the Big Bang to the Present Day
    by Claire Vallance

Atmospheric Chemistry: From the Surface to the Stratosphere
    by Grant Ritchie

Principles of Nuclear Chemistry
    by Peter A C McPherson

Orbitals: With Applications in Atomic Spectra
    by Charles Stuart McCaw

The Essence of Crystallography
    by Mark Ladd

Essential Textbooks in Chemistry

The Essence of
Crystallography

Mark Ladd

Formerly Head of Chemical Physics, University of Surrey, UK

World Scientific

NEW JERSEY · LONDON · SINGAPORE · BEIJING · SHANGHAI · HONG KONG · TAIPEI · CHENNAI · TOKYO

Published by

World Scientific Publishing Europe Ltd.

57 Shelton Street, Covent Garden, London WC2H 9HE

Head office: 5 Toh Tuck Link, Singapore 596224

USA office: 27 Warren Street, Suite 401-402, Hackensack, NJ 07601

Library of Congress Cataloging-in-Publication Data

Names: Ladd, M. F. C. (Marcus Frederick Charles), author.
Title: The essence of crystallography / by Mark Ladd (Formerly Head of Chemical Physics,
    University of Surrey, UK).
Description: New Jersey : World Scientific, 2019. | Series: Essential textbooks in chemistry |
    Includes bibliographical references.
Identifiers: LCCN 2018043883 | ISBN 9781786346315 (hc : alk. paper)
Subjects: LCSH: Crystallography--Textbooks.
Classification: LCC QD905.2 .L3295 2019 | DDC 548--dc23
LC record available at https://lccn.loc.gov/2018043883

British Library Cataloguing-in-Publication Data
A catalogue record for this book is available from the British Library.

For any available supplementary material, please visit
https://www.worldscientific.com/worldscibooks/10.1142/Q0188#t=suppl

Desk Editors: V. Vishnu Mohan/Jennifer Brough/Shi Ying Koe

Typeset by Stallion Press
Email: enquiries@stallionpress.com

An expanded version of the Periodic Table is available with the Web Materials on the publisher's website at (https://www.worldscientific.com/worldscibooks/10.1142/Q0188#t=suppl).

Foreword

How does one become a crystallographer? I, for example, studied physics, then took and gave classes in classical crystallography (stereographic projection and matrices in non-Cartesian space), specialized in physical crystallography, got a job at a chemistry department and was asked to do structures for the department because 'crystallography' showed up in my application documents. There we go, now what? I had done a few X-ray structures, guided by someone 'who knew', and was suddenly facing impatient students and faculty for getting their structures done 'subito' and flawlessly as in: 'you got the final answer regarding the molecule' — no pressure, right?

If you are reading this and find this history strangely familiar or got thrown at solving your own structures with little help at hand, don't put this book away: you need it because you do not want to walk through lengthy proofs of physical equations or overly poetic descriptions of the matter you are dealing with. You want to get information quickly when you need it. So, what is this book about? Would I myself have used it when I started dealing with X-ray structures? Yes and no: because I had the benefit of that classical crystallography training, and part of Mark Ladd's book does help you catch-up on symmetry and visualization via stereographic projection. Without being introduced into 'three-dimensional thinking', how would one ever be able to understand crystallographic symmetries? Does this help? Of course, just imagine you got three atoms while refining a structure, but nobody told you that this is a half benzene, to be completed by symmetry, and you have to expand (in your mind) so to place the aromatic protons correctly. Are you ready to imagine a disordered solvent molecule on a special position where all atoms are collapsed into one-half of the total mass? Are you capable of identifying an additional symmetry in your structure that was missed by the automatic space-group finding routine?

You will find this book quite helpful for preparing you to deal with these kinds of troubles. And following through, yes, I would have found the book helpful and bought it, if it had been available right when I took over the X-ray lab of that chemistry department. It is time now for an up-to-date monograph on the fundamentals of structural crystallography. The book by Mark Ladd moves you from those symmetries to symmetry operations in Chapter 2 (dry math, makes an Egyptian mummy look moist, but that is exactly as intended, and no ink gets wasted), to crystal lattices, space groups, and then to X-ray diffraction itself in Chapter 5. Each chapter is accompanied by a set of problems for self-testing your knowledge. If, on top of your lab duties, you have to give classes in crystallography, this helps to keep your students busy. Chapter 6 provides background and talking stuff on how to solve a structure. Admittingly, for most people, it is more important to find a structure solution than how the process actually works, but you will impress your peers, students, or facility users if you do know those details.

Chapter 7 makes the book's intentions even more clear: now it is time to explain the crystallographic jargon, aka: what's behind the button captions in your X-ray structure software (scaling, weighting schemes, displacement parameters, torsion angles, bond parameters, etc.). Besides useful references to databases for crystal structures, the book also gives examples of the author's software for investigating the principles of solving structures, free to download. I will have to take a look at those e-tools some time, curious if they are better than what I was using for years.

Werner Kaminsky
Professor of Crystallography
University of Washington
Seattle

Preface

The origin of the subject of crystallography lies in the description and classification of crystals and minerals, and included a study of properties such as colour, shape, hardness and cleavage. Today, the thrust of crystallography almost always lies in the experimental investigations of crystalline materials, particularly macromolecules, by means of the diffraction of X-rays or neutrons from them in order to determine their complete structures: this is the essence of crystallography.

Crystallographic structure-solving computer program systems are now so advanced and well documented that it is open to all physical scientists, armed with the appropriate diffraction data, to solve structures pertinent to their own interests. Although it is not essential to be trained in X-ray crystallography in order to engage with these facilities, their best use is gained with, at least, a basic knowledge of the various procedures and of the mathematics involved in them. This book is not addressed to the trained crystallographer. Its aim is to discuss the basics of the crystallography of structure solving for those who wish to undertake and understand this activity, and to increase the awareness the processes taking place, whether starting from a set of corrected intensity data or from the crystal itself. There can be little doubt that the procedures of crystal structure analysis are best understood through repeated practical involvement with them. The simulation of that experience is one of the features attempted in this book by means of the interactive XRAY program system supplied with the book as the 'Program Suite'; two-dimensional data are used in these applications, for obvious reasons.

The early chapters discuss the geometry and symmetry of crystals, noting particularly the structural information that may be deduced from space group data. Symmetry is treated in some detail as the author has found that this topic often presents a little difficulty to those meeting it for the first time. The book then addresses X-ray diffraction and the collection and correction of intensity data, and some important statistical properties of intensities are discussed, all leading up to the point that structure solving properly begins.

There exist today a number of different procedures that can be applied for solving crystal structures from diffraction data. Three of them in particular are chosen for description herein as they are responsible for the majority of crystal structures that have been solved. They comprise the following: the Patterson synthesis and electron density calculations, often termed the 'method of successive Fourier synthesis'; the application of probability equations, also known as 'direct methods'; and the 'isomorphous replacement procedure', which is important in solving the structures of non-centrosymmetric crystals containing macromolecules. However, although proteins are included in this class of substances, the 'molecular replacement' technique is also used for their structure analyses. A solved structure is always subjected to a refinement procedure, based normally on an application of the method of least squares. This procedure is treated in the penultimate chapter together with the calculations of molecular geometry and its precision, and noting the many accessible databases that have recorded solved crystal structures. Indeed, one of the first questions when anticipating a crystal structure analysis should be 'has it been done already'.

All stages of a crystal structure analysis involve computational procedures of varying degrees of complexity that are discussed herein. Throughout the book, reference has been made to the Program Suite; it is available on the publisher's website which can be accessed at (https://www.worldscientific.com/worldscibooks/10. 1142/Q0188#t=suppl). Of particular significance in the Suite is the

XRAY program. It is interactive, and allows many of the basic X-ray analysis procedures to be followed through on the monitor screen, albeit in two-dimensional projection. The accompanying data sets for this practice have been selected such that the molecular species chosen are well resolved in two dimensions. Other programs in the Suite apply to various crystallographic calculations and to the indexing analysis of X-ray powder patterns. The majority of these programs are deliberately unsophisticated since this feature allows more interaction with the actual operations involved in the computations; this chapter also explains the use of the Python interpreter for x-y graph plotting.

Each chapter, other than Chapter 8, contains a set of problems designed to assist in the understanding of the textual material, and detailed tutorial solutions are provided; there are also self-assessment exercises running through the text. The reader is encouraged to engage with these problems in order to become familiar with the procedures on which they are based. In this context, it may be helpful to consider some aspects of this final chapter at an early stage.

I am pleased to acknowledge Dr Neil Bailey of the University of Sheffield for the original XRAY program, which has been enhanced here by the present author with plotting facilities and additional crystallographic routines. I thank those publishers for permission to reproduce the illustrations that carry appropriate acknowledgements. I am most grateful to my colleague Dr Rex Palmer, Emeritus Reader, Birkbeck College, University of London for reading the manuscript in its draft form and for his timely comments, and also to Professor Werner Kaminsky of the University of Washington, Seattle for writing the Foreword. Finally, I thank Dr Merlin Fox and Mr Vishnu Mohan of World Scientific for encouragement and for assistance in bringing the book to publication.

Disclaimer

Every effort has been made to ensure the correct functioning of the software associated with this book. However, the reader planning to use the software should note that, from a legal point of view, there is no warranty, expressed or implied, that the programs are free from error or will prove suitable for a particular application. By using the software, the reader accepts full responsibility for all the results produced, and the author and publisher disclaim all liability from any consequences arising from the use of the software. The software should not be relied upon for solving a problem, the incorrect solution of which could lead to injury to a person or loss of property. If you do use the programs in such a manner, it is at your own risk. The author and publisher disclaim all liability for direct or consequential damages resulting from your use of the programs.

Contents

Physical Data, Notation and Online Materials

The physical constants tabulated hereunder are the CODATA (2014) recommended values; the majority of these constants are available with precisions greater than those indicated here. Where a unit is expressed other than in any of the seven basic SI units (kg, m, s, A, K, mol, cd) the corresponding kg_m_s components (and some useful equivalents) are added in parentheses.

A current review of the SI system is under consideration in which exact values will be set for the basic units, excluding the Kelvin but including the hyperfine splitting frequency for ^{133}Cs; the Kelvin will be set by defining an exact value for the Boltzmann constant. Such changes, if agreed, will not come into force before 2018 and, in any case, would not affect the use of the values of these constants employed in this book.

Physical constants[a]

Atomic mass unit	m_u	1.66054×10^{-27} kg
Avogadro constant	L	6.02214×10^{23} mol^{-1}
Boltzmann constant	k	1.38065×10^{-23} J K^{-1} (kg m^2 s^{-2})
Electron mass	m_e	9.10938×10^{-31} kg
Electron radius	r_e	2.81794×10^{-15} m
Elementary charge	e	1.60218×10^{-19} C (A s)
Faraday	F	9.64853×10^4 C mol^{-1} (A s mol^{-1})
Molar gas constant	R	8.31446 J K^{-1} mol^{-1} (kg m^2 s^{-2})
Molar mass of carbon-12	$M(^{12}C)$ (exact)	12×10^{-3} kg mol^{-1}

(Continued)

Neutron mass	m_0	$1.67493 \times 10^{-27}\,\mathrm{kg}$
Planck constant	h	$6.62607 \times 10^{-34}\,\mathrm{J\,s\,(kg\,m^2\,s^{-2})}$
Proton mass	m_p	$1.67262 \times 10^{-27}\,\mathrm{kg}$
Speed of light in a vacuum	c	$2.99792 \times 10^{8}\,\mathrm{m\,s^{-1}}$

[a]http://physics.nist.gov/cuu/Constants.

Prefixes to units[a]

atto	femto	pico	nano	micro	milli	centi	deci	kilo	mega	giga	tera	peta	exa
a	f	p	n	μ	m	c	d	K	M	G	T	P	E
10^{-18}	10^{-15}	10^{-12}	10^{-9}	10^{-6}	10^{-3}	10^{-2}	10^{-1}	10^{3}	10^{6}	10^{9}	10^{12}	10^{15}	10^{18}

[a]The prefixes deca (10^1) and hecto (10^2) are used only rarely.

Notation

These notes indicate the main symbols used throughout the book. Inevitably, some of them have more than one application, partly from general usage, and partly from a desire to preserve a mnemonic character in the notation wherever possible. Two or more uses of one and the same symbol are separated by a semicolon in the presentation hereunder.

Å	Ångström unit ($1\,\text{Å} = 10^{-10}\,\mathrm{m}$)
A', B'	$|F|\cos\phi$ and $|F|\sin\phi$, respectively
A, iB	Real and imaginary geometrical structure factor amplitudes
a, b, c	Unit-cell edge lengths along the x-, y- and z-axes, respectively; glide plane symbols
a, b, c	Unit-cell translation vectors along the x-, y- and z-axes, respectively

(Continued)

a^*, b^*, c^*	Reciprocal unit-cell edge lengths along the x^*-, y^*- and z^*-axes, respectively
$\mathbf{a^*, b^*, c^*}$	Reciprocal unit-cell translation vectors along the x^*-, y^*-, and z^*-axes, respectively
a_b	Plane containing a and b
c	Speed of light
calc	(as a subscript) Calculated
d	Interplanar spacing of a family of planes; diamond glide plane symbol
d^*	Distance in reciprocal space
$\mathbf{E}$	Normalized structure factor
e	Electron charge; double glide plane symmetry symbol
^{e}, exp	Exponential function
esd	Estimated standard deviation
f, f_θ	Atomic scattering factor
$\mathbf{F}$	Structure factor
$|\mathbf{F}|$	Amplitude of structure factor; often just F in a centrosymmetric situation
g	Glide line symmetry symbol
h	Miller index parallel to the x-axis; Planck's constant; order of a group
(hkl)	Miller indices of planes associated with the x-, y- and z-axes, respectively
$(hkil)$	Miller–Bravais indices of planes (hexagonal system)
hkl	Reciprocal lattice point corresponding to the (hkl) family of planes
$\mathbf{i, j, k}$	Unit vectors parallel to the x-, y- and z-axes, respectively
i	$\sqrt{-1}$
L	Avogadro constant
M_r	Relative molecular mass
$M(^{12}C)$	Molar mass of carbon (^{12}C isotope)

(*Continued*)

m	Mirror (reflection) symmetry symbol
$P(uvw)$	Patterson function at the point u, v, w
m_e	Mass of electron
m_n	Mass of neutron
m_p	Mass of proton
m_u	Atomic mass unit
n	Diagonal glide plane symbol
P, R, A, B, C, F	Unit-cell symbols in three dimensions
p, c; p	Primitive and centred two-dimensional unit-cell symbol; primitive one-dimensional
R	Symmetry element; point group; rotation axis of degree R; gas constant
$\bar{R}$	Inversion symmetry symbol of degree R
$\boldsymbol{R}$	Symmetry operation or operator
$\mathbf{R}$	Symmetry operation (or operator); symmetry matrix
$\mathbf{r}$	Vector distance, as in $\mathbf{r} = \mathbf{i}x + \mathbf{j}y + \mathbf{k}z$, for example
rms	Root mean square
$[uvw]$	Direction in a lattice
$[UVW]$	Zone symbol
V_c	Volume of a crystal unit cell
V^*	Volume of reciprocal unit cell
X, Y, Z	Coordinates of points in absolute measure
x, y, z	Labels for reference axes; fractional coordinates in the unit cell
x, y, u, z	Labels for reference axes (hexagonal system)
x-y and similar	Plane containing the directions x and y
Z	Atomic number
Z_c	Number of formula entities of relative molecular mass M_r per unit cell
α	Interaxial angle $\widehat{y\,z}$ in real space
α^*	Interaxial angle $\widehat{y^*\,z^*}$ inreciprocal space

(*Continued*)

β	Interaxial angle $\widehat{z\,x}$ in real space
β^*	Interaxial angle $\widehat{z^*\,x^*}$ in reciprocal space
$\cancel{c}$	Not constrained by symmetry to equal
ϕ	Phase angle
γ	Interaxial angle $\widehat{x\,y}$ in real space
γ^*	Interaxial angle $\widehat{x^*\,y^*}$ in reciprocal space
λ	Wavelength
π	Pi (ratio of the circumference of a circle to its diameter), equal to $4\tan^{-1}(1)$
θ	Bragg angle
ρ_o	Observed (experimental) density
ρ_c	Calculated density
$\rho(xyz)$	Electron density function at the point x, y, z
$^\mathrm{o}$	Superscript: degree, as in 90°
$\langle\,\rangle$	Average value, as in $\langle X \rangle$

Online materials

Computer programs (Program Suite) relevant to the text and the problems have been devised and are available from the publisher's website: https://www.worldscientific.com/worldscibooks/10.1142/Q0188#t=suppl.

(1) Web Appendices;
   WA1 Stereographic projection;
   WA2 Reciprocity of F and I unit cells;
   WA3 Orthogonal functions;
(2) Programs (.EXE files) and data (.TXT files).

Chapter 1

Crystals and Crystal Geometry

*And there is a certain facility for learning
all other subjects in which we know that those who
have studied geometry lead the field.*
Plato

Key Topics

- Crystals and crystalline substances
- Crystal form and crystal habit
- Periodicity and aperiodicity
- Reference axes for crystals
- Miller indices
- Equation of a plane
- Zones
- Stereographic projection (stereogram)

1.1.   Introduction

Crystallography is the study of the crystalline state of matter in its widest sense and application. The word *crystal* is derived from the Greek word κρυσταλλοσ, applied originally to the mineral quartz, which was thought to be water permanently congealed by intense cold. Together with γραπο, meaning *write*, the word κρυσταλλογραπια, or *crystallography* emerges; it is an established

1

field in its own right. The essence of crystallography is the determination of the arrangements of atoms and molecules in crystalline substances by experimental methods that are based upon data obtained mainly by the diffraction of X-rays, neutrons or other appropriate radiations from crystalline materials.

Crystals, with their plane faces, sharp angles and varied colours, have excited interest from the earliest of times. Indeed, these properties formed the first definition of a crystal; most other naturally occurring substances exhibit rounded or curved outlines. The external shape of a crystal is a manifestation of its internal structural regularity. Ideally, it comprises a three-dimensional, regular stacking of identical building blocks. Figure 1.1(a) illustrates the crystal form of an octahedron obtained by a three-dimensional stacking of minute, identical, cubical blocks of structure, as envisaged by the French mineralogist Haüy [1]. One form of the crystalline mineral fluorite CaF_2 exhibits such an octahedral development, as illustrated in Fig. 1.1(b). A crystal *form* is a set of crystal faces that are related to each other by the symmetry of the crystal, while a description of its general shape together with its crystallographic forms is known as the crystal *habit*.

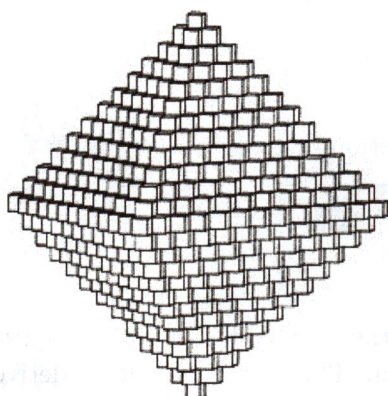

(a)   (b)

Fig. 1.1.   (a) A regular octahedron formed by stacking cube-shaped building blocks of *ca.* 5 Å dimensions. (b) An example of a crystal of fluorite CaF_2, known as Blue John, showing the {111} form that leads to its octahedral habit.

An important property of crystals is their ability to produce discrete spot patterns by the diffraction of X-rays or other appropriate radiation from them. A periodic crystal, like the fluorite example, has structural regularity in three dimensions. However, a material such as DNA exhibits a diffraction pattern but does not possess three-dimensional regularity: It is crystalline, but it is aperiodic and may be termed a *quasicrystal* [2, 3]. This feature led to the modern definition of a crystalline substance as *a material capable of producing a clear-cut diffraction pattern, with ordering that is either periodic or aperiodic.*

The concept of aperiodicity in a crystalline material was introduced by Schrödinger in 1944 in an explanation of the storage of hereditary material. Molecules were deemed too small for this purpose, while amorphous solids were chaotic. Therefore, he argued, it had to be a kind of crystal: A periodic structure could not encode sufficient information, so the pattern had to be of another type, namely, aperiodic [4]. Later, DNA was discovered and shown to possess properties similar to those described by Schrödinger — an ordered regular but non-periodic crystalline structure. An interesting summary exists online that gives an outline of the DNA story from 1869, when Miescher extracted DNA from the nuclei of white blood cells, to 1953 when the structure of DNA was determined [5]. Samples of DNA are essentially fibrous and partially ordered along the fibre direction; this ordering can be increased by stretching in the fibre direction.

This book will be concerned with crystals (sometimes called classical crystals) and crystalline materials that possess three-dimensional regularity and will make use of the words 'crystal' and 'crystalline' with this understanding.

1.2. Crystal Geometry

An inspection of any well-developed crystal, such as that in Fig. 1.2(a), indicates the desirability of a method for describing the geometry of the crystal and its faces. As a three-dimensional body, it may be referred to three non-coplanar, right-handed, *reference axes*

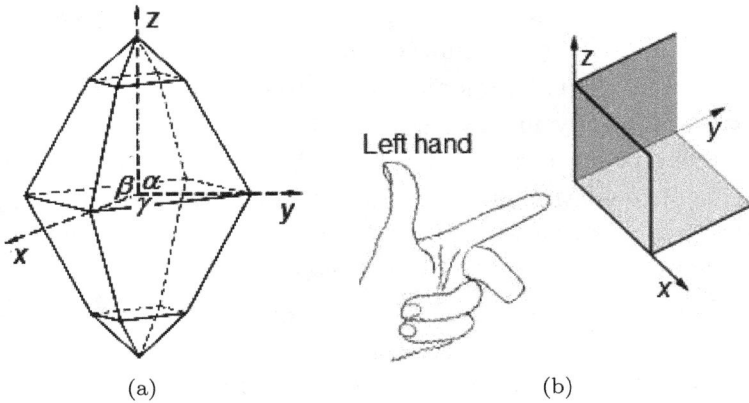

(a)   (b)

Fig. 1.2.   (a) Idealized crystal, showing the conventional, right-handed x-, y- and z-axes and the interaxial angles $\alpha = \widehat{yz}$, $\beta = \widehat{zx}$ and $\gamma = \widehat{xy}$. (b) Using the left hand as shown, the conventional, *right-handed* axes correspond to the middle finger as x, the index finger as y and the thumb as z; interchanging any pair of axial labels would produce left-handed axes.

designated x, y and z, which are illustrated further in Fig. 1.2(b). The interaxial angles are symbolized as α between y and z, β between z and x, and γ between x and y, as shown in Fig. 1.2(a). The crystallographic reference axes are set always along or in close relation to the symmetry directions of the crystal, but they are not always orthogonal: two lines are *orthogonal* if they are mutually perpendicular, that is, there is no component of any one on to another.

1.3.   Miller Indices of a Crystal Plane

Once the reference axes are defined for a crystal and one face that intersects all three axes has been chosen as the *parametral* plane, any other face can be described in terms of three numbers related to that plane and known as its *Miller indices* (hkl) [6]. If, as is usual, the parametral plane is labelled (111), then the indices of all other crystal faces are small integers; generally, h, k and l are numbers less than 6 [6], with no common factor among the three numbers (see also Table 5.1 and associated text).

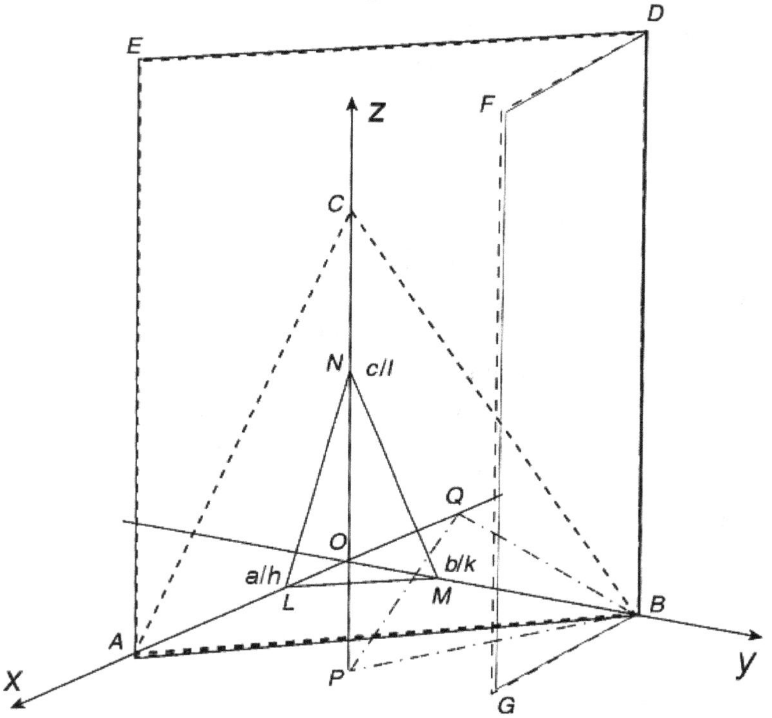

Fig. 1.3.   Illustration of Miller indices of crystal planes: ABC is defined as the parametral plane (111); then, plane LMN is (432), plane $ABDE$ is (110), plane $GBDF$ is (010) and plane PBQ is ($\bar{2}1\bar{3}$).

Figure 1.3 shows a plane ABC which intercepts the x-, y- and z-axes at distances that will be termed a, b and c, respectively, and is the chosen parametral plane. Another plane LMN makes intercepts a/h, b/k and c/l on the x-, y- and z-axes, where h, k and l are integers. Its Miller indices are the ratios of the intercepts of the parametral plane to the corresponding intercepts of the plane LMN, that is, $a/(a/h)$, $b/(b/k)$ and $c/(c/l)$, or h, k and l. The diagram has been drawn such that $OL = a/4$, $OM = b/3$ and $ON = c/2$, whereupon the plane LMN is (432). If fractions remain after the divisions, they are cleared by multiplying throughout by the lowest common denominator.

Self-assessment 1.1. If in Fig. 1.3 the plane LMN is chosen as the parametral plane, so that OL, OM and ON are then a, b and c, respectively, what would be the Miller indices of the plane ABC?[a]

The plane $ABDE$ is parallel to the z-axis so that its intercept is effectively ∞; thus, the plane $ABDE$ is (110). Similarly, the plane $GBDF$ is (010). A plane that intercepts an axis on its negative side has a corresponding negative Miller index; thus, if the plane PBQ has intercepts $-a/2$, b and $-c/4$, then it is designated as $(\bar{2}1\bar{4})$. Note the neat format of placing the negative sign associated with a number above that number; conventionally, a number $\bar{x}$ is referred to as 'bar-x' ('x-bar' in the USA).

In crystals based on hexagonal symmetry, a fourth index may be defined. Thus, with the hexagonal prismatic form in Fig. 1.4, three symmetry-equivalent axes, x, y and u, lie in one plane at 120° to one another, with z normal to that plane. The corresponding indices are known as *Miller–Bravais* indices $(hkil)$, and it is a straightforward matter to show that $i = -(h + k)$.

Self-assessment 1.2. Show that for the Miller–Bravais indices of a plane $(hkil)$, $i = -(h + k)$.

1.4. Equation of a Plane

In Fig. 1.5, $P(X, Y, Z)$ is a point in the ABC plane, ON of length d is the normal to that plane from the origin O and has the direction cosines $\cos\chi$, $\cos\psi$ and $\cos\omega$, with respect to x, y and z reference axes; $PK \parallel OC$ and $KM \parallel OB$. Then, the lengths OM, MK and KP are equal to X, Y and Z, and their vector sum is OP. The length d is the projection of OP on to ON, and it comprises the sum of the

---

[a]Answers to all self-assessments are given at the end of the chapter.

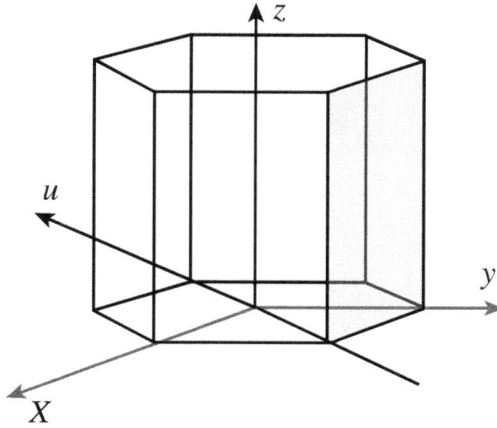

Fig. 1.4. Hexagonal prismatic crystal showing the Miller–Bravais axes x, y, u and z. The angles $\widehat{xy} = \widehat{yu} = \widehat{ux} = 120°$, and z is normal to the x_y_u plane. The forms shown are the hexagonal pinacoid $\{0001\}$ and the hexagonal prism $\{10\bar{1}0\}$; the symbol u is useful but not essential.

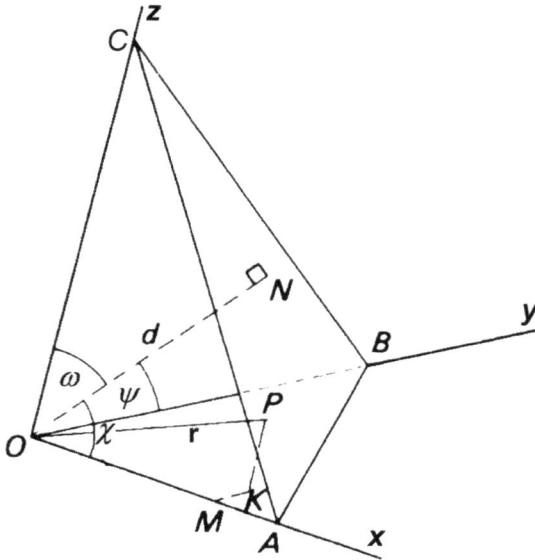

Fig. 1.5. A plane ABC; the normal ON to the plane from the origin O has the length d.

projections of *OM*, *MK* and *KP* all on to *ON*. Hence,

$$d = X\cos\chi + Y\cos\psi + Z\cos\omega \qquad (1.1)$$

In $\triangle OAN$, $d = OA\cos\chi$; similarly, $d = OB\cos\psi = OC\cos\omega$. If OA, OB and OC are written as a, b and c, respectively, then Eq. (1.1) becomes

$$\frac{X}{a} + \frac{Y}{b} + \frac{Z}{c} = 1 \qquad (1.2)$$

which is the *intercept equation* for the plane ABC. If the plane is moved in parallel orientation to pass through the origin O, then the equation becomes

$$\frac{X}{a} + \frac{Y}{b} + \frac{Z}{c} = 0 \qquad (1.3)$$

Although developed for orthogonal axes, Eqs. (1.2) and (1.3) are applicable generally, provided that a, b and c are the intercepts of the parametral plane on the x-, y- and z-axes [7].

In an alternative, vector treatment, OP is the vector $\mathbf{r}$, and $\mathbf{n}$ is a unit vector along ON and normal to the plane ABC. Then, the scalar product $\mathbf{r}\cdot\mathbf{n}$ is given by

$$\mathbf{r}\cdot\mathbf{n} = rn\cos\widehat{PON} = r\cos\widehat{PON} = d \qquad (1.4)$$

If the reference axes are orthogonal, then resolving $\mathbf{r}$ and $\mathbf{n}$ leads to the following equation:

$$\mathbf{r}\cdot\mathbf{n} = (\mathbf{i}X + \mathbf{j}Y + \mathbf{k}Z)\cdot(\mathbf{i}n_X + \mathbf{j}n_Y + \mathbf{k}n_Z)$$
$$= (n_X X + n_Y Y + n_Z Z) = d \qquad (1.5)$$

which is an expression of the plane in the form

$$A'X + B'Y + C'Z = d \qquad (1.6)$$

and may be recast conveniently as

$$AX + BY + CZ = 1 \qquad (1.7)$$

Since $n_X = d/a$, and similarly for n_Y and n_Z *mutatis mutandis*,[b] substitution into Eq. (1.5) leads back to Eq. (1.2). If the reference axes are not orthogonal, the cross terms arising from Eq. (1.5) must be included.

If fractional coordinates are used (Section 3.5), then Eq. (1.2) becomes

$$hx + ky + lz = 1 \qquad (1.8)$$

1.5. Zones

The examination of a well-formed crystal, such as that of zircon illustrated in Fig. 1.6, shows that the faces are arranged in sets with respect to certain directions in the crystal. In other words, the crystal exhibits symmetry, which is an external manifestation of the regular ordering of the atomic and molecular components of the crystal.

A set of faces with a common direction are said to lie in one and the same *zone*, and the common direction is the *zone axis*. Two faces $(h_1k_1l_1)$ and $(h_2k_2l_2)$ are sufficient to define a zone, and the zone axis may be given by the line of intersection of two planes of the type of Eq. (1.3):

$$\frac{h_1X}{a} + \frac{k_1Y}{b} + \frac{l_1Z}{c} = 0$$
$$\frac{h_2X}{a} + \frac{k_2Y}{b} + \frac{l_2Z}{c} = 0 \qquad (1.9)$$

The solution of these equations is the line

$$\frac{X}{a(k_1l_2 - k_2l_1)} + \frac{Y}{a(l_1h_2 - l_2h_1)} + \frac{Z}{a(h_kk_2 - h_2k_1)} \qquad (1.10)$$

---

[b]A Latin phrase meaning 'the necessary changes having been made'.

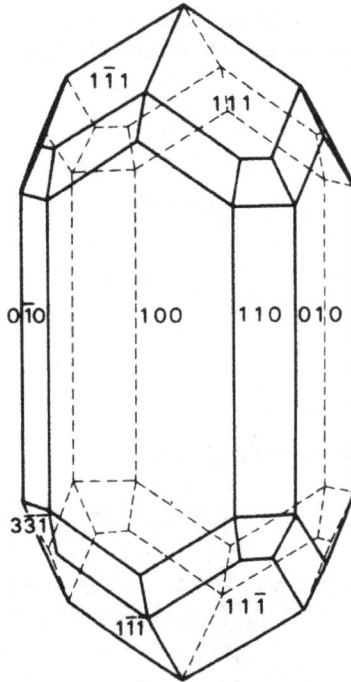

Fig. 1.6.   Zircon ZrSiO$_4$, a highly symmetrical crystal showing Miller indices of faces. The zone common to (1 1 1) and (11$\bar{1}$) is [$\bar{1}$ 1 0]. Faces such as (331) and ($\bar{3}\,\bar{3}\,\bar{1}$) lie in this zone: $3 \times (-1) + (3 \times 1) + (1 \times 0) = 0$. [*Structure Determination by X-ray Crystallography*, Mark Ladd and Rex Palmer, 5th edn. (2013). Reproduced by permission of Springer Science+Business Media, NY.]

which also passes through the origin and may be written as

$$\frac{X}{aU} = \frac{Y}{bV} = \frac{Z}{cW} \qquad (1.11)$$

where $[UVW]$ is the zone symbol. If another face (hkl) is *tautozonal*, that is, it lies in the same zone, then

$$hU + kV + lW = 0 \qquad (1.12)$$

which is the *Weiss zone law* [8].

Example 1.1. The Weiss zone law may be applied in the following manner. Let two tautozonal planes be (123) and ($2\bar{3}1$). The indices are written twice, and the first and last columns are neglected. Then, by cross-multiplication, as with a 2×2 determinant, U, V and W are obtained:

$$
\begin{array}{c|cccc|c}
1 & 2 & 3 & 1 & 2 & 3 \\
& & \times & \times & \times & \\
2 & \bar{3} & 1 & 2 & \bar{3} & 1
\end{array}
$$

$$U = 2 - (-9) = 11, \ V = 6 - 1 = 5, \ W = -3 - 4 = -7$$

The zone symbol is $[11, 5\bar{7}]$.

Two zone symbols $[U_1 V_1 W_1]$ and $[U_2 V_2 W_2]$ define a plane that is common to both zones. Then, from Eq. (1.12),

$$
\begin{aligned}
hU_1 + kV_1 + lW_1 = 0 \\
hU_2 + kV_2 + lW_2 = 0
\end{aligned}
\tag{1.13}
$$

and these equations can be solved in the manner of Example 1.1.

Self-assessment 1.3. Determine the face that is common to the zones $[1\bar{3}2]$ and $[02\bar{1}]$.

The program ZONE in the Program Suite accompanying this book solves the equations discussed above to give either (hkl) or $[UVW]$ values according to the data presented. The following standard conventions should be noted:

- (hkl), a plane;
- {hkl}, a form of planes;
- $[UVW]$, a zone symbol[c];
- $\langle UVW \rangle$, a form of zone symbols;

---

[c]The notation $[uvw]$ is used for a direction in a lattice.

- A Miller index, or a zone symbol, of two digits is followed by a comma unless it is the end number in the triplet.

A study of the external features of crystals, that is, its *morphology*, does not distinguish between (hkl) and (nh, nk, nl), where n is an integer; indeed, in a morphological description, such a distinction is not required. In X-ray crystallographic studies however, general indices nh, nk and nl $(n = 1, 2, 3, \ldots)$ for planes in a lattice are meaningful and necessary: they represent a *family* of parallel, equidistant planes, but $1/n$ times the perpendicular distance of the plane (hkl) from the origin.

1.6. Crystal Representation by Stereographic Projection

It is desirable to be able to represent a crystal and other three-dimensional features, such as bond directions from a central atom, by means of a plane drawing. The *stereographic projection*, or *stereogram*, may be used for this purpose as it is an angle-true projection. The development of the stereogram of a crystal is discussed in detail elsewhere [9, 10]; here, its meaning will be shown by means of diagrams.

Consider the crystal in Fig. 1.7(a) placed at the centre of a sphere of arbitrary radius. From a point within the crystal, conveniently its centre, lines are drawn normal to each crystal face and extended to cut the sphere, as shown in the *spherical projection* of Fig. 1.7(b). The reference axes x, y and z shown in Fig. 1.7(a) are Ob, Oe and Od on Fig. 1.7(b); a' is a face parallel to a, but not shown in the crystal drawing, and there are 12 other such faces.

The plane of the stereographic projection, or *primitive plane* (aka primitive), is $ABCD$ in Fig. 1.7(c), and the *primitive circle* encloses the stereogram details. A point of intersection of a normal with the surface of the sphere, such as r on the upper hemisphere, is joined to the lowest point P on the spherical projection. Its intersection with the primitive plane is the *pole*, R; it is marked with a dot • on the primitive plane, as in Fig. 1.7(c). In order to not increase the

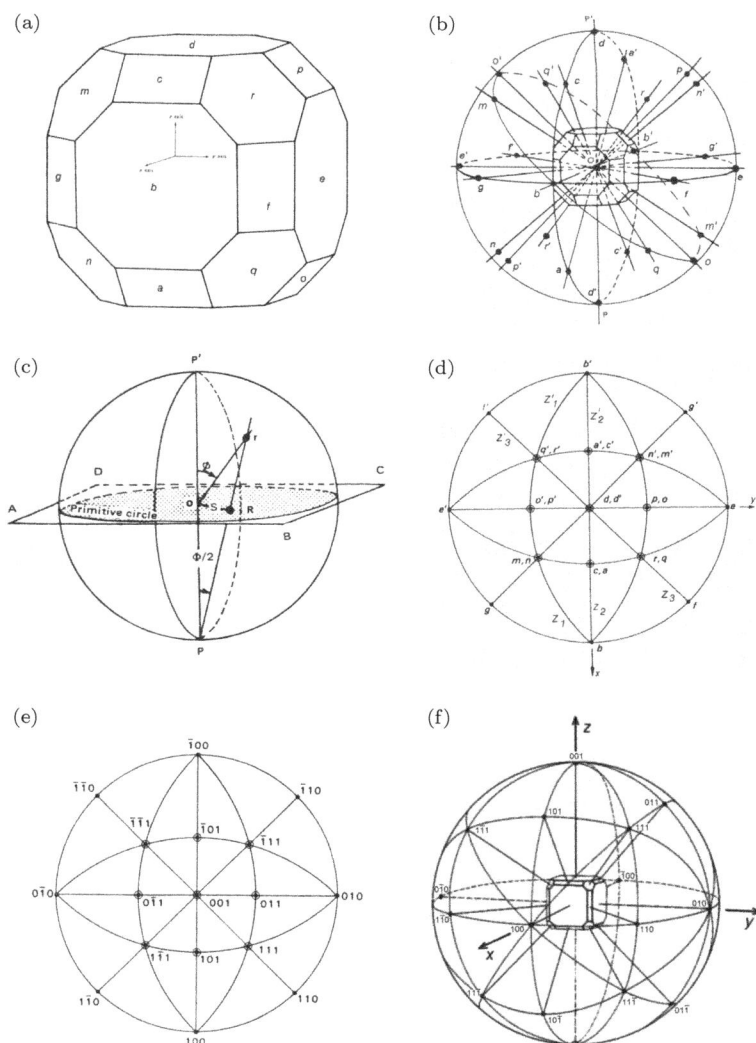

Fig. 1.7.   Formation of a stereographic projection (stereogram). (a) Crystal (of cubic symmetry) showing three forms: cube, b, e and d; octahedron, r, m, n and q; rhombic dodecahedron, f, c, g, a, o and p. Twelve parallel faces also exist, such as b', e', d' parallel to b, e, d, and similarly for the other two forms. (b) Spherical projection of the crystal; each of the 26 lines radiating from O, such as a_a', is normal to the face through which it passes. (c) Development of the stereogram from the spherical projection; the intersection at r forms the pole at R. (d) Completed stereogram: the *zone circles* $Z_1 Z_1'$, $Z_2 Z_2'$ and $Z_3 Z_3'$ have the zone symbols [011], [010] and [1$\bar{1}$0]. (e) Indexed stereogram. (f) Zone circles of the poles and Miller indices of the faces in Fig. 1.7(a). [*Structure Determination by X-ray Crystallography*, Mark Ladd and Rex Palmer, 5th edn. (2013). Reproduced by permission of Springer Science+Business Media, NY.]

size of the stereogram to unmanageable proportions, intersections on the lower hemisphere are joined to the uppermost point P' on the spherical projection and their poles are marked with circles o on the primitive — invert the diagram to get the picture.

The stereogram, completed in this way, is shown in Fig. 1.7(d). The parametral plane is r with indices 111 (no parentheses on a stereogram). On account of the chosen orientation of the crystal within the spherical projection, pole d, representing the face (001), lies at the centre of the stereogram and has the indices 001; similarly, face b is the intersection of the x-axis and the primitive circle. The fully indexed stereogram is shown in Fig. 1.7(e). There is no need to index any point marked o lying around the dot symbol •; if the symbol • represents hkl, then, in the given orientation, the o around it is $hk\bar{l}$; Fig. 1.7(f) shows zone circles and the Miller indices of the poles of the faces in Fig. 1.7(a).

While the stereogram has many applications in morphology [10], its use in this book is confined to illustrating the symmetry of crystals and bonds radiating from a central atom in a chemical species. Notwithstanding that a stereogram is a two-dimensional figure, it is helpful to indicate the provenance of the pole from either the upper or the lower hemisphere of the spherical projection. This can be done with a modified notation, devised by the author, and based on space-group notation; it is discussed in Section 2.4.2. Further notes on the stereographic projection can be found in the Web Materials, Appendix WA1.

Problems

1.1. Write Miller indices for planes which make the following intercepts on the x-, y- and z-axes: (a) $a/2$, $-3b/2$, c; (b) $-a$, ∥ $b, c/3$; (c) $a/2, b/4, -3c$.

1.2. Evaluate zone symbols for each of the following pairs of planes: (a) $(001)_-(111)$; (b) $(0\bar{1}3)_-(100)$; (c) $(1\bar{2}1)_-(\bar{3}\bar{2}1)$.

1.3. In which of the zones evaluated in Problem 1.2 does each of the following faces lie? Some faces may lie in more than one zone. (a) $(\bar{1}\bar{1}2)$; (b) $(11\bar{3})$; (c) (021).

1.4. What face lies at the intersection of each of the following pairs of zones?

(a) $[001]_-[111]$; (b) $[120]_-[021]$; (c) $[113]_-[21\bar{4}]$.

1.5. What are two possible planes in the zone $[1\bar{2}3]$? There is more than one answer, and a check may be made by cross-multiplication.

1.6. Given a zone defined by the planes $(h_1 k_1 l_1)$ and $(h_2 k_2 l_2)$, show that the plane $(mh_1 + nh_2, mk_1 + nk_2, ml_1 + nl_2)$ lies in the same zone; m and n are integers.

1.7. Assign hkl indices to the poles b–g on the stereogram shown below; pole a is 210. **Hint:** Begin with pole c.

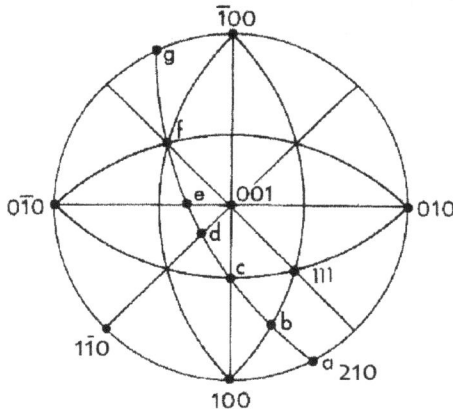

1.8. Use the program PLANE from the Program Suite to obtain the best-fit plane to the following six data points (given in absolute measure), and list the deviations of the points from the plane together with the rms value.

X	Y	Z	X	Y	Z	X	Y	Z
1	2	3	2	3	−4	−1	2	3
1	1	2	3	1	−2	−2	2	1

1.9. (a) Find the zone axis for the pair of planes $(121)\_(231)$. (b) Show that the (011) plane lies in that zone. (c) Find the values of p and q that enable (011) to be written as $(ph_1 + qh_2, pk_1 + qk_2, pl_1 + ql_2)$, where the subscripts refer to (121) and (231), respectively.

1.10. A molecule in a given crystal contains two chlorine atoms with the orthogonal, integer coordinates $\mathbf{r}_1(2, 6, 8)$Å and $\mathbf{r}_2(4, 12, 10)$Å with respect to an origin O. Calculate the $Cl(1) \cdots Cl(2)$ distance d and the angle θ formed by O and the two chlorine atoms.

Answers to Self-Assessments

1.1. $h = a/4a$, $k = b/3b$ and $l = c/2c$, so that (hkl) is $(1/4, 1/3, 1/2)$ or (346).

1.2. The following diagram shows three vectors (see Section A1ff) $\mathbf{h}, \mathbf{k}$ and $\mathbf{i}$ of equal magnitudes h, k and i along the three coplanar hexagonal axes x, y and u, respectively. It is evident from the diagram that $\mathbf{h} + \mathbf{k} + \mathbf{i} = 0$ whereupon it follows, in scalar terms, that $i = -(h + k)$.

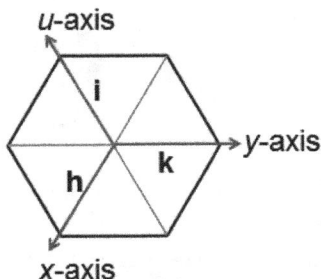

1.3.

$$
\begin{array}{ccccc}
1\,|\,\bar{3} & 2 & 1 & \bar{3}\,|\,2 \\
 & \times & \times & \times & \\
0\,|\,2 & \bar{1} & 0 & 2\,|\,\bar{1}
\end{array}
$$

The plane (hkl) is $(\bar{1}12)$.

References

[1] Whitlock HP, *Amer. Min.* **3**, 92 (1918). Online at ⟨http://www.minsocam. org/msa/collectors_corner/arc/hauyiv.htm⟩.

[2] Janot C, *Quasicrystals: A Primer*. Oxford University Press (2012).

[3] Weber S, *Quasicrystals*. Online at ⟨http://www.jcrystal.com/steffenweber/ qc.html⟩.

[4] Schrödinger E, *What is life*. Cambridge University Press (1944).

[5] Aldridge S, *The DNA Story* (2003). Online at ⟨https://www.chemistryworld. com/news/the-dna-story/3003946.article⟩.

[6] Miller WH, *Treatise on Crystallography*, Cambridge University Press (1839).

[7] Ladd M and Palmer R, *Structure Determination by X-ray Crystallography*, 5th edn. Springer (2013).

[8] Weiss CS, *Abhandlungen der Königlichen Akademie der Wissenschaften*, Berlin (1815).

[9] Ladd M, *Symmetry of Crystals and Molecules*. Oxford University Press (2014).

[10] Phillips FC, *Introduction to Crystallography*. Longmans Higher Education_New edition (1977).

Chapter 2

Symmetry Operations and Point Groups

A thing is symmetrical if there is something
you can do to it so that after you have finished doing it,
it looks the same as before.
Hermann Weyl

Key Topics

- Symmetry defined
- Groups
- Symmetry elements and symmetry operations
- Crystal systems and classes
- Crystallographic point groups and their recognition
- Enantiomorphism
- Non-crystallographic point groups
- Subgroups
- Matrix representation of point groups

2.1. Introduction

Life is surrounded by symmetry: two of the objects in Fig. 2.1 are well known in everyday life, and the molecular structure (c) in this illustration is known to chemists as 2,4,6-triazidotriazine. All three entities share the common feature of three-fold rotational

(a) (b) (c)

Fig. 2.1. Three objects exhibiting three-fold rotational symmetry about their centres: (a) Logo of NatWest plc; (b) Emblem of Mercedes-Benz; (c) Molecular structure of 2,4,6-triazidotriazine.

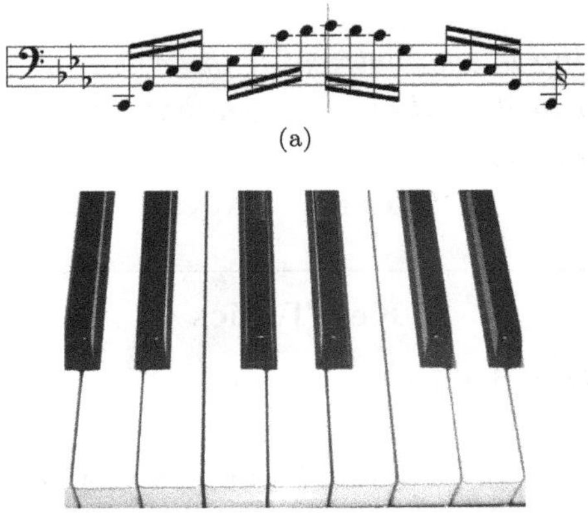

(a)

(b)

Fig. 2.2. Reflection symmetry in music. (a) An extract from the Étude for piano Op. 10 No. 12 by Chopin; the thin, vertical line marked is a reflection symmetry line. (b) Symmetry across a D key of a piano keyboard.

symmetry about their centres. Symmetry is evident in the music of all ages. Figure 2.2(a) shows a portion of the bass line from Chopin's Étude Op. 10 No. 12; even the piano keyboard itself has symmetry (Fig. 2.2(b)). In architecture and plants, the symmetry of their design and construction is evident as exemplified in Fig. 2.3, where (a) depicts the well-known Taj Mahal, while (b) is a bloom of the orchid *Cattleya walkeriana*, variation *Alba*, that shows bilateral symmetry.

One aspect of the symmetry of the illustrations in Figs. 2.2 and 2.3 is that of reflection across a line or across a plane; Fig. 2.4 is

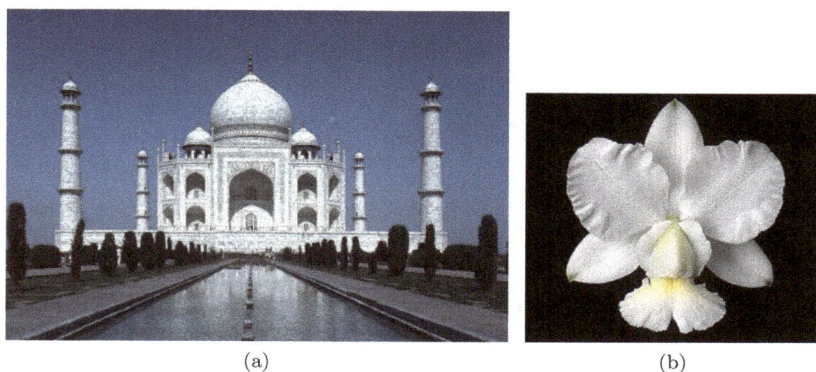

(a)   (b)

Fig. 2.3. Symmetry in plants and in architecture. (a) The Taj Mahal: An imagined vertical plane through its centre divides it into reflective halves. (Note that the symmetry of the picture as a whole is degraded by the unsymmetrical horticultural state of the bushes.) (b) A bloom of the orchid *Cattleya walkeriana*, variation *Alba* showing bilateral reflection symmetry. [Reproduced by courtesy of Greg Allikas.]

Fig. 2.4.   Model of a crystal of gypsum $CaSO_4 \cdot 2H_2O$; reflection symmetry exists across an imagined vertical plane through the centre of the crystal.

a model crystal of gypsum $CaSO_4 \cdot 2H_2O$ that shows *inter alia* the same symmetry across a conceptual vertical plane through the centre of the model. The symmetry of macroscopic crystals and other objects, as well as the symmetry of the internal structure of crystals will be studied in this book. Figure 2.5 shows the crystal structure of oxalic acid dihydrate $(CO_2H)_2 \cdot 2H_2O$ as a stereoscopic pair of images.

The symmetry of plane objects, such as those illustrated in Fig. 2.1, is easily recognized: all parts of a two-dimensional object are

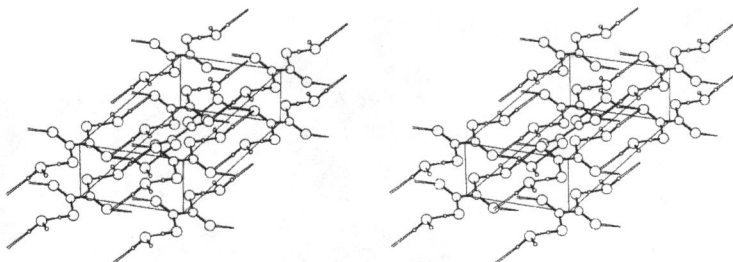

Fig. 2.5.   Stereoview of the crystal structure of oxalic acid dihydrate, $(CO_2H)_2.2H_2O$; the circles in decreasing order of size represent O, C and H atoms. The double, open-ended lines indicate hydrogen bonds with an average H–O $\cdots$ H bond distance of 2.50 Å. The sum of the van der Waals radii for hydrogen and oxygen is 2.72 Å, which is evidence of strong hydrogen bonding in this structure. Appendix A3 discusses stereoviews and stereoviewing.

seen simultaneously, so that the relationship of each to the whole is seen at one and the same time. In three dimensions, the total symmetry is not always so obvious. Objects like flowers, pencils and building bricks are simple enough to be examined and rotated physically, but the manipulation of more complex objects may require assistance in order to relate its parts to the whole object.

Section 1.6 described how the external forms of crystals may be represented by stereographic projections. Another method, which is particularly suited to crystal structures, is that of representing the object by a pair of stereoscopic images, such as that shown in Fig. 2.5, and Appendix A3 provides notes on stereoviewers and stereoviewing.

2.2.   Defining Symmetry

The symmetry description of a body may depend upon the method of its examination, that is, on the examining probe. For example, the structure of elemental chromium as determined by the X-ray diffraction has the body-centred cubic structural unit as shown in Fig. 2.6(a); the atom at the centre of the cubic unit is identical to those at its corners. Neutrons interact with the magnetic moments of substances. Elemental chromium has the electronic configuration

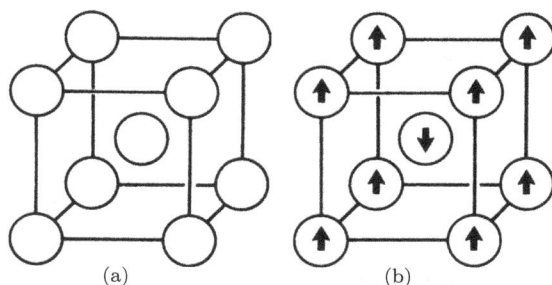

Fig. 2.6. A unit cell of the crystal structure of elemental chromium as seen by (a) X-rays and (b) neutrons. The arrows in (b) represent the directions of the magnetic moment vectors of the chromium atoms.

$(\text{Ar})(4\text{s})^1(3\text{d})^5$ and possesses a magnetic moment by virtue of its unpaired d-electrons. The structure of chromium with respect to neutron diffraction has been shown to be based on a primitive cubic structural unit: the atom at the centre is different from those at the corners on account of the direction of its magnetic moment vector, as shown in Fig. 2.6(b).

In this book, *symmetry* will refer to the positional distributions of the parts of an object or pattern as revealed by visual inspection, by diffraction techniques or by microscopic examination. A definition of symmetry is *that spatial property of an assemblage by which it can be brought from an initial state to another indistinguishable state by means of a certain operation—a symmetry operation.* The term 'assemblage' is useful because it can describe the distributions of faces on a crystal, of bonds radiating from a central atom, and of the spatial arrangement of diffraction spectra from crystalline materials.

2.3. Groups

A group G is a set of members $G\{\mathbf{1}, \mathbf{A}, \mathbf{B}, \mathbf{C}, \ldots\}$ that are interrelated according to well-defined rules:

(1) The product of any two members of the set is also a member of the set. The term *product* refers to a *law of combination* and is

not always algebraic multiplication. Using A, B and C as typical members, the combination $AB = C$ would be an example of the rule, and this property is termed *closure*. Note that AB is not necessarily equal to BA, that is, the combinations need not *commute*.

(2) The *associative law* of combination must hold, that is, $A(BC) = (AB)C$.

(3) The group contains the *identity* member 1 (aka I), such that $A1 = A$, and similarly for all members in the set.

(4) Each member A has an inverse A^{-1} that is also a member of the set, so that $AA^{-1} = 1$; the member 1 is its own inverse.

Example 2.1. An interesting corollary of rule (4) is that the inverse of a product of two or more members is the product of the individual inverse members in the reverse order: Let $ABC = D$ be the product. Then,

$$(ABC)C^{-1}B^{-1}A^{-1} = DC^{-1}B^{-1}A^{-1}$$

and

$$DC^{-1}B^{-1}A^{-1} \equiv A(B1B^{-1})A^{-1} = A1A^{-1} = 1$$

Thus, writing the last equation as

$$1 = DC^{-1}B^{-1}A^{-1}$$

and pre-multiplying by D^{-1}, it follows that

$$D^{-1}1 = D^{-1}DC^{-1}B^{-1}A^{-1}$$

or

$$D^{-1} = C^{-1}B^{-1}A^{-1}$$

showing that

$$(ABC)^{-1} = C^{-1}B^{-1}A^{-1}$$

2.3.1. Examples of groups

Groups may be finite or infinite. Groups that describe the symmetries of molecules are finite except for linear molecules such as hydrogen chloride HCl or carbon dioxide CO_2. An example of an infinite group is the series

$$-\infty, \ldots - n, \ldots, -3, -2, -1, 0, 1, 2, 3, \ldots, n, \ldots, \infty$$

With addition as the law of combination, the group is *Abelian*, that is, its members commute, $1 + 2 = 2 + 1$; it is associative, $1 + (2 + 3) = (1 + 2) + 3$; the identity member is 0, since $0 + n = n$; and each member has an inverse, for $-n + n = 0$.

2.3.2. Group multiplication tables

A group comprising h members is defined when its h^2 products are known; they can be displayed by a *Cayley table* within which each row and each column list each member once only. Consider a group $G_3\{A, B, C\}$ for which $A^2 = B$ and $AB = 1$; its group multiplication table is as follows:

G_3	1	A	B
1	1	A	B
A	A	B	1
B	B	1	A

The top line is read first, so that the combination $AB = 1$ is read as B followed by A, and the table shows that $AB = 1$. Other multiplicative relationships are $B^2 = A$, $A^2B = A(AB) = A$. The order of operation is not important in this group; the members commute, so that $A(BA) = A$ and the group is Abelian.

One of the simplest non-Abelian crystallographic point groups is 32; the mineral quartz exhibits this symmetry. In the following, a crystal of quartz is illustrated together with its Cayley table; the

x, y and u two-fold axes lie in one plane, with the three-fold z-axis normal to this plane:

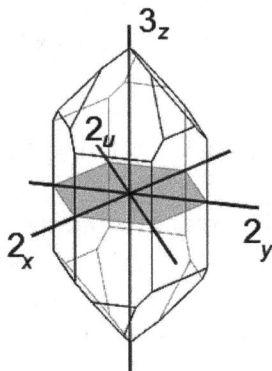

G_{32}	1	3	3^{-1}	2_x	2_y	2_u
1	1	3	3^{-1}	2_x	2_y	2_u
3	3	3^{-1}	1	2_u	2_x	2_y
3^{-1}	3^{-1}	1	3	2_y	2_u	2_z
2_x	2_x	2_y	2_u	1	3^{-1}	3
2_y	z_y	2_u	2_x	3	1	3^{-1}
2_u	2_u	2_x	2_y	3^{-1}	3	1

The table shows, for example, that the combination $32_x = 2_u$, whereas $2_x 3 = 2_y$. Thus, a Cayley table of the symmetry operations of a given point group shows how the operations of that group combine. If a group contains h symmetry operations, then the group will contain h^2 products; h is the *order* of the group. Note that the inverse operation 3^{-1} may be written as 3^2, that is, two successive three-fold operations ($33 = 3^2 = 3^{-1}$); 2 is its own inverse ($2_y 2_y^{-1} = 2_y 2_y = 1$).

The mathematics of groups will not be pursued except insofar as it is related to the crystallographic point groups and space groups. The reader wishing to study the topic more generally should have recourse to the literature [1–3].

Self-assessment 2.1. Consider the following questions in relation to the Cayley table for the group G. (a) Which of the members is the identity member? (b) For which member(s) is the relationship $\boldsymbol{X} = \boldsymbol{X}^{-1}$ true? (c) Are the group operations commutative?[a]

G	A	B	C	D
A	C	A	D	B
B	A	B	C	D
C	D	C	B	A
D	B	D	A	C

2.4. Symmetry Elements, Symmetry Operations and Symmetry Operators

A *symmetry element* is a conceptual geometrical entity, a point, a line or a plane, combined with a set of symmetry operations that has the geometrical entity in common. Although only conceptual, it can be helpful to accord it a sense of reality. The symmetry element associates all parts of the assemblage (to which it refers) into a number of symmetrically related sets. Thus, a four-fold axis is a line about which the operation of four-fold rotational symmetry can be performed. The concept of symmetry element has been reviewed in detail in the literature [4, 5].

A *symmetry operation* when applied to a body transforms it to a state that is indistinguishable from its state before the operation, thus revealing the symmetry inherent in the body. A single symmetry element can be associated with more than one symmetry operation. Thus, in a four-fold symmetry, operations 4^2 and 4^3 may be regarded either as multiple steps, 2 and 3 steps, respectively, of the operation 4

---

[a] Answers to all self-assesments are given at the end of the chapter.

(which itself could be called 4^1) or as a single operation each in its own right. Note that the terms symmetry element and symmetry operation are not synonyms.

Example 2.2. Consider orthogonal axes with a four-fold symmetry axis along z. A four-fold, *right-handed* rotational operation transforms (x, y, z) to $(\bar{y}, x, z)$:

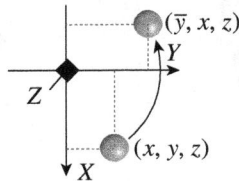

The operation on a crystal plane (hkl) can be represented on a stereogram:

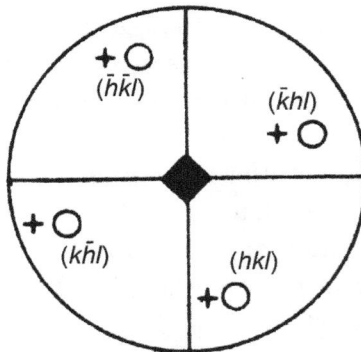

$(hkl) \xrightarrow{\;4\;} (\bar{k}\,hl); (hkl) \xrightarrow{\;4^2(\equiv 2)\;} (\bar{h}\,\bar{k}l); (hkl) \xrightarrow{\;4^3(\equiv 4^{-1})\;}$
$(k\bar{h}\,l).$

A *symmetry operator* may be regarded as a mathematical function that generates a symmetry operation in a definite manner and orientation. It is represented in a matrix form, which is discussed in Section 2.13 and Appendix A1. The notation used in this context is, for example, m (italic font) for a mirror plane symmetry element, 4 (italic bold font) for a symmetry operation or symmetry operator, and **4** (Roman bold font) for a matrix *representing* a symmetry operation or an operator, or normal font otherwise. In addition, a notation 'the plane x_y', for example, is used to imply a plane containing the axes/directions x and y.

Self-assessment 2.2. The above diagram is a stereoscopic pair for the chlorobenzene molecule. Identify the symmetry elements in type and position. With the point-group recognition program, SYMM, the appropriate model number is **16**.

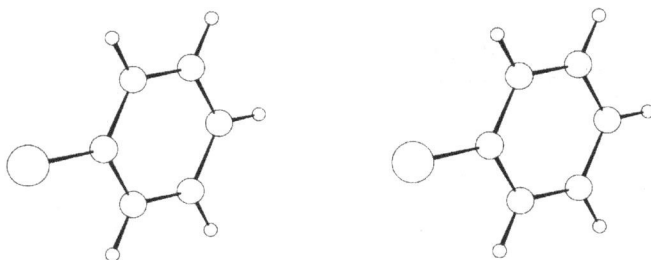

2.4.1. Point group

In finite bodies, symmetry elements and symmetry operations occur singly or in combinations, and may be defined for one-, two- or three-dimensional space. A set of associated symmetry operations or just one such operation is known as a point group. A *point group* may be defined as a set of symmetry operations any or all of which leave at least one point unmoved. All symmetry operations pass through this point, which is also the origin for the reference axes. It follows that translational symmetry is not permitted in a point group. In crystals, it is the three-dimensional point groups that are of importance. Point groups are discussed in more detail in Section 2.7ff, and also in the literature [6, 7].

Symmetry operations in finite bodies are of three types — rotation, reflection and inversion — and exhibit similar characters

Table 2.1.   Symmetry operations in one, two and three dimensions.

Symmetry	One dimension	Geometrical extension	
	One dimension	Two dimensions	Three dimensions
Reflection	Across a point	Across a line	Across a plane
Rotation	–	About a point	About a line
Inversion	–	–	Through a point

whether in one, two and three dimensions; they differ in geometrical extension, as Table 2.1 shows.

2.4.2.   Reflection symmetry

If an object can be brought into self-coincidence by reflection across an imaginary plane within it, it possesses reflection symmetry and the plane is termed a *mirror* plane, symbol m. The object and its mirror image are *enantiomorphs*; they are related as a right hand is to a left hand. Several examples of reflection symmetry have been shown in Figs. 2.2–2.4. Unlike rotation, a mirror symmetry operation cannot be performed physically on a finite body, but its presence can be appreciated visually and also from the stereogram of a body containing an m-symmetry relationship.

Figure 2.7 is a stereogram of the point group of gypsum (Fig. 2.4). A stereogram of a point group needs to show a *general* form of planes in order to reveal the true point-group symmetry; a form of planes lying on symmetry elements is termed as a *special* form. An example of special forms is shown in Fig. 2.8, which is a diagram of a crystal of N-iodosuccinimide $C_4H_4O_2NI$. The apparent point group is $4mm$; the four-fold axis lies in the apparent, vertical mirror planes. Only special forms are present in this example and the true point group, deduced from an X-ray diffraction study, is 4.

2.4.3.   Rotation symmetry

If an object can be brought from an initial state to another, indistinguishable state for every rotation of $(360/R)°$ around a line

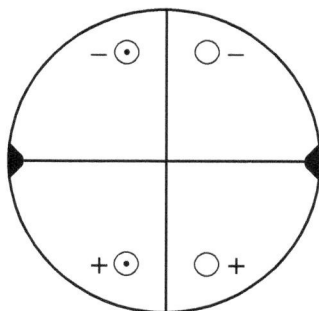

Fig. 2.7. Stereogram of the point group of gypsum showing a general form {*hkl*}: The thick line represents a vertical *m* plane; the filled-in semi-oval represents a two-fold rotation axis (not a standard symbol) lying the plane of the diagram, and the open circle at the centre represents a centre of symmetry, at the point where the *y*-axis cuts the mirror plane.

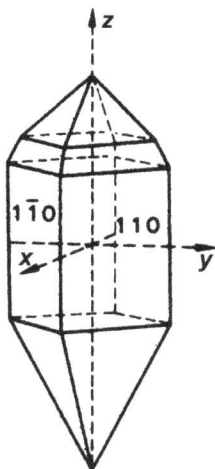

Fig. 2.8. A crystal of *N*-iodosuccinimide showing only special forms, such as {1 1 0}; the crystal symmetry is apparently 4*mm*, but the true point group (determined by X-ray diffraction) is 4.

(rotation axis), then that line is a symmetry axis of degree R, or an R-fold rotation axis. In principle, R can take any value from 1 to ∞, but in (classical) crystals, R is restricted to the values 1, 2, 3, 4 and 6. Identical solids each with one of these rotational symmetries can be packed to fill a region of space completely. Figure 2.9 illustrates the

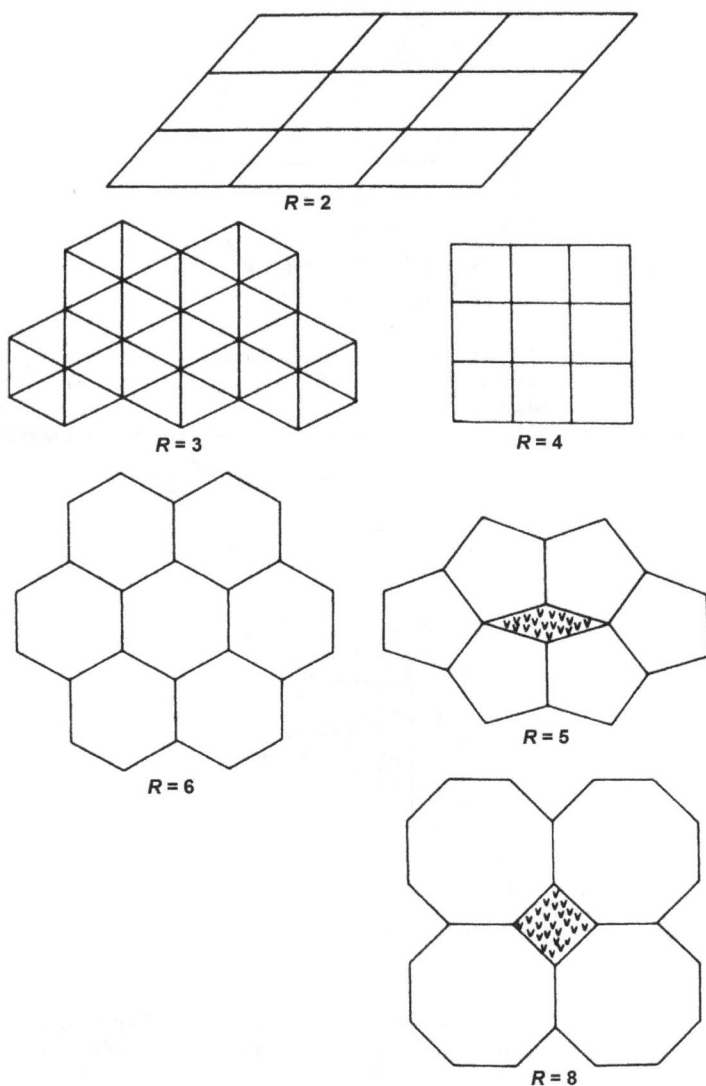

Fig. 2.9.   The packed parallelepipeda are shown in projection along their principal symmetry axes R. Only those parallelepipeda with $R = 2, 3, 4$ and 6 can fill space completely.
With $R = 5$ or 8, voids v remain within the packing. [*Structure Determination by X-ray
Crystallography*, Mark Ladd and Rex Palmer, 5th edn. (2013). Reproduced by permission
of Springer Science + Business Media, NY.]

packing, in projection, of polyhedra that are based on the rotation degrees 2, 3, 4, 5, 6 or 8; in the figures with $R = 5$ and $R = 8$, for example, voids remain within the packed polyhedra.

All operations of rotational symmetry involve *congruent* aspects of the initial and rotated positions of the object under examination, whereas the reflection symmetry involves *enantiomorphic* aspects.

2.4.4. Roto-inversion symmetry

An object possesses roto-inversion symmetry (aka inversion symmetry) $\overline{R}$ if it can be brought from an initial state to an indistinguishable final state for every rotation of $(360/R)°$ around a line (inversion axis) combined with inversion through about a point on the $\overline{R}$ inversion axes; in crystals, this point is the unmoved point in the definition of point group. Like mirror symmetry, this operation cannot be performed physically on a body. It can be represented on a stereogram, and Section A3.2.3 gives instructions for constructing a model having $\overline{4}$ (four-fold inversion) symmetry, together with a stereogram of that operation.

2.4.5. Roto-reflection symmetry

Roto-reflection symmetry (aka alternating symmetry) can be used in place of inversion symmetry in the description of point groups. A crystal exhibits alternating symmetry of degree R around a line (alternating axis) if it can be brought from an initial state to another indistinguishable state for every rotation of $(360/R)°$ around the line combined with reflection across a plane normal to the alternating axis. It should be noted that this plane need not be a reflection plane of the point group containing the alternating axis. Compare the stereograms for point groups $\overline{4}$ and $\frac{4}{m}$ (Fig. 2.11); if the plane normal to $\overline{4}$ *is* a mirror plane, then the point group $\frac{4}{m}$ is obtained.

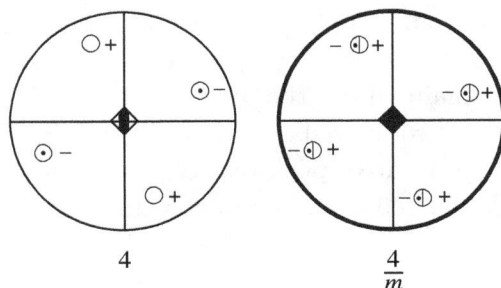

Roto-reflection symmetry is a feature of the Schönflies point-group symmetry notation [8], whereas in crystallography the Hermann–Mauguin notation [9, 10] is employed. The operations of roto-inversion, roto-reflection and mirror reflection[b] are known also as *improper rotations* or *operations of the second kind*, and with these operations, the object and its symmetry-related aspect are enantiomorphic; operations of rotation are known as *proper rotations* or *operations of the first kind*.

2.5. Crystal Systems

The crystallographic point groups are divided unequally among seven *crystal systems*: they form a broad classification of crystals based on their *characteristic* symmetry, that is, the symmetry that is essential in order to allocate a crystal to its system. A given crystal will display frequently more than its characteristic symmetry. The symmetry of the crystal determines special relationships between the intercepts a, b, c of the parametral plane and between the interaxial angles α, β, γ; Table 2.2 lists these relationships together with the characteristic symmetry for the seven crystal system.

A detailed derivation of crystal systems is given in the literature [7]. As an example, consider type $R2$ in the orthorhombic system,

---

[b] $m \equiv \bar{2}$.

Table 2.2. The seven crystal systems and their characteristic properties.

System	Characteristics	Relationships between a, b, c and α, β, γ[a,b]
Triclinic	None	$a \neq b \neq c$; $\alpha \neq \beta \neq \gamma \neq 90°$, $120°$
Monoclinic	One 2 or $\bar{2}$[c] along y	$a \neq b \neq c$; $\alpha = \gamma = 90°$; $\beta \neq 90°$, $120°$
Orthorhombic	Three mutually perpendicular 2 or $\bar{2}$-axes along x, y, z	$a \neq b \neq c$; $\alpha = \beta = \gamma = 90°$
Tetragonal	One 4- or $\bar{4}$-axis along z	$a = b \neq c$; $\alpha = \beta = \gamma = 90°$
Cubic	Four 3-axes inclined at $54.74°\,(\cos^{-1} 1/\sqrt{3})$ to x, y, z	$a = b = c$; $\alpha = \beta = \gamma = 90°$
Trigonal[d]	One 3-axis along z	$a = b \neq c$; $\alpha = \beta = 90°$; $\gamma = 120°$
Trigonal[e]	One 3-axis along $\mathbf{a} + \mathbf{b} + \mathbf{c}$	$a = b = c$; $\alpha = \beta = \gamma \neq 90°$, $< 120°$
Hexagonal[f]	One 6- or $\bar{6}$-axis along z	$a = b \neq c$; $\alpha = \beta = 90°$; $\gamma = 120°$

[a]The same relationships apply also to the conventional unit cells of the Bravais lattices, as will be discussed shortly.
[b]The symbol $\neq$ means 'not constrained by symmetry to equal'.
[c]$\bar{2}$ is equivalent to m if m is perpendicular to it; conventionally, 2 is along y and m is the x–z plane.
[d]Refers to hexagonal axes.
[e]Refers to rhombohedral axes; $\mathbf{a} + \mathbf{b} + \mathbf{c}$ is the direction $[1\,1\,1]$ in the unit cell.
[f]In relation to the axis denoted by u, the vector $-\mathbf{u} = \mathbf{a} + \mathbf{b}$.

where $R = 2$, and let the two-fold axes lie along x and y. Then, for a two-fold symmetry operation **2** along x:

$$x, y, z \xrightarrow{\;2_x\;} x, -y, -z \qquad (2.1)$$

and for symmetry **2** along y:

$$x, y, z \xrightarrow{\;2_y\;} -x, y, -z \qquad (2.2)$$

The combination of these two operations is given by

$$x, y, z \xrightarrow{\;2_y 2_x\;} -x, -y, z \qquad (2.3)$$

which implies symmetry **2** along z. It follows from the sign changes in Eqs. (2.1) and (2.2) that $x \perp y$ and z, and that $y \perp x$ and z. Thus, the x-, y- and z-axes, and the corresponding symmetry operators, are mutually orthogonal, which is the characteristic symmetry of the orthorhombic system.

2.6. Geometric Crystal Classes

Each of the 32 *geometric crystal classes* (crystal classes) may be
characterized by the name of its general form, which has the full
symmetry (holosymmetry) of the crystal under consideration. Crystals frequently do not display their general form, so the true symmetry is not immediately apparent, as with the crystal example in
Fig. 2.8. The general form for the class of point group 4 is the tetragonal pyramid whereas that for $4mm$ is the ditetragonal pyramid. The
morphological names of general forms may be found in the literature [11, 12], but it is not the purpose of this book to consider them
further. The term crystal class is not exactly synonymous with point
group: it is a classificatory term. Crystals of a given class exhibit the
same point group.

2.7. The Crystallographic Point Groups

The derivation of the crystallographic point groups has been given
in the literature [6, 7, 13, 14] and need not be repeated here. Table 2.3
lists a scheme for writing the crystallographic point-group symbols
under seven type-headings that will suffice for present purposes. The
table indicates how the point groups are assembled as the characteristic crystal rotational symmetry R increases.

2.7.1. Point-group symbols

In order to understand point-group and space-group symbols, their
full meaning must be known. Symmetry operations themselves are
straightforward, but it is essential to appreciate both the relative
orientations of the symmetry elements in a point group, and how
these orientations change according to the characteristic symmetry
of each system.

A point-group symbol comprises a combination of the elements
R, $\overline{R}$ and m, which form *three positions* in a symbol that are related
to the reference axes x, y and z, which are themselves related to

Table 2.3. Crystallographic point-group scheme: $R = 1, 2, 3, 4, 6$.[a]

System	R	$\bar{R}$	$\overline{R}\bar{1}$	$R2$	Rm	$\overline{R}m$	$R2\,\&^{\text{b}}\,\bar{1}$
				Type			
Triclinic	1	$\bar{1}$	–				
Monoclinic	2	m^{c}	$\frac{2}{m}$				
Orthorhombic				222	$mm2$	–	mmm
Tetragonal	4	$\bar{4}$	$\frac{4}{m}$	422	$4mm$	$\bar{4}2m$	$\frac{4}{m}mm$
Cubic[d,e]	23	$m\bar{3}$	–	432	–	$\bar{4}3m$	$m\bar{3}m$
Trigonal	3	$\bar{3}$	–	32	$3m$	$\bar{3}m$	–
Hexagonal	6	$\bar{6}$	$\frac{6}{m}$	622	$6mm$	$\bar{6}m2$	$\frac{6}{m}mm$

[a]The reader may wish to consider the implication of the spaces marked – in the table.
[b]The symbol & implies here a combination of the operations $R2$ and $\bar{1}$.
[c]m is equivalent to $\bar{2}$ perpendicular to it.
[d]The cubic system is characterized by four three-fold axes along $\langle 111 \rangle$.
[e]Point groups $m\bar{3}$ and $m\bar{3}m$ may be found as $m3$ and $m3m$ in older literature.

symmetry directions in the crystal. Not all three positions are listed in every point-group symbol because to do so would be trivial in some cases. For example, the monoclinic symbol $\frac{2}{m}$ implies a two-fold axis (by convention along y) with a mirror plane perpendicular to it; $\frac{R}{m}$ is always a single direction with $m \perp R$. Thus, $\frac{2}{m}$ represents symmetry with respect to the direction y, and this description is sufficient. In full, the point group would be written as $1\frac{2}{m}1$, with the three symbols 1, $\frac{2}{m}$ and 1 representing the symmetry with respect to x, y and z. Since there is only the trivial symmetry 1 in the x- and z-directions, the symbol 1 is generally omitted.

In the same way, the point group $4mm$ can be written adequately as $4m$ because the interaction of 4 and m generates a second form of m planes. The best practice is to use the three symbols with the exception of trivialities to which one soon becomes accustomed. The argument is illustrated in Figs. 2.10(a–d) and legend.

The meanings of the point-group symbols are listed in Table 2.4; they are given in the Hermann–Mauguin symmetry notation, while Fig. 2.11 displays stereograms of the crystallographic point groups.

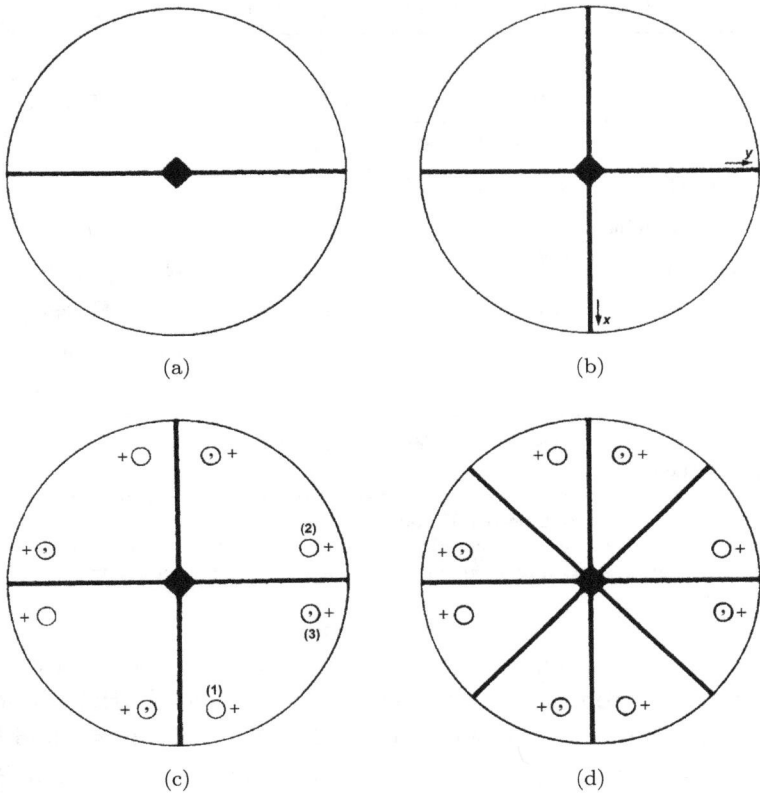

Fig. 2.10.   Point group 4*mm*. (a) A single mirror plane is inconsistent with the presence of the four-fold symmetry along z (the normal through the centre of the plane); a second mirror plane is generated as shown in (b). (b) The x and y axial directions are traces of the vertical m planes; these two planes constitute a form of m planes. (c) Symmetry 4*m*: a general point is added to (b), marked (1); the four-fold rotation generates point (2), and reflection across an m plane generates point (3). (d) The stereogram is then completed by adding a second form of symmetry lines: points such as (1) and (3) are related by m symmetry; the second form of m planes lies at 45° to the first form, and the full point-group symbol is 4*mm*.

Note that, for convenience, the symbol $\bar{2}$ is used in Table 2.4; more generally, it is represented by m, the mirror plane perpendicular to it.

The stereogram and its representation were discussed in Section 1.6 and Fig. 2.12; Fig. 2.13 illustrates further aspects of the notation used herein. Whereas in point group 2, the symmetry-related

Table 2.4. Two-dimensional and three-dimensional point-group symbols.

System	Point groups[a]	First position	Second position	Third position
Two-dimensional				
Oblique	1, 2	Rotation 1, 2 about a point	–	–
Rectangular	$1m$, $2mm$	Rotation 1, 2 about a point	$m\perp x$	$m\perp y$[b]
Square	4	Rotation 4 about a point	$m\perp x, y$	–
	$4mm$	Rotation 4 about a point	$m\perp x, y$	m 45° to x, y
Hexagonal	3, 6	Rotation 3, 6 about a point		
	$3m$	Rotation 3 about a point	$m\perp x, y, u$	–
	$6mm$	Rotation 6 about a point	$m\perp x, y, u$	m 60° to x, y, u
Three-dimensional				
Triclinic	$1, \bar{1}$	All directions in crystal	–	–
Monoclinic[c]	$2, m, 2/m$	2 and/or $\bar{2}$ along y	–	–
Orthorhombic	222, $mm2$, mmm	2 and/or $\bar{2}$ along x	2 and/or $\bar{2}$ along y	2 and/or $\bar{2}$ along z
Tetragonal	$4, \bar{4}, 4/m$	4 and/or $\bar{4}$ along z	–	–
	422, $4mm$, $\bar{4}2m$, $\frac{4}{m}mm$	4 and/or $\bar{4}$ along z	2 and/or $\bar{2}$ along x, y	2 and/or $\bar{2}$ at 45° to x, y and in the x, y plane, i.e. along $\langle 110 \rangle$
Cubic[d]	23, $m\bar{3}$	2 and/or $\bar{2}$ along x, y, z	3 and/or $\bar{3}$ at[e] 54.73° to x, y, z, i.e. along $\langle 111 \rangle$	–
	432, $\bar{4}3m$, $m\bar{3}m$	4 and/or $\bar{4}$ along x, y, z	3 and/or $\bar{3}$ at[e] 54.73° to x, y, z, i.e. along $\langle 111 \rangle$	2 and/or $\bar{2}$ at 45° to x, y, z, i.e. along $\langle 110 \rangle$
Trigonal[f]	$3, \bar{3}$	3 and/or $\bar{3}$ along z	–	–
	$32, 3m, \bar{3}m$	3 and/or $\bar{3}$ along z	2 and/or $\bar{2}$ along x, y, u	–

(*Continued*)

Table 2.4.   (*Continued*)

System	Point groups[a]	First position	Second position	Third position
Hexagonal	$6, \bar{6}, \frac{6}{m}$	6 and/or $\bar{6}$ along z	–	–
	622, 6mm, $\bar{6}m2$, $\frac{6}{m}mm$	6 and/or $\bar{6}$ along z	2 and/or $\bar{2}$ along x, y, u	2 and/or $\bar{2} \perp x, y, u$ and in the x, y, u plane

[a] $\frac{R}{m}$ occupies a single position in a point-group symbol because only one direction is involved (m is perpendicular to R; for convenience, $\frac{R}{m}$ is often written as R/m.
[b] In plane group 2mm.
[c] In the monoclinic system, the 2- or $\bar{2}$-axis is along y by convention; hence, an m plane, if present, is normal to it, and in other crystal systems *mutatis mutandis*.
[d] Actually, $\cos^{-1}(1/\sqrt{3})$.
[e] Earlier notation used $m3$ and $m\bar{3}$ for $m\bar{3}$ and $m\bar{3}m$, respectively.
[f] For convenience, the trigonal system is referred to hexagonal axes, on rhombohedral axes, the orientations of the three positions are, in order, [111], [1$\bar{1}$0] and [11$\bar{2}$].

objects are congruent, point group $\bar{4}$ exhibits enantiomorphous relationships on account of the inversion symmetry, but the difference is not clear on the traditional stereograms. However, the difference becomes clear by the notation of Fig. 2.13; furthermore, the $\pm$ signs serve usefully to indicate symmetry-related positions with respect to the plane of the diagram.

As an example from Table 2.4, consider the stereogram for point group $4mm$ in Fig. 2.11. Table 2.4 lists symmetry 4 along z for position 1, symmetry $\bar{2}$ along x *and* y for position 2 ($\equiv m \perp x$ and y), and symmetry $\bar{2}$ along $\langle 110 \rangle$ for position 3 ($\equiv m$ at 45° to x and y), thus specifying the orientation of the symmetry elements unambiguously. The reader should find it helpful to study Table 2.4 and Fig. 2.11 together with examples of crystal or crystal models (Appendix A3.3).

The scheme in Table 2.5 develops the above discussion of point groups and lists also the *Laue classes* (aka Laue groups), which are 11 crystal classes that possess *inter alia* a centre of symmetry and are written in bold font in the table. The arrows show how the addition

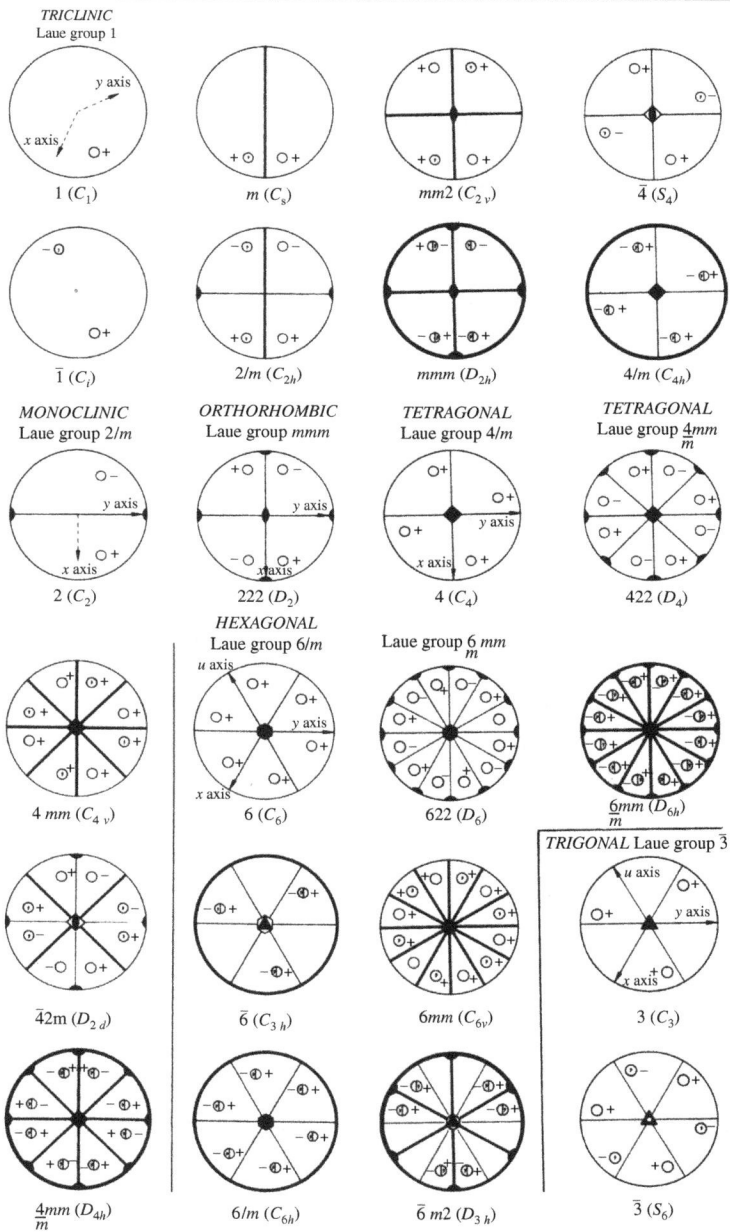

Fig. 2.11 (*Continued*).

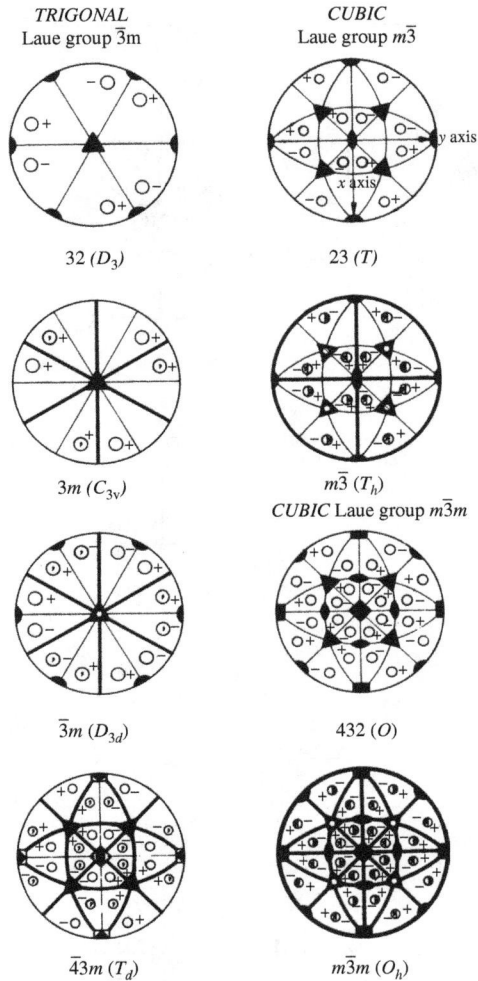

Fig. 2.11. Stereograms of the crystallographic point groups arranged by crystal system and Laue class or classes. Each stereogram shows the symmetry elements and a general form {*hkl*}. The reference axes are shown once for each system. An axis terminating within the primitive implies that that axis makes an oblique angle with the *z*-axis, the latter being always normal to the diagram. The symbols in parentheses are the equivalent Schönflies symbols. [*Structure Determination by X-ray Crystallography*, Mark Ladd and Rex Palmer, 5th edn. (2013). Reproduced by permission of Springer Science + Business Media, NY.]

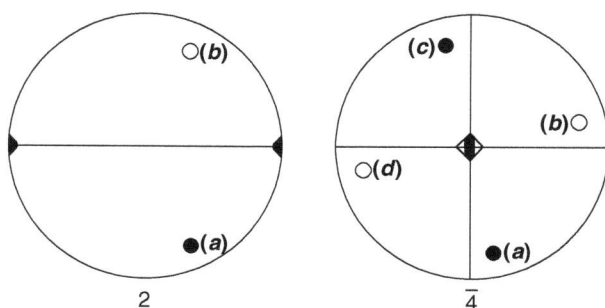

Fig. 2.12. Traditional stereogram drawings of a general form in point groups 2 and $\bar{4}$. In point group 2, points (a) and (b) imply a congruent relationship, whereas in point group $\bar{4}$, points (a) and (b) imply an enantiomorphic relationship on account of the inversion symmetry, but the notation does not reveal this difference.

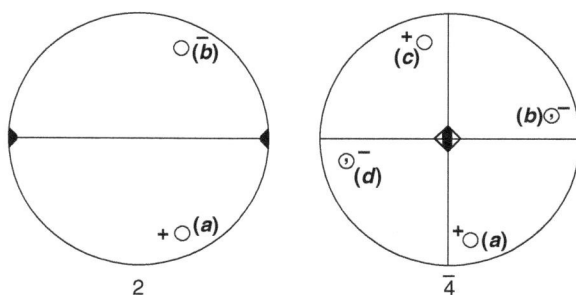

Fig. 2.13. The same stereograms in the revised notation; the congruent and enantiomorphic aspects are now indicated clearly.

of a centre of symmetry to non-centrosymmetric groups leads to the Laue classes of the systems.

Crystallography has well-defined, generally accepted conventions of notation, some of which have been described already. There are standard symbols for all symmetry axes and reflection planes, and Tables 2.6 and 2.7 illustrate them. Screw axes and glide planes are included as they will arise in the study of space groups. There may seem to be a wealth of notation, but it is essential for clarification and consistent communication in crystallography.

Table 2.5. Crystallographic point-group symbols with Laue groups in bold font.

System				Type			
	R	$\overline{R}$	$R\overline{1}$	$R2$	Rm	$\overline{R}m$	R & $\overline{1}$
Triclinic	1	$\overline{\mathbf{1}}$	–				
Monoclinic	2	m	$\frac{\mathbf{2}}{m}$				
Orthorhombic				222	$mm2$	–	$\boldsymbol{mmm}$
Tetragonal	4	$\overline{4}$	$\frac{\mathbf{4}}{m}$	422	$4mm$	$\overline{4}2m$	$\frac{\mathbf{4}}{m}\boldsymbol{mm}$
Cubic	23	$\boldsymbol{m\overline{3}}$	–	432	–	$\overline{4}3m$	$\boldsymbol{m\overline{3}m}$
Trigonal	3	$\overline{\mathbf{3}}$	–	32	$3m$	$\overline{\mathbf{3}}\boldsymbol{m}$	–
Hexagonal	6	$\overline{6}$	$\frac{\mathbf{6}}{m}$	622	$6mm$	$\overline{6}m2$	$\frac{\mathbf{6}}{m}\boldsymbol{mm}$

2.8. Point-Group Recognition

There are several methods of point-group recognition. An interactive method [15] divides crystals into four types depending upon the presence of

(1) neither a mirror plane nor a centre of symmetry
(2) a mirror plane only
(3) a centre of symmetry only
(4) a mirror plane and a centre of symmetry

Table 2.6.   Point-group and space-group axial symmetry symbols.

Axis and symbol	Graphical symbol	Translation
1		None
2 $\perp$ diagram plane		None
2 in diagram plane		None[a]
$2_1 \perp$ diagram plane		$c/2$
2_1 in diagram plane		$a/2$ or $b/2$
3		None
3_1		$c/3$
3_2		$2c/3$
4		None[a]
4_1		$c/4$
4_2		$2c/4$ ($\equiv c/2$)
4_3		$3c/4$
6		None
6_1		$c/6$
6_2		$2c/6$ ($\equiv c/3$)
6_3		$3c/6$ ($\equiv c/2$)
6_4		$4c/6$ ($\equiv 2c/3$)
6_5		$5c/6$
$\bar{1}$		None
$\bar{3}$		None
$\bar{4}$		None
$\bar{6}$		None

[a]In stereograms, the half-oval and half-square are used for this 2 and 4 lying in the primitive.

Table 2.7.   Point-group and space-group reflection symmetry symbols.

Plane ‖ diagram	Symbol	Graphical symbol placed around the unit cell	Graphical symbol placed within the unit cell	Translation
Mirror	m			None
Glide	a or b			$a/2$ or $b/2$
Glide	c			$c/2$
Diagonal glide	n			$(a+b)/2$ or $(b+c)/2$ or $(c+a)/2$
Double glide	e			$a/2$ and $b/2$ or $b/2$ and $c/2$ or $c/2$ and $a/2$
Diamond glide	d	3/8 · 1/8		$(b\pm c)/4$ and $(c\pm a)/4$ or $(a\pm b)/4$ and $(b\pm c)/4$ and $(c\pm a)/4$ or $(a\pm b\pm c)/4$

so that the first step is a search for one of these four types for the crystal under examination.

A centre of symmetry in a crystal model may be demonstrated by placing it in any orientation on a flat surface. In the presence of a centre of symmetry, the uppermost face will be parallel to the supporting surface. A mirror plane divides an object into enantiomorphic halves, showing a right-hand–left-hand relationship. Some models may be constructed readily, as described in Appendix A3.

Next, the principal rotation axis R, the rotation axis of highest degree, must be identified, together with the number of such axes, the presence and number of mirror planes m and two-fold rotation axes.

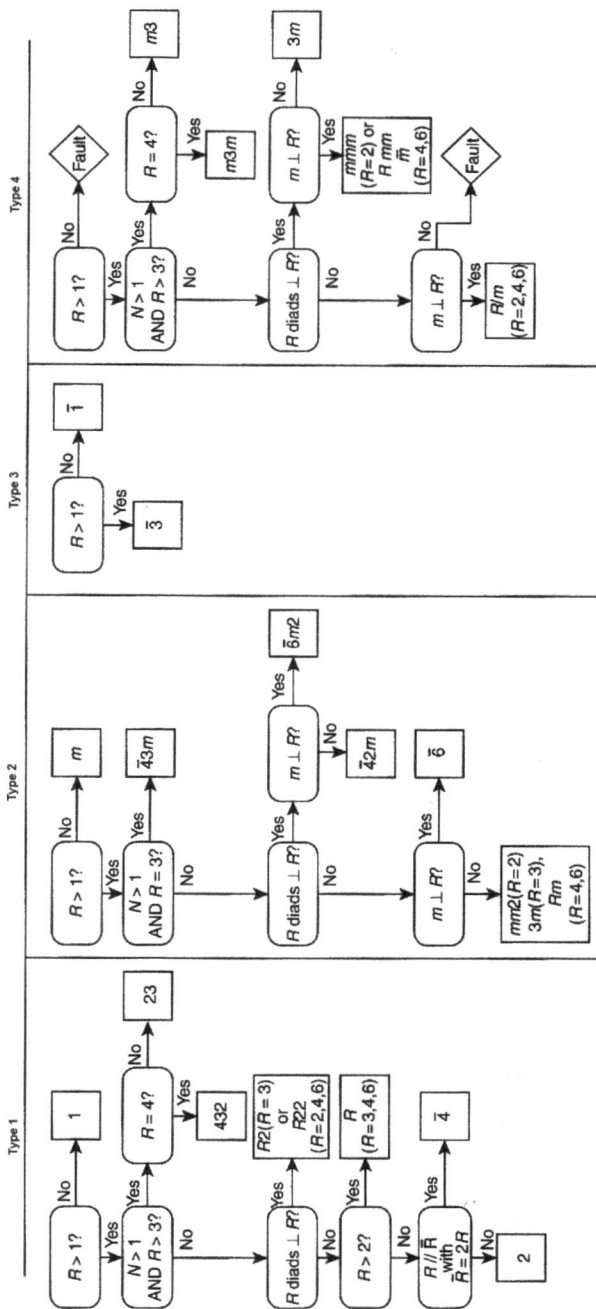

Fig. 2.14. Flow diagram of the point-group recognition program SYMM. Point groups ∞m and ∞/m evolve in columns 4 and 9, respectively, with $R = 0$ (zero).

Table 2.8.   Point groups in the four types for point-group recognition.

Type 1 neither m nor $\bar{1}$	Type 2[a] m and no $\bar{1}$	Type 3 $\bar{1}$ and no m	Type 4[a] m and $\bar{1}$
1, 2, 3, 4, $\bar{4}$, 6	m, $mm2$, $3m$, $4mm$	$\bar{1}$, $\bar{3}$	$2/m$, mmm, $\bar{3}m$, $4/m$
222, 32, 422	$\bar{4}2m$, $\bar{6}$, $6mm$		$4/mmm$, $6/m$, $6/mmm$
622, 23, 432	$\bar{6}m2$, $\bar{4}3m$, ∞m		$m\bar{3}$, $m\bar{3}m$, ∞/m

[a]∞m and ∞/m, although non-crystallographic, are included in program SYMM; they relate to linear molecules.

The program SYMM in the Program Suite enables a deduction of a point group to be carried out and confirmed. An incorrect response to a question directs the user back to the point of error. The program also identifies certain non-crystallographic symmetry elements that are found in molecules. A flow diagram of the program is given in Fig. 2.14. The models described in Appendix A3.2.3 and that in Self-assessment 2.2 can be examined by this program. Table 2.8 lists the crystallographic point groups in terms of the types 1–4 described above. Practice with this interactive program will assist the user in deducing point groups of solid objects or models of chemical species. Other examples using the program SYMM can be found in the literature [6].

2.9.   Enantiomorphism and Chirality

Species that exhibit *enantiomorphism* are said to be *chiral*; they have no form of inversion symmetry; in the earlier literature, chiral species were termed dissymmetric. They exist as non-superposable mirror images or enantiomers, and their symmetry comprises only proper rotations; objects with no symmetry (point group 1) are termed asymmetric.

Organic enantiomorphic molecular species frequently crystallize as *racemates* or *racemic mixtures*, that is, they are mixtures of both enantiomers. A well-known example is lactic acid; its enantiomers are non-superposable:

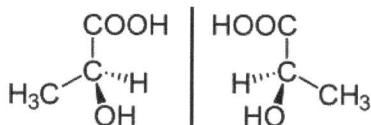

Enantiomorphs of lactic acid

Sodium chlorate of point group 23 is also chiral, but only in the solid state; the chirality arises from the screw symmetry adopted when this substance crystallizes from aqueous solution (see also Section 4.11.3.1).

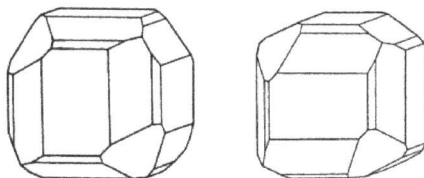

Enantiomorphs of sodium chlorate

2.10. Non-crystallographic Point Groups

Certain molecular species exhibit degrees of rotational symmetry other than those discussed so far. Some examples are degree 5 in (η^5-cyclopentadienyl)nickel nitrosyl, degree 7 in uranium heptafluoride, degree 8 inoctacyanotungstate(VI), and degree ∞ or ∞/m in linear molecules; Fig. 2.15 is stereoscopic illustration of a nickel compound. Other forms of non-crystallographic symmetry in crystals are discussed under pseudosymmetry in Section 6.6.5.

2.11. Subgroups

A subgroup is a subset of members of a group that satisfies the four group rules. It must contain the identity member as one of the subsets. Thus, point group 422 has subgroups 4, 2 and 1.

The subgroups of the crystallographic point groups are illustrated in Fig. 2.16.

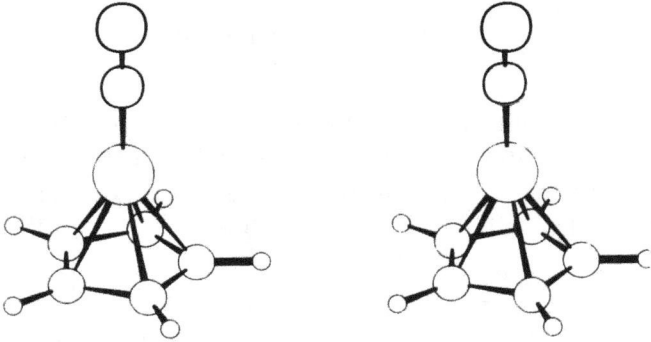

Fig. 2.15. Stereoview of the molecule of (η^5-cyclopentadienyl) nickel nitrosyl C_5H_5 NiNO, an example of a non-crystallographic point group.

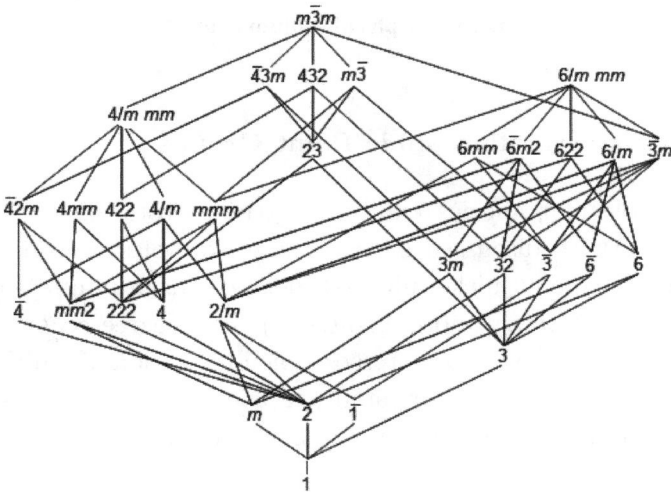

Fig. 2.16. Diagrammatic representation of the subgroups of the crystallographic point groups.

Table 2.9.   Plane groups and their subgroups.

Plane group (short symbol)	Subgroup	Plane group (short symbol)	Subgroup
1	None	2mm	1, 2, m
2	1	3m	1, 3, m
3	1	4mm	1, 2, 4, m, 2mm
4	1, 2	6mm	1, 2, 3, 6, m
6	1, 2, 3		2mm, 3m
m	1		

2.12.   Plane Point Groups

There are only two one-dimensional point groups, namely, 1 and m. In two dimensions, there are ten two-dimensional point groups (aka plane groups); they are listed, together with their subgroups, in Table 2.9. They are easily understood, and will appear again when studying space groups; they need no further elaboration at this stage.

2.13.   Matrix Representation of Point-Group Symmetry Operations

This section makes use of some of the results developed in Appendix A1. A symmetry operation $\boldsymbol{R}$ acting on a crystal plane of Miller indices (hkl) transforms it to indices $(h'k'l')$; for conciseness, the symbols $\mathbf{h}$ and $\mathbf{h}'$, respectively, will be used for these example planes.

A symmetry operation may be written in the most general manner as

$$\boldsymbol{R}\mathbf{h} + \mathbf{t} = \mathbf{h}' \tag{2.4}$$

where $\boldsymbol{R}$ is a symmetry operation and $\mathbf{t}$ is a translation vector. By definition, $\mathbf{t}$ is identically zero in a point group; this condition is obtained because all the symmetry elements in the group pass through one and the same point. If it were not so, then two parallel

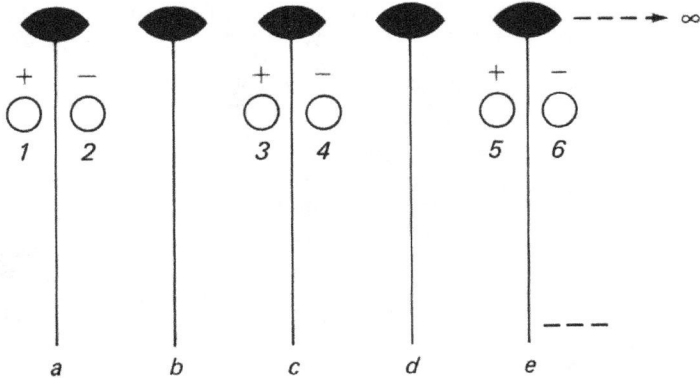

Fig. 2.17.   The repetition a, b, c, ... of a symmetry element by translation leads to a succession of symmetry axes and points 1, 2, 3, ... that are wholly incompatible with a point group. The infinite repetition of symmetry axes and related entities is a feature of space groups, as will be discussed shortly.

two-fold axes, for example, would generate additional axes and points *ad infinitum*, as Fig. 2.17 shows. Thus, in point groups, a symmetry operation is given by

$$\boldsymbol{R}\mathbf{h} = \mathbf{h}' \tag{2.5}$$

Consider the stereogram for two-fold symmetry in Fig. 2.12 with the two-fold axis along y. The position (a) may be likened to a plane (hkl). Rotation of this plane about the axis changes the signs of h and l. Thus, the matrix for this operation and its action are given by

$$\underbrace{\begin{pmatrix} \bar{1} & 0 & 0 \\ 0 & 1 & 0 \\ 0 & 0 & \bar{1} \end{pmatrix}}_{\mathbf{2}} \underbrace{\begin{pmatrix} h \\ k \\ l \end{pmatrix}}_{\mathbf{h}} = \underbrace{\begin{pmatrix} \bar{h} \\ k \\ \bar{l} \end{pmatrix}}_{\mathbf{h}'} \tag{2.6}$$

multiplying by the usual rule for matrices, as described in Appendix A1.4.7. If this operation is followed by that of reflection across the x–z plane, then only the y-coordinate changes sign, and the matrix equation becomes

$$
\begin{pmatrix} 1 & 0 & 0 \\ 0 & \bar{1} & 0 \\ 0 & 0 & 1 \end{pmatrix} \begin{pmatrix} \bar{h} \\ k \\ \bar{l} \end{pmatrix} = \begin{pmatrix} \bar{h} \\ \bar{k} \\ \bar{l} \end{pmatrix} \tag{2.7}
$$

$\qquad\quad$ **m** $\qquad$ **h′** $\qquad$ **h″**

The result is a plane $(\bar{h}\,\bar{k}\,\bar{l})$ centrosymmetrically related to (hkl), and the complete process of symmetry operations may be written for the given orientations as

$$
\boldsymbol{m\ 2 = \bar{1}} \tag{2.8}
$$

which is read as **2** *followed by* **m** is equivalent to $\bar{1}$. What is the result of **m** followed by **2**?

Self-assessment 2.3. What is the result in the above example, Eq. (2.7), if the second operation **m** is replaced by two-fold symmetry along a z-axial direction normal to the $x\text{-}y$ plane m?

As a second example, consider the stages in Fig. 2.10(c). The four-fold anticlockwise (right-handed) rotation (1) to (2) relates (hkl) to $(\bar{k}hl)$; then reflection (2) to (3) generates the plane (khl). In the matrix notation, the two matrices may be multiplied first as **m4**, and then the resulting matrix operated on the plane (khl): It is permissible first to multiply the two matrices and then operate on the plane (hkl) (note the order of multiplication with **m4**):

$$
\begin{pmatrix} \bar{1} & 0 & 0 \\ 0 & 1 & 0 \\ 0 & 0 & 1 \end{pmatrix} \begin{pmatrix} 0 & \bar{1} & 0 \\ 1 & 0 & 0 \\ 0 & 0 & 1 \end{pmatrix} = \begin{pmatrix} 0 & 1 & 0 \\ 1 & 0 & 0 \\ 0 & 0 & 1 \end{pmatrix}
$$

$\quad$ **m⊥x** $\qquad$ **4 along z** $\qquad$ **m 45° to** $\qquad$ (2.9)

$\qquad\qquad\qquad\qquad\qquad\qquad$ **x and y**

$$\begin{pmatrix} 0 & 1 & 0 \\ 1 & 0 & 0 \\ 0 & 0 & 1 \end{pmatrix} \begin{pmatrix} h \\ k \\ l \end{pmatrix} = \begin{pmatrix} k \\ h \\ l \end{pmatrix}$$

$$\mathbf{m}\ 45°\ \text{to} \qquad \mathbf{h} \qquad \mathbf{h''} \tag{2.10}$$

$$x\ \text{and}\ y$$

Self-assessment 2.4. What is the result in the above example, Eq. (2.9), if the mirror plane normal to y is used for the second symmetry operation?

All symmetry combination operations can be constructed in similar manner. Comprehensive lists of matrices representing all crystallographic operations are provided in the literature [6, 7, 14]. Matrix operations will be addressed further in Section 4.13 and Appendix A1.

2.13.1.   Useful rotation matrices

The rotation matrix $\boldsymbol{\Psi}$ that was derived in Appendix A4 applies to those situations in which the rotation axis is along z and normal to a plane containing the x- and y-axes.

$$\boldsymbol{\Psi} = \begin{pmatrix} (\cos\phi - \cos\gamma\sin\phi/\sin\gamma) & (-\sin\gamma\sin\phi - \cos^2\gamma\sin\phi/\sin\gamma) & 0 \\ (\sin\phi/\sin\gamma) & (\cos\phi + \cos\gamma\sin\phi/\sin\gamma) & 0 \\ 0 & 0 & 1 \end{pmatrix} \tag{2.11}$$

In this matrix, the angle of rotation about the symmetry axis is ϕ and the angle between the x and y axes is γ. A point $\mathbf{X}(x, y, z)$ may be transformed to $\mathbf{X}'(x', y', z')$ according to the equation

$$\boldsymbol{\Psi}\mathbf{X} = \mathbf{X}' \tag{2.12}$$

where $\mathbf{X}$ and $\mathbf{X}'$ are columns vectors.

Example 2.3. What are the coordinates of a point $\mathbf{X}(x, y, z)$ after a right-handed six-fold rotation about z in the hexagonal system?

In the hexagonal system, $\gamma = 120°$, and the angle of rotation ϕ is given as $60°$. Then,

$$\boldsymbol{\Psi} = \begin{pmatrix} 1 & \bar{1} & 0 \\ 1 & 0 & 0 \\ 0 & 0 & 1 \end{pmatrix} \quad \text{and} \quad \mathbf{X}' = \begin{pmatrix} 1 & \bar{1} & 0 \\ 1 & 0 & 0 \\ 0 & 0 & 1 \end{pmatrix} \begin{bmatrix} x \\ y \\ z \end{bmatrix} = \begin{pmatrix} x - y \\ x \\ z \end{pmatrix}$$

that is, $x, y, z \xrightarrow{\ 6\ } x - y, x, z$.

Problems

2.1. What is the full, conventional meaning that is conveyed by the point-group symbols (a) m, (b) $mm2$, (c) $\bar{4}2m$, (d) 32 (on hexagonal axes), (e) 622, (f) 23.

2.2. From the symbol 422, write symbols of all possible combinations of rotation and inversion axes. Which of these symbols represent crystallographic point groups, and what are their conventional symbols?

2.3. What are the point groups of the following chemical species? (a) CCl_4; (b) $CHCl_3$; (c) CH_2Cl_2 (two forms); (d) CH_3Cl; (e) CH_4.

2.4. What are the point groups of the following chemical species, shown in stereoviews? The numbers in bold font refer to the model numbers for the program SYMM:
(a) Dibenzyl (**78**); (b) 2,4,6-Triazidotriazine (**89**); (c) Pentafluoroantimonate(III)ion (**55**); (d) Cyclohexane (chair form) (**38**); (e) 1,4-Dichlorobenzene (**65**); (f) E-1,2-Dichloroethene (**68**).

(a)

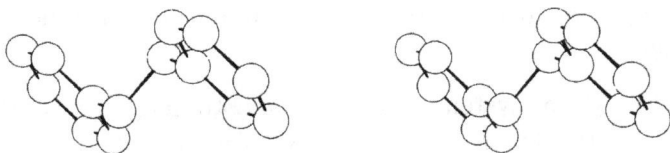

(b)

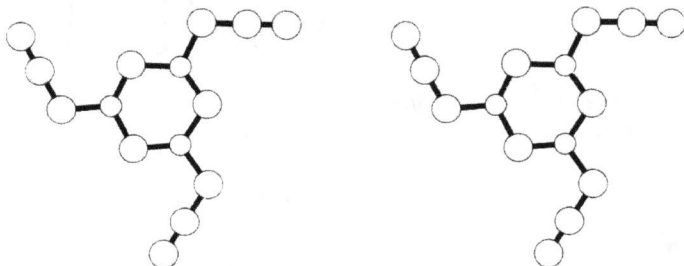

(c)

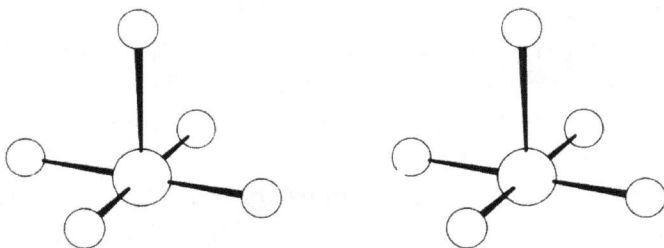

(d)

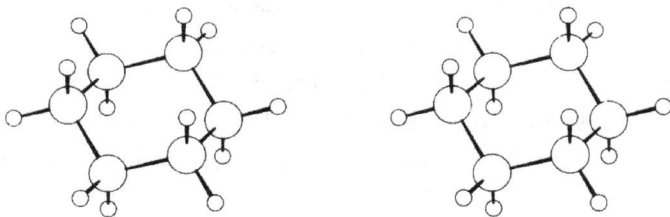

(e)

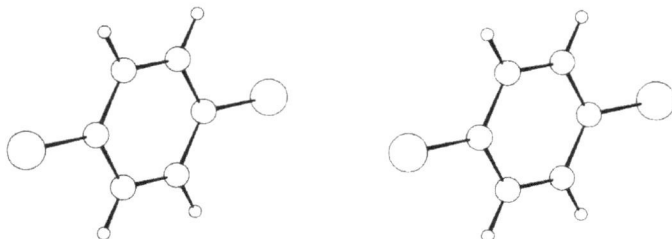

(f)

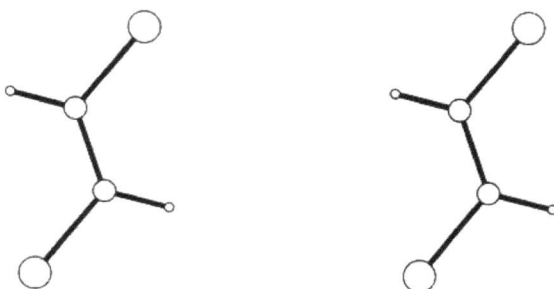

2.5. (a) Set up matrices for the symmetry operations $\overline{4}$ along z and m normal to x appropriate to the tetragonal system. (b) Use the matrices to determine the nature and position of the symmetry operation of $\overline{4}$ followed by m. (c) What is the result of $\overline{4}m$ and how is it related to the operation in (b)?

2.6. What symmetry is formed from the combination of symmetry operations $\overline{4}_{[010]}\overline{4}_{[100]}$?

2.7. What are the point groups of benzene and its 13 possible chloro-derivatives shown in the diagram below?

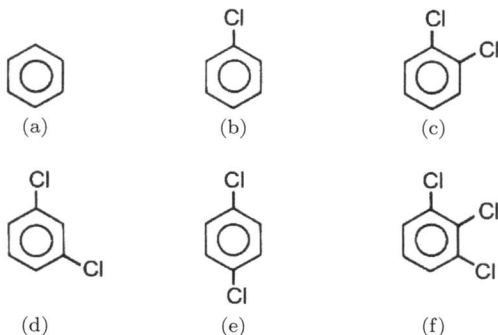

(a) (b) (c)

(d) (e) (f)

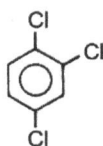

(g)

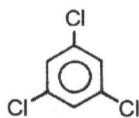

(h)

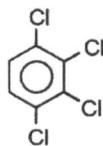

(i)

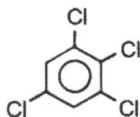

(j)

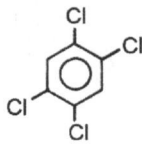

(k)

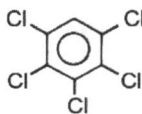

(l)

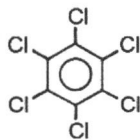

(m)

2.8. Name the Platonic solids (a)–(e) hereunder, and write the numbers of faces F, edges E and vertices V that each of solids exhibits. What is the relationship between F, E and V?

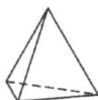

(a)

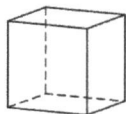

(b)

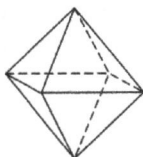

(c)

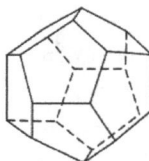

(d)

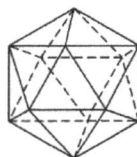

(e)

2.9. List the subgroups of point groups 32, $\bar{4}2m$ and $m\bar{3}$.

2.10. The full symbol for point group $6m$ is $6mm$. Why is $3m$ not written as $3mm$?

2.11. (a) What are the symmetry operations of a water molecule? (b) What pairs of these operations commute?

2.12. In the cubic system, how many points are generated from x, y, z by the symmetry operations $\bar{4}_{[001]}$ followed by $\bar{3}_{[111]}$

followed by $\overline{2}_{[100]}$ followed by $\overline{1}$, and what is the point group at each stage?

2.13. What symmetries are revealed by packing together (a) two, and (b) four, irregular but identical quadrilaterals?

2.14. The diagram below is a stereoview of the trichlorome-thane $CHCl_3$ molecule. Construct a Cayley table for this species.

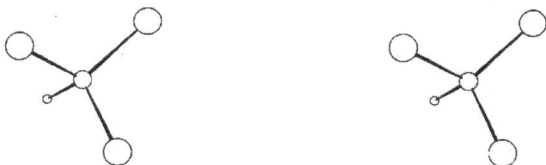

Answers to Self-Assessments

2.1. (a) $\boldsymbol{B}$ is the identity member since $\boldsymbol{BX} = \boldsymbol{X}$ for $\boldsymbol{X} = \boldsymbol{A}, \boldsymbol{B}, \boldsymbol{C}, \boldsymbol{D}$.

(b) $\boldsymbol{B}$ and $\boldsymbol{C}$, since $\boldsymbol{BB} = \boldsymbol{B}$ and $\boldsymbol{CC} = \boldsymbol{B}$ ($\boldsymbol{B}$ is the identity member) whereas $\boldsymbol{AA} = \boldsymbol{C}$ and $\boldsymbol{DD} = \boldsymbol{C}$.

(c) $\boldsymbol{AB} = \boldsymbol{BA}$ and similarly for all other pairs; thus, the combinations commute. The group is commutative since $\boldsymbol{XY} = \boldsymbol{YX}$ for $\boldsymbol{X}/\boldsymbol{Y} = \boldsymbol{A}, \boldsymbol{B}, \boldsymbol{C}, \boldsymbol{D}$.

2.2. Two-fold rotation axis through the Cl and C(4) atoms, a vertical m-plane through the atoms Cl and C(4), and an m-plane as the plane of the molecule.

2.3. On orthogonal axes, two-fold symmetry along z changes the signs of h and k. Hence, for the second operation

$$\begin{pmatrix} \overline{1} & 0 & 0 \\ 0 & \overline{1} & 0 \\ 0 & 0 & 1 \end{pmatrix} \begin{pmatrix} \overline{h} \\ k \\ \overline{l} \end{pmatrix} = \begin{pmatrix} h \\ \overline{k} \\ \overline{l} \end{pmatrix}$$

which implies two-fold rotational symmetry along x.

2.4.

$$
\begin{pmatrix} 1 & 0 & 0 \\ 0 & \bar{1} & 0 \\ 0 & 0 & 1 \end{pmatrix}
\begin{pmatrix} 0 & \bar{1} & 0 \\ 1 & 0 & 0 \\ 0 & 0 & 1 \end{pmatrix}
=
\begin{pmatrix} 0 & \bar{1} & 0 \\ \bar{1} & 0 & 0 \\ 0 & 0 & 1 \end{pmatrix}
\quad \text{and}
$$

$$\mathbf{m} \perp y \qquad \mathbf{4} \text{ along } z \qquad \mathbf{m}'45° \text{to}$$

$$x \text{ and } y$$

$$
\begin{pmatrix} 0 & \bar{1} & 0 \\ \bar{1} & 0 & 0 \\ 0 & 0 & 1 \end{pmatrix}
\begin{pmatrix} h \\ k \\ l \end{pmatrix}
=
\begin{pmatrix} \bar{k} \\ \bar{h} \\ l \end{pmatrix}
$$

$$\mathbf{m}'45° \text{to}$$

$$x \text{ and } y$$

which is a second m-plane, $\mathbf{m}''$ at 45° to x and y.

References

[1] McWeeny R, *Symmetry*. Dover Publications Ltd. (2003).
[2] Burns G, *Introduction to Group Theory with Applications*. Elsevier (1977).
[3] IUCr, *International Tables for Crystallography A*. 6th edn. Wiley (2016).
[4] de Wolff PM *et al.*, *Acta Crystallogr. A* **45**, 494 (1989).
[5] Flack HD *et al.*, *Acta Crystallogr. A* **56**, 96 (2000).
[6] Ladd M, *Symmetry of Crystals and Molecules*. Oxford University Press (2014).
[7] Burns G and Glazer AM, *Space groups for Solid State Scientists*. Elsevier (2013).
[8] Schönflies AM, *Kristallsysteme und Kristallstruktur*. Teubner (1891).
[9] Hermann C, *Z. Kristallogr.* **68**, 257, *ibid.* **69**, 226, *ibid.* **69**, 533 (1928); *ibid.* **76**, 559 (1931).
[10] Mauguin C-V, *Z. Kristallogr.* **76**, 542 (1931).
[11] Phillips FC, *An Introduction to Crystallography*, 2nd edn. Longmans (1956).

[12] Howard M, *Crystallography and Minerals Arranged by Crystal Form*. Online at ⟨http://www.webmineral..com/crystal/Tetragonal.shtml{#}.WT Li9Te1uM8⟩.

[13] IUCr, *Derivation of Point Groups*. Online at ⟨http://www.mx.iucr.org/ iucr-top/comm/cteach/pamphlets/10/node7.html⟩.

[14] Rigault G, *Metric Tensor and Symmetry Operations in Crystallography*. Online at ⟨http://www.iucr.org/education/pamphlets/10/full-text⟩.

[15] Ladd MFC, *Int. J. Math. Educ. Sci. Tech.* **7**, 395 (1976).

Chapter 3

Lattices

*X-ray crystallography is nowadays an accurate
and rapid method of determining conformation in the
crystal lattice, which conformation usually corresponds
to the preferred conformation in solution.*
D. H. R. Barton

Key Topics

- Lattice defined
- Unit cell
- Bravais lattices
- Lattice directions
- Reciprocal lattice
- Unit-cell transformations
- Metric tensor
- Bilbao Crystallographic Server

3.1. Introduction

The next step in the study of crystal structure determination is the
consideration of the internal symmetry of crystals, which brings with
it the concept of the lattice. Lattices that arise in the study of crystallography can be defined in one, two or three dimensions. The idea
of a lattice is the same in each of these dimensions, the variation

lies in its spatial extent; crystals are based on one of several types of three-dimensional lattices.

3.2. Definition of Lattice

A crystal lattice may be defined as an infinite, regular, three-dimensional arrangement of points[a] in space, such that each point has the same environment as all other points. A lattice can be generated by three non-coplanar *translation vectors* along the crystallographic reference axes x, y and z. In each of these directions, the minimum distances of separation are denoted by the vectors $\mathbf{a}$, $\mathbf{b}$ and $\mathbf{c}$, respectively. A lattice is, then, an infinite, regular, linearly independent three-dimensional array of basic translation vectors.

A distance t between points in a lattice is defined in terms of the three basic translations. Thus, taking any lattice point as an origin, any other point can be reached by a translation vector $\mathbf{t}$, given with respect to that origin by

$$\mathbf{t} = u\mathbf{a} + v\mathbf{b} + w\mathbf{c} \qquad (3.1)$$

where u, v and w are the integer coordinates at the head of the vector, the 'other' lattice point.

In this discussion, the angles between the translation vectors are governed by the characteristic symmetry of the crystal. Referring to Table 2.2, the quantities a, b, c can now be seen to have an additional significance.

It may be noted that common terminologies such as 'sodium chloride lattice' and 'body-centred lattice' are misnomers. The first of these terms refers to a *crystal structure*, that is, a three-dimensional, periodic array of chemical entities, whereas the second term refers

---

[a] A lattice point may be regarded as a zero-dimensional object.

to the type of *unit cell* that has been chosen from within the given lattice to represent it. However, the usage is common, and may be applied with the above understanding.

3.3. Unit Cell

A *unit cell* is the smallest, three-dimensional repeating unit of its lattice. It is a region of space bounded by the basic translation vectors **a**, **b** and **c**, and one lattice point is associated with its volume; such a unit cell is termed *primitive* P (or p in two dimensions). Unit cells stacked together in three dimensions build up an entire lattice. Figure 3.1 illustrates the most general type of unit cell. It belongs to the triclinic system and is a primitive unit cell. It is necessary to remember that a unit cell is a representative portion of an infinite array of such cells; each point shown here contributes one-eighth of a point to the unit cell. A unit cell may contain more than one lattice point. In general, regular arrangements of points may

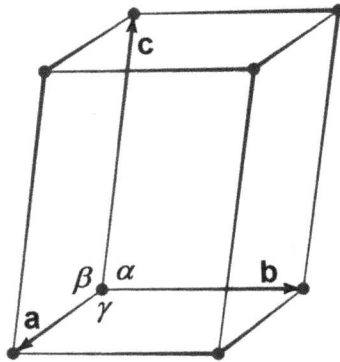

Fig. 3.1.   A unit cell of the most general (triclinic) lattice. Reiterating the conditions given in Table 2.2, $a \neq b \neq c$; $\alpha \neq \beta \neq \gamma \neq 90°$ or $120°$. Angles of $90°$ and $120°$ are associated with rotational symmetries 4, 3 and 6, and would not imply the most general case. One or more of these angles could, by chance, have a value of $90°$ or $120°$ in a triclinic crystal, within experimental error; but the symmetry of the structure governs the crystal system and it would remain triclinic.

imply lattice points at positions other than the corners of a cell, in which case their contributions to a single unit cell depend on their location:

Corners	1/8 each
Edges	1/4 each
Faces	1/2 each
Within the cell	1

It is possible to choose a unit cell larger than primitive, as Fig. 3.2 shows, and this is true for all lattices, but it would be unconventional.

A three-dimensional lattice representation can be imagined to be generated in stages. In Fig. 3.3(a), a line of identical bricks of infinitesimal size can be represented by a set of equally spaced

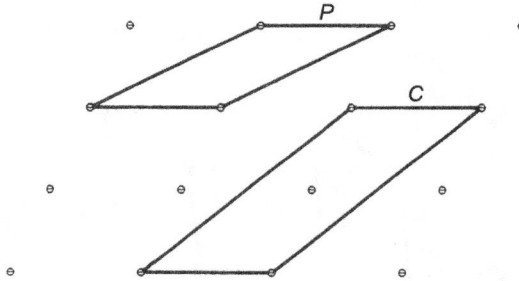

Fig. 3.2. Two unit cells in one and the same lattice, as seen in projection; P primitive (p in two dimensions) and centred C (c in two dimensions). The C cell has twice as many points per unit cell as does the P cell; the P unit cell is the conventional choice for this lattice.

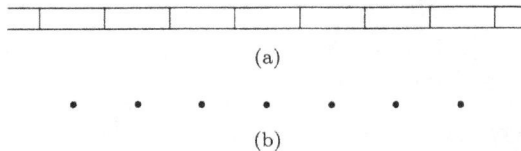

(a)

(b)

Fig. 3.3. One-dimensional lattice: (a) Line of identical, infinitesimal bricks; (b) Line of equally spaced points representing the bricks and illustrating a one-dimensional lattice or *row* of spacing, say b. The simplest unit cell contains one lattice point (brick).

(a)

(b)

Fig. 3.4. Two-dimensional lattice: (a) Wall of identical, infinitesimal bricks; (b) Array of points regularly arranged in two dimensions representing the bricks and forming a two-dimensional lattice or *net* of spacings a and b. The smallest unit cell is a primitive rhombus, $a = b$, $\gamma \neq 90°$ or $120°$. Conventionally, the preferred crystallographic unit cell is a centred, rectangular cell, $a \neq b$, $\gamma = 90°$.

Fig. 3.5. Stereoscopic view of a three-dimensional lattice: Nets of points, as shown in Fig. 3.4(b), are arranged regularly at a spacing c, non-coplanar with a and b, to form the *lattice*. The smallest unit cell is primitive, as in Fig. 3.1. (This diagram is best seen through a stereoviewer.)

points that constitute a one-dimensional lattice or *row*, as shown in Fig. 3.3(b).

A number of similar bricks arranged regularly in two dimensions, as in Fig. 3.4(a), would build a wall, which may be represented by a two-dimensional lattice or *net*, as illustrated in Fig. 3.4(b).

Finally, stacking a number of such nets regularly in a third dimension, non-collinear with the two dimensions already defined, leads to a *lattice* on which a crystal may be based. Figure 3.5 is a stereoview of the most general lattice; it belongs to the triclinic crystal system.

3.4. Bravais Lattices and Conventional Unit Cells

Bravais determined that there are 14 different arrangements of points that can be defined in space within the allowed crystal symmetries [1]. It is conventional, and desirable, to select the basic translation vectors **a**, **b** and **c** in a close relationship with the crystal symmetry. This principle is evident from Table 2.2; the lattices so devised are termed *Bravais lattices*, and they include certain *centred* (non-primitive) arrangements. The centring conditions for most of the lattices present no particular difficulties, and the unit cells of the Bravais lattices are shown in Fig. 3.6.

The centring of a unit cell must comply with the definition of a lattice. A unit cell such as that shown in projection in Fig. 3.7 is described as primitive, with two lattice points per unit cell: a vector equivalent to **OP** set with its tail at the point P does not terminate at another lattice point; thus, it is not any form of centred arrangement.

The interrelationship of the hexagonal and trigonal systems makes their lattices a little more complicated than those in the other crystal systems. A pattern based on six-fold rotational symmetry introduces three-fold rotations, as the following diagram shows; it is a projection of a hexagonal lattice on to the x-y plane of its unit cell; the arrays of six representative points highlight the positions of three-fold and six-fold symmetry axes. Thus, the hexagonal lattice can be used for certain trigonal crystals.

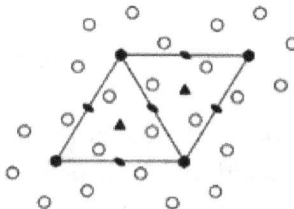

Six-fold and three-fold rotational symmetry

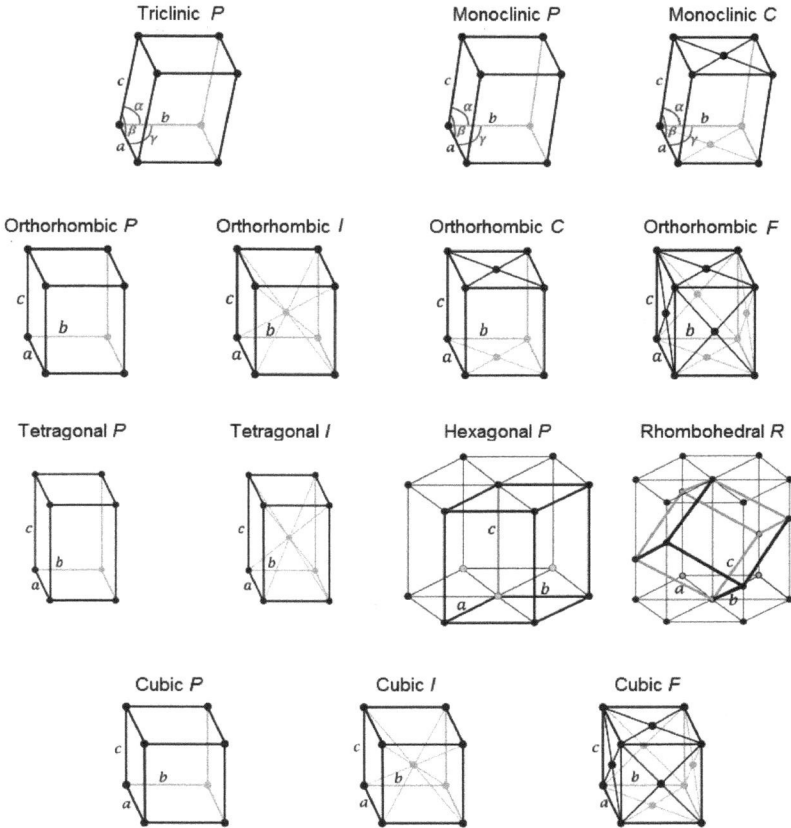

Fig. 3.6. The 14 Bravais lattices. Lattices in the trigonal system can be referred to either a hexagonal P unit cell, or a rhombohedral R unit cell. If the trigonal system is referred to hexagonal axes, the R_{hex} unit cell contains three lattice points.

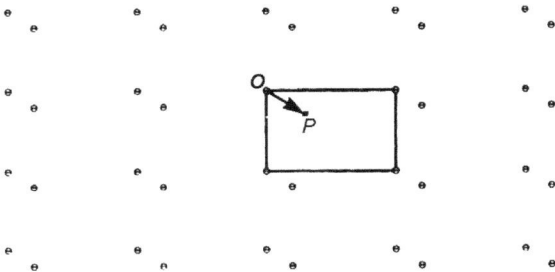

Fig. 3.7. A rectangular unit cell with two lattice points per unit cell. A true lattice has all lattice points in identical environments.

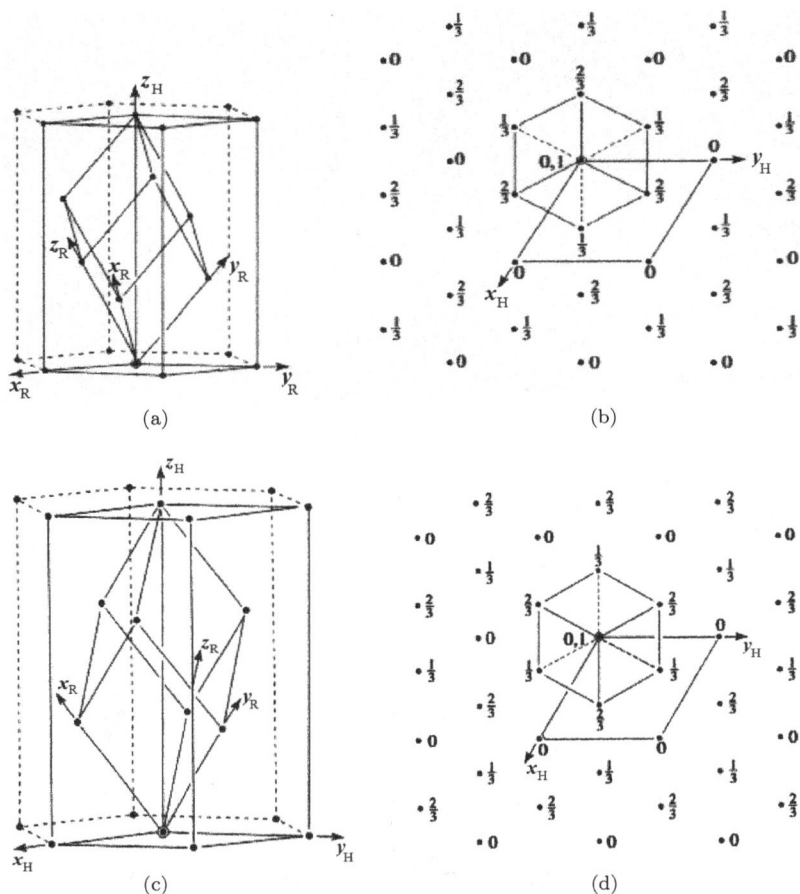

Fig. 3.8. Triply-primitive hexagonal unit cells in the rhombohedral lattice. (a) The *obverse* (standard) setting of the R unit cell as developed from a triply primitive hexagonal H unit cell. (b) The obverse setting as seen along the z_H-axis; the fractions refer to values of c_{hex}. (c) The *reverse* setting; the rhombohedral lattice is rotated clockwise about [111] by 60°. (d) The obverse setting as seen along the z_H-axis. The ratio of the volumes of two cells in one and the same lattice is equal to the ratio of the number of lattice points per unit cell.

The rhombohedral unit cell can be oriented within a hexagonal unit cell in two ways. The standard setting is the *obverse*, as shown in Figs. 3.8(a) and 3.8(b). The alternative, *reverse* setting, is obtained by rotating the rhombohedral unit cell clockwise by 60° about the vertical axis to give the orientation shown in Figs. 3.8(c) and 3.8(d).

When hexagonal axes are employed for a trigonal crystal, the smallest hexagonal unit cell is not primitive; the unique lattice points are at 0, 0, 0; 1/3, 2/3, 2/3; 2/3, 1/3, 1/3. The situation may be clarified by considering points related by three-fold symmetry in the three situations. In a P unit cell in a hexagonal lattice, the unique coordinates related by three-fold symmetry are

$$x, y, z; \ \bar{y}, x - y, z; \ y - x, \bar{x}, z$$

whereas in a rhombohedral R unit cell, the same symmetry delivers the unique points

$$x, y, z; \ z, y, x; \ y, z, x$$

The rhombohedral R unit cell referred to hexagonal axes, which may be termed R_{hex}, has the unique coordinates of the P unit cell together with two additional points obtained by adding the translations 1/3, 2/3, 2/3 and 2/3, 1/3, 1/3 to the coordinates of the P unit cell, leading to the triply primitive hexagonal unit cell.

The R unit cell may be thought of as a cubic unit cell extended along [111]; it is the only unit cell that contains more than one lattice point lying completely within it. Figure 3.9 shows a stereoscopic view of the rhombohedral unit cell within a hexagonal unit cell.

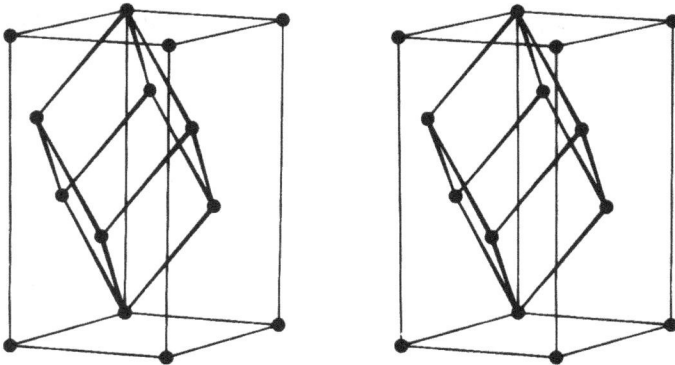

Fig. 3.9. Stereoview of the obverse rhombohedral R unit cell.

Self-assessment 3.1. (a) This figure can be found in some
literature as an example of a hexagonal lattice in projection on
to the x-y plane, with lattice points at each intersection. Is this
a satisfactory description? Comment.

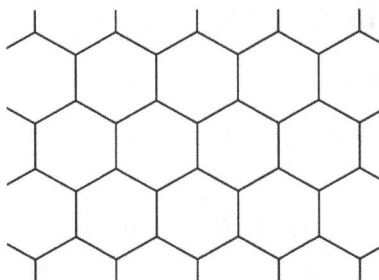

'Hexagonal' lattice

(b) What is the result of centring the basal (a-b) planes of a
close-packed array of primitive (P) hexagonal unit cells?[b]

A crystal can be considered as a three-dimensional stacking of
identical unit cells. Each of the seven crystal systems has its own
unique values of the six parameters a, b, c, α, β and γ. The conditions
attached to the nets and the lattices are governed by the symmetry
of the crystal. Table 3.1 lists unit-cell data for the nets and Table 3.2
gives the corresponding details for the Bravais lattices. The volume

Table 3.1. The five nets.

System name	Unit-cell symbol	Symmetry at each lattice point	Unit-cell parameters
Oblique	p	2	$a \neq b$; $\gamma \neq 90°, 120°$
Rectangular	p, c	$2\,mm$	$a \neq b$; $\gamma = 90°$
Square	p	$4\,mm$	$a = b$; $\gamma = 90°$
Hexagonal	p	$6\,mm$	$a = b$; $\gamma = 120°$

---

[b] Answers to all self-assessments are given at the end of the chapter.

Table 3.2.   The Bravais lattices.[a]

Unit-cell centring sites	Unit-cell symbol and name	h, k, l indices of centred faces	Unit-cell translations dependent on centring
None	P, primitive	—	0, 0, 0
None	R, primitive (rhombohedral)	—	0, 0, 0
	R_{hex}^{b}	—	0, 0, 0; 1/3, 2/3, 2/3; 2/3, 1/3, 1/3
b_c faces	A, A-centred	(100)	0, 0, 0; 0, 1/2, 1/2
c_a faces	B, B-centred	(010)	0, 0, 0; 1/2, 0, 1/2
a_b faces	C, C-centred	(001)	0, 0, 0; 1/2, 1/2, 0
Body	I, body centred	—	0, 0, 0; 1/2, 1/2, 1/2
All faces	F, face centred	(100), (010) and (001)	0, 0, 0; 0, 1/2, 1/2; 1/2, 0, 1/2; 1/2, 1/2, 0

[a]See Table 2.2 for the relationships between a, b and c, and between α, β and γ.
[b]Trigonal crystal referred to hexagonal axes; pseudo-centring points arise at 1/3, 2/3, 2/3 and 2/3, 1/3, 1/3.

of a unit cell is given by the vector product $\mathbf{a} \cdot (\mathbf{b} \times \mathbf{c})$, as discussed in Appendix A1.3; in two dimensions, it is $ab \sin \gamma$. The most prominent faces of a crystal are those parallel to internal planes having the greatest density of lattice points, a rule stated by Bravais. Further discussions on the Bravais lattices can be found in the literature [2–4].

3.4.1.   Wigner–Seitz cells

In certain theoretical studies and in spectroscopy, it is often desirable to work with a primitive unit cell whatever be the crystal system. A primitive, space-filling unit cell for each of the 14 Bravais unit cells is defined by a Wigner–Seitz cell. However, there is a total of 24 Wigner–Seitz cells because in some crystal systems, the shape of the Wigner–Seitz cell is governed by the axial ratios, a/b and c/b [2]. Figure 3.10 illustrates the Wigner–Seitz cell obtained from a cubic I unit cell. It is obtained by first drawing lines from a lattice point, conveniently that at the centre of the unit cell, to its nearest neighbour lattice points. Then, planes are constructed to bisect these lines perpendicularly, and then extended, if necessary, to form a closed polyhedron. The cell contains a single lattice point and is, therefore,

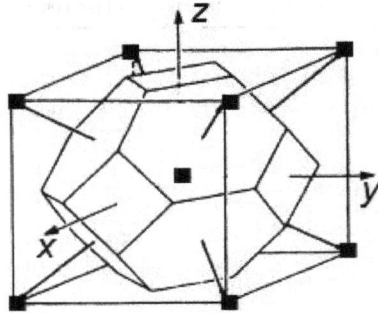

Fig. 3.10. Wigner–Seitz cell developed from a body-centred cubic unit cell; it has the shape of a truncated octahedron and is a space-filling solid.

of the smallest volume for that lattice, and it can be packed so as to fill space completely.

3.5. Lattice Directions

A *direction*, or *directed line*, in a lattice between two lattice points was introduced earlier by Eq. (3.1). In this equation, the triplet of integers $[uvw]$ may be termed the coordinates of the lattice point and can be positive or negative. The notation is similar to that used for a zone symbol: in general, a direction $[uvw]$ is not normal to the plane of those indices. A form of directions is represented as $\langle uvw \rangle$. The distance d from the origin of the lattice to any general point x, y, z in the unit cell is given by

$$\mathbf{d} = x\mathbf{a} + y\mathbf{b} + z\mathbf{c} \tag{3.2}$$

which, in vector notation, is written as

$$\mathbf{d} = (\mathbf{a}\ \mathbf{b}\ \mathbf{c}) \begin{pmatrix} x \\ y \\ z \end{pmatrix} \tag{3.3}$$

In these equations, x, y and z are *fractional coordinates*, that is, $x = X/|\mathbf{a}|$, where X is in absolute measure and $|\mathbf{a}|$ is the unit-cell dimension along X; d is evaluated as the scalar product $\mathbf{d} \cdot \mathbf{d}$.

Example 3.1. A monoclinic unit cell has the dimensions $a = 10.00\,\text{Å}$, $b = 5.000\,\text{Å}$, $c = 15.00\,\text{Å}$ and $\beta = 110.0°$. What is the length of the direction $[31\bar{2}]$? From Eq. (3.1), the distance may be evaluated as $\mathbf{d}\cdot\mathbf{d}$ for the monoclinic system giving $d^2 = a^2 + b^2 + c^2 + 2ca\cos\beta$, since $\alpha = \gamma = 90°$. Then,

$$d^2[31\bar{2}] = (3 \times 10.00\,\text{Å})^2 + (1 \times 5.000\,\text{Å})^2 + (-2 \times 15.00\,\text{Å})^2$$
$$+ 2(-2 \times 15.00\,\text{Å} \times 3 \times 10.00\,\text{Å})$$
$$\times (-0.34202) = 2440.6\,\text{Å}^2$$

so that $d = 49.40\,\text{Å}$.

3.6. Rotational Symmetries of Crystal Lattices

In Section 2.4.3, the permitted crystal rotational symmetries were discussed in terms of the space-filling properties of solids. An alternative proof of the *crystallographic restriction theorem* can be given in the following terms.

Consider the counterclockwise rotation about the z-axis of a vector $\mathbf{r}$ through an angle θ as shown in the diagram, assuming Cartesian axes.

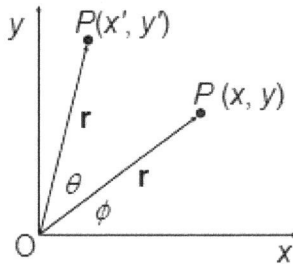

Then, $x = r\cos\phi$ and $y = r\sin\phi$. After the rotation through an angle, $\theta, x' = r\cos(\phi + \theta) = r\cos\phi\cos\theta - r\sin\phi\sin\theta = x\cos\theta - y\sin\theta$. Similarly, the y-coordinate after rotation is given by $y = x\sin\theta + y\cos\theta$; z is unchanged throughout. Thus, the matrix for this right-handed rotation through an angle θ acting on a point

x, y, z may be written as

$$(x' \; y' \; z') = \begin{pmatrix} \cos\theta & -\sin\theta & 0 \\ \sin\theta & \cos\theta & 0 \\ 0 & 0 & 1 \end{pmatrix} \begin{pmatrix} x \\ y \\ z \end{pmatrix} \qquad (3.4)$$

The trace of the matrix, following Appendix A1.7, is $2\cos\theta + 1$, and in the case of a symmetry operation, the trace is always integral [5]. Hence, $2\cos\theta + 1 = m$, where m is an integer, or $2\cos\theta = M$, where M is another integer. Since $|\cos\theta| \leq 1$, the analysis leads to the following results:

M	-2	-1	0	1	2
θ (deg)	180	120	90	60	0 (360)
R	2	3	4	6	1

3.7. Reciprocal Lattice

It is appropriate to consider the reciprocal lattice at this stage as it will be needed when discussing X-ray diffraction in the subsequent chapters. The idea of reciprocal vectors was introduced by Gibbs (*ca.* 1881). He derived *inter alia* a quantity $\mathbf{r}$ given by

$$\mathbf{r} = \frac{\mathbf{b} \times \mathbf{c}}{\mathbf{a} \cdot (\mathbf{b} \times \mathbf{c})} \qquad (3.5)$$

which is a vector perpendicular to the $b\text{-}c$ plane [6] and may be termed a *reciprocal vector*. The application of reciprocal vectors and the *reciprocal lattice* in X-ray crystallography was devised by Bernal [7] for the interpretation of X-ray diffraction records and developed in detail by Ewald [8].

The reciprocal lattice will be considered first geometrically. Figure 3.11(a) shows the projection of a monoclinic lattice on to the $x\text{-}z$ plane; $\mathbf{a}$ and $\mathbf{c}$ are the basic translation vectors of a primitive unit cell of the lattice. Three families of planes are illustrated by the planes (100), (001) and (101) and lines OP, OR and OQ

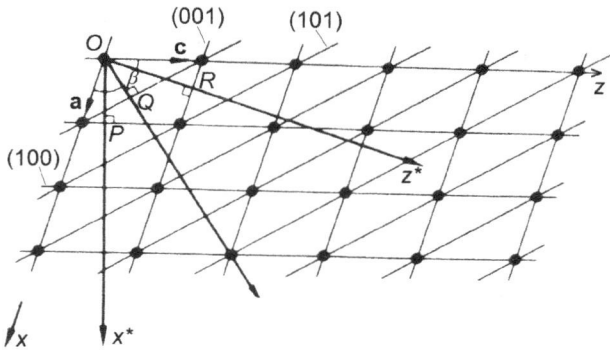

(a)

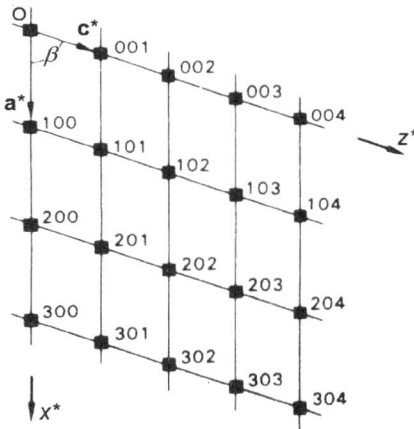

(b)

Fig. 3.11. Reciprocal lattice. (a) Monoclinic lattice in projection on to the $x\text{-}y$ plane; primitive unit cells are shown together with traces of the planes (100), (001) and (101). (b) The reciprocal lattice developed from (a); $a^* = d^*_{100} = 1/d_{(100)} = 1/(a \sin \beta)$, $b^* = 1/b$, $c^* = d^*_{001} = 1/d_{(001)} = 1/(c \sin \beta)$.

are drawn from the origin O normal to these planes. Along each of these lines, distances are defined that are inversely proportional to the corresponding interplanar distances in the Bravais (real) lattice, thus forming the points of the reciprocal lattice shown in Fig. 3.11(b). The general relationship is

$$d^*(hkl) = K/d(hkl) \tag{3.6}$$

where K is a constant and equal to unity in this discussion. The vectors $\mathbf{d}^*(100)$, $\mathbf{d}^*(010)$ and $\mathbf{d}^*(001)$ define the translation vectors for a unit cell in the reciprocal lattice.

From Eq. (3.6), $d^*(100) = 1/d(100) = a^*$; then, since $d(100) = a \sin \beta$, it follows that

$$a^* = \frac{1}{a \sin \beta} \tag{3.7}$$

By a similar argument,

$$c^* = \frac{1}{c \sin \beta} \tag{3.8}$$

But

$$b^* = \frac{1}{b} \tag{3.9}$$

because d^*_{010} is normal to the a–c plane. The β^* angle is given by

$$\beta^* = 180° - \beta \tag{3.10}$$

Also, following Example 3.1, the scalar product

$$\mathbf{a} \cdot \mathbf{a}^* = a\,a^* \cos \mathbf{a}\,\mathbf{a}^* = a \left(\frac{1}{a \sin \beta} \right) \cos(\beta - 90°) = 1 \tag{3.11}$$

since $\cos(\beta - 90) = \sin \beta$, whereas the scalar product

$$\mathbf{a} \cdot \mathbf{b}^* = ab^* \cos \mathbf{a}\,\mathbf{b}^* = 0 \tag{3.12}$$

with similar results *mutatis mutandis* for all other such products. While Eqs. (3.7)–(3.10) apply to the monoclinic example, Eqs. (3.11) and (3.12) are valid for all crystal systems.

A more general treatment of the reciprocal lattice follows from Fig. 3.12. In this figure, the z^*-axis is normal to the a–b plane. Then, $c^* = |\mathbf{c}^*| = \frac{1}{c \cos ROC}$. Since $\mathbf{c}^*$ is normal to both $\mathbf{a}$ and $\mathbf{b}$, its direction is that of the vector product $(\mathbf{a} \times \mathbf{b})$, and may be written as

$$\mathbf{c}^* = \kappa(\mathbf{a} \times \mathbf{b}) \tag{3.13}$$

where κ is a constant — not the K of Eq. (3.6). From Appendix A1.3, a unit-cell volume V_c may be written as $\mathbf{c} \cdot (\mathbf{a} \times \mathbf{b})$, and using

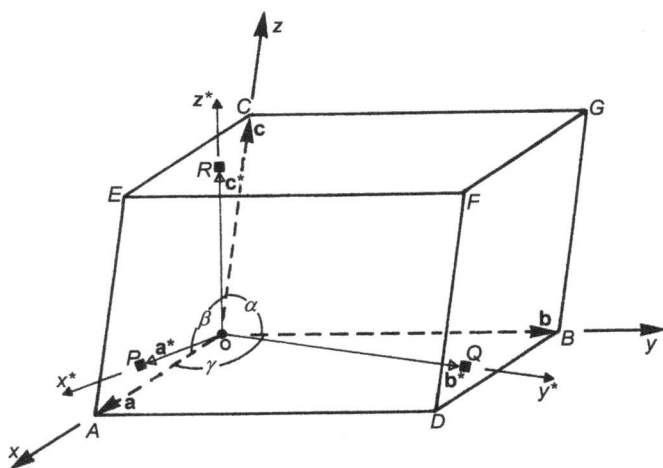

Fig. 3.12.   General reciprocal lattice: a primitive unit cell of a real space triclinic lattice defined by the parameters $\mathbf{a}$, $\mathbf{b}$, $\mathbf{c}$, α, β and γ. The axes x^*, y^* and z^* are normal to the b_c, c_a and a_b planes, respectively; $\mathbf{a}^*$, $\mathbf{b}^*$ and $\mathbf{c}^*$ are the basic translation vectors of the reciprocal lattice.

Eq. (3.11), the scalar product $\mathbf{c} \cdot \mathbf{c}^*$ becomes

$$\mathbf{c} \cdot \mathbf{c}^* = \kappa \mathbf{c} \cdot (\mathbf{a} \times \mathbf{b}) = \kappa V_c = 1 \qquad (3.14)$$

Thus, $\kappa = 1/V_c$, so that it follows from Eq. (3.14) that $\mathbf{c}^* = (\mathbf{a} \times \mathbf{b})/V_c$ which in vector form, and with similar relationships for $\mathbf{a}^*$ and $\mathbf{b}^*$, leads to

$$
\begin{aligned}
\mathbf{c}^* &= \frac{(\mathbf{a} \times \mathbf{b})}{\mathbf{c} \cdot (\mathbf{a} \times \mathbf{b})} = \frac{ab \sin \gamma}{V_c} \\[1mm]
\mathbf{a}^* &= \frac{(\mathbf{b} \times \mathbf{c})}{\mathbf{a} \times (\mathbf{b} \times \mathbf{c})} = \frac{bc \sin \alpha}{V_c} \\[1mm]
\mathbf{b}^* &= \frac{(\mathbf{c} \times \mathbf{a})}{\mathbf{b} \cdot (\mathbf{c} \times \mathbf{a})} = \frac{ca \sin \beta}{V_c}
\end{aligned}
\qquad (3.15)
$$

The angles α^*, β^*, γ^* of the reciprocal unit cell may be obtained by the equations of spherical trigonometry (Appendix A5). Alternatively, in Fig. 3.13, the angle β^*, for example, is the angle between $\mathbf{c}^*$ and $\mathbf{a}^*$. In the spherical triangle, ABC, OA, OB and OC are lengths of the translation vectors of the real space unit cell; its sides are the

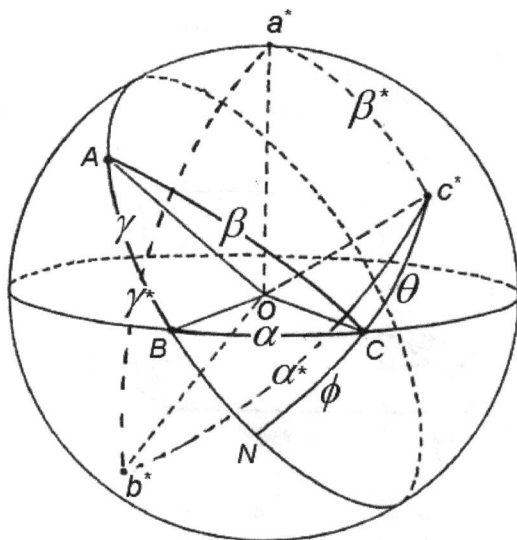

Fig. 3.13.   Derivation of the angles α^*, β^* and γ^* of the reciprocal unit cell.

interaxial angles α, β, γ in real space. Similarly, OA^*, OB^* and OC^* are the lengths of the translation vectors of the reciprocal unit cell. Now, OC^* is normal to the inclined great circle through A and B, and similarly with OA^* and OB^*. Hence, $A^*B^*C^*$ is a spherical triangle polar to triangle ABC, so that β^* is the angle between the great circles through AB and BC, or $\pi - B$, where B is the angle of the spherical triangle ABC at B. Then, from the formulae of spherical trigonometry,

$$\cos \alpha^* = -\cos A = (\cos \beta \cos \gamma - \cos \alpha)/\sin \beta \sin \gamma$$
$$\cos \beta^* = -\cos B = (\cos \gamma \cos \alpha - \cos \beta)/\sin \gamma \sin \alpha \quad (3.16)$$
$$\cos \gamma^* = -\cos C = (\cos \alpha \cos \beta - \cos \gamma)/\sin \alpha \sin \beta$$

Simpler expressions are obtained when the crystal symmetry is higher than triclinic. An alternative derivation of these angles is given in Appendix A6.

It is shown next that the reciprocal lattice is a true lattice. From Eq. (A1.40), the vector $\mathbf{n}$ normal to a plane of Miller indices h, k and l in the real space unit cell is given by

$$\mathbf{n} = h(\mathbf{b} \times \mathbf{c}) + k(\mathbf{c} \times \mathbf{a}) + l(\mathbf{a} \times \mathbf{b}) \qquad (3.17)$$

Dividing throughout by the unit-cell volume V_c and from (3.15), the resulting vector $\mathbf{d}^*(hkl)$ is

$$\mathbf{d}^*_{hkl} = \left[h\frac{(\mathbf{b} \times \mathbf{c})}{\mathbf{a} \times (\mathbf{b} \times \mathbf{c})} + k\frac{(\mathbf{c} \times \mathbf{a})}{\mathbf{b} \cdot (\mathbf{c} \times \mathbf{a})} + l\frac{(\mathbf{a} \times \mathbf{b})}{\mathbf{c} \cdot (\mathbf{a} \times \mathbf{b})} \right]$$

$$= h\mathbf{a}^* + k\mathbf{b}^* + l\mathbf{c}^* \qquad (3.18)$$

Note that reciprocal lattice points are denoted by the h, k and l indices of the family of planes from which they are derived, but written without parentheses.

Self-assessment 3.2. A unit cell has the dimensions $a = 5\,\text{Å}$, $b = 6\,\text{Å}$, $c = 7\,\text{Å}$, $\alpha = 90°$, $\beta = 100°$, $\gamma = 110°$. Calculate (a) the unit-cell volume and (b) the dimensions and volume of the reciprocal unit cell, all to four significant figures.

3.8.   Unit-Cell Transformations

In practice, a transformation of the reference axes may be found to be desirable for reasons such as convention or convenience. In general, one or more of the following parameters relating to the setting of the axes may need to be transformed: $\mathbf{a}$, $\mathbf{b}$, $\mathbf{c}$; h, k, l; $\mathbf{a}^*$, $\mathbf{b}^*$, $\mathbf{c}^*$; U, V, W; u, v, w; x, y, z; x^*, y^*, z^*.

A general transformation comprises two parts: a *linear* part and an origin-shift part. In this discussion, the linear part is a 3×3 matrix $\mathbf{S}$ of elements s_{ij} that represents a change in the length or the orientation of the basic vectors $\mathbf{a}$, $\mathbf{b}$ and $\mathbf{c}$; the change of origin procedure will be discussed at a later stage.

The position vector $\mathbf{r}$ defined by Eq. (3.2) can be written in the form

$$\mathbf{r} = (\mathbf{a}\ \mathbf{b}\ \mathbf{c}) \begin{pmatrix} x \\ y \\ z \end{pmatrix} \tag{3.19}$$

Let $\mathbf{a}$, $\mathbf{b}$ and $\mathbf{c}$ be transformed to $\mathbf{a}'$, $\mathbf{b}'$ and $\mathbf{c}'$, respectively, such that

$$\begin{aligned} \mathbf{a}' &= s_{11}\mathbf{a} + s_{12}\mathbf{b} + s_{13}\mathbf{c} \\ \mathbf{b}' &= s_{21}\mathbf{a} + s_{22}\mathbf{b} + s_{23}\mathbf{c} \\ \mathbf{c}' &= s_{31}\mathbf{a} + s_{32}\mathbf{b} + s_{33}\mathbf{c} \end{aligned} \tag{3.20}$$

The transformation may be written generally, using row vectors, as

$$(\mathbf{a}\ \mathbf{b}\ \mathbf{c}) \begin{pmatrix} s_{11} & s_{21} & s_{31} \\ s_{12} & s_{22} & s_{32} \\ s_{13} & s_{23} & s_{33} \end{pmatrix} = (\mathbf{a}'\ \mathbf{b}'\ \mathbf{c}') \tag{3.21}$$

or concisely as

$$(\mathbf{a}\ \mathbf{b}\ \mathbf{c})\mathbf{S}^{\mathrm{T}} = (\mathbf{a}'\ \mathbf{b}'\ \mathbf{c}') \tag{3.22}$$

where $\mathbf{S}^{\mathrm{T}}$ is the transpose of $\mathbf{S}$, from Eq. (3.20). A row matrix may be thought of as a 3×3 matrix in which each of the second and third *column* elements is zero.

Some sources write the transformation, using column vectors, as

$$\begin{pmatrix} \mathbf{a}' \\ \mathbf{b}' \\ \mathbf{c}' \end{pmatrix} = \begin{pmatrix} s_{11} & s_{21} & s_{31} \\ s_{12} & s_{22} & s_{32} \\ s_{13} & s_{23} & s_{33} \end{pmatrix} \begin{pmatrix} \mathbf{a} \\ \mathbf{b} \\ \mathbf{c} \end{pmatrix} \tag{3.23}$$

or

$$\begin{pmatrix} \mathbf{a}' \\ \mathbf{b}' \\ \mathbf{c}' \end{pmatrix} = \mathbf{S} \begin{pmatrix} \mathbf{a} \\ \mathbf{b} \\ \mathbf{c} \end{pmatrix} \tag{3.24}$$

It is clear that both procedures give identical results, but the procedure of Eq. (3.22) is conventional practice.

3.8.1. Metric tensor

The metric tensor provides for a neat computational method for obtaining *inter alia* the distance between two points, the angle between two vectors, and related properties. Some of these calculations have been carried out earlier by geometrical arguments.

For a unit cell with translation vectors **a**, **b** and **c** and angles $\alpha = \widehat{|\mathbf{b}|\,|\mathbf{c}|}$, $\beta = \widehat{|\mathbf{c}|\,|\mathbf{a}|}$, $\gamma = \widehat{|\mathbf{a}|\,|\mathbf{b}|}$, the *metric tensor* G (aka metric matrix) is given by

$$G = \begin{pmatrix} a^2 & ab\cos\gamma & ca\cos\beta \\ ba\cos\gamma & b^2 & bc\cos\alpha \\ ca\cos\beta & cb\cos\alpha & c^2 \end{pmatrix} \quad (3.25)$$

The distance from a vector **p** to a vector **q** in a unit cell with translation vectors a_i, for $i = 1-3$ ($\equiv a$, b and c), is given by $\mathbf{q} - \mathbf{p}$, where

$$\mathbf{p} = \sum_{i=1}^{3} p_i a_i \quad (3.26)$$

and

$$\mathbf{q} = \sum_{j=1}^{3} q_j a_j \quad (3.27)$$

Then, the distance $\mathbf{d} = \mathbf{q} - \mathbf{p}$ is given by

$$\mathbf{d} = \sum_{i,j=1}^{3} p_i a_i G q_j a_j \quad (3.28)$$

Example 3.2. A monoclinic unit cell has vectors **a**, **b** and **c** of length 5.000 Å, 10.00 Å and 15.00 Å, respectively, and the β-angle is 110.0°. What is the length d of the body diagonal of the cell? The vector **p** may be taken to start at the origin with **q** at the

Example 3.2. (*Continued*) other end of the vector; then $\mathbf{d} = \mathbf{q} - \mathbf{p}$ defines the direction [111]. Hence, from the foregoing,

$$d^2 = (\mathbf{q} - \mathbf{p}) \cdot (\mathbf{q} - \mathbf{p})$$

$$= [1\ 1\ 1] \begin{pmatrix} 25 & 0 & -25.6515 \\ 0 & 100 & 0 \\ -25.6515 & 0 & 225 \end{pmatrix} \begin{bmatrix} 1 \\ 1 \\ 1 \end{bmatrix}$$

$$= [1\ 1\ 1] \begin{pmatrix} -0.6515 \\ 100 \\ 199.3485 \end{pmatrix} = 298.697\,\text{Å}^2$$

so that $d = 17.28\,\text{Å}$ (all quantities in the matrix tensor G have the dimensions of Å^2).

3.8.2.   Centred unit cells

The conventional (Bravais) unit cells are shown in Fig. 3.6. Could there be other centred unit cells? Figure 3.14 shows two adjacent B-centred unit cells. A smaller unit cell can be delineated with the transformation

$$\mathbf{a}' = \mathbf{a} \tag{3.29}$$
$$\mathbf{b}' = \mathbf{b} \tag{3.30}$$
$$\mathbf{c}' = -\mathbf{a}/2 + \mathbf{c}/2 \tag{3.31}$$

Example 3.3. Let a monoclinic B unit cell have the exact dimensions $a = 6\,\text{Å}$, $b = 7\,\text{Å}$, $c = 8\,\text{Å}$, $\beta = 110°$. It is evident that a and b are unchanged by the above transformation; new values are required for c' and β'. The unit-cell transformation is given by $(a', b', c') = (a, b, c)\mathbf{S}^{\mathrm{T}}$:

$$(a'\ b'\ c') = (a\ b\ c) \underbrace{\begin{pmatrix} 1 & 0 & \overline{1/2} \\ 0 & 1 & 0 \\ 0 & 0 & 1/2 \end{pmatrix}}_{\mathbf{S}^{\mathrm{T}}} \begin{pmatrix} a \\ b \\ c \end{pmatrix}$$

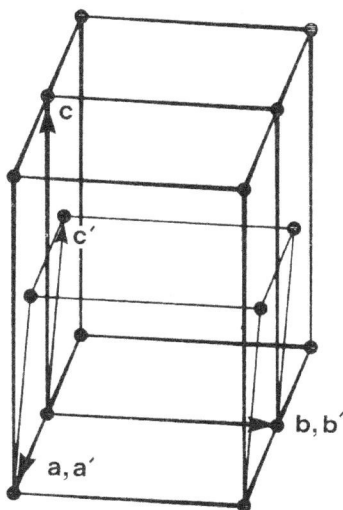

Fig. 3.14. A *B*-centred unit cell in a monoclinic lattice can be transformed to a *P* unit cell within one and the same lattice.

Example 3.3. (*Continued*) (a) The new value of c' is given by $(-0.5\ 0\ 0.5)\, G (-0.5\ 0\ 0.5)^{\mathrm{T}}$, where T implies the transpose of the row vector. Calculating c' with the program METTENS gives

$$c'^2 = (-0.5\ \ 0\ \ 0.5) \begin{pmatrix} 36.00 & 0 & -16.417 \\ 0 & 49.00 & 0 \\ -16.417 & 0 & 64.00 \end{pmatrix} \begin{pmatrix} -0.5 \\ 0 \\ 0.5 \end{pmatrix}$$

$$= (-0.5\ \ 0\ \ 0.5) \begin{pmatrix} -26.2085 \\ 0 \\ 40.2085 \end{pmatrix} \begin{pmatrix} -0.5 \\ 0 \\ 0.5 \end{pmatrix} = 33.2085\,\text{Å}^2$$

hence, $c' = 5.7627\,\text{Å}$.

(b)

$$\cos \beta' = \frac{\mathbf{a}' \cdot \mathbf{c}'}{a'c'} = \frac{1}{ac'}\, \mathbf{a} \cdot (-\mathbf{a}/2 + \mathbf{c}/2) = \frac{1}{ac'} \left(-\frac{a^2}{2} + \frac{ac\cos\beta}{2} \right)$$

$$= \frac{-18\,\text{Å}^2 - 8.20085\,\text{Å}^2}{6.000\,\text{Å} \times 5.7627\,\text{Å}} = -0.75799,$$

so that $\beta' = 139.29°$.

This procedure for calculating c' may seem unduly elaborate. Indeed, the answer can be seen to be $\sqrt{(\frac{-\mathbf{a}}{2} + \frac{\mathbf{c}}{2}) \cdot (\frac{-\mathbf{a}}{2} + \frac{\mathbf{c}}{2})} =$ $(\frac{a^2}{4} + \frac{c^2}{4} - \frac{ac\cos\beta}{2})^{1/2}$. However, the value of the metric tensor is that it lends itself to straightforward programming for unit-cell transformations in all crystal systems. This calculation may be carried out by the program METTENS in the Program Suite.

In considering the possibility of a centred cell in a crystal system giving rise to a new lattice, the questions relating to the centred cell, in the order of priority, are as follows:

- Does the centred lattice represent a true lattice?
- If it is a lattice, is the symmetry of the centred cell different from that which is characteristic of the crystal system?
- If the symmetry is unchanged, does the centred cell represent a new lattice, which is equivalent to asking if the unit-cell parameters obey the relationships in Table 2.3 for the given system?

Self-assessment 3.3. (a) Centre a primitive orthorhombic unit cell on the B faces. Does it represent a new lattice? (b) Carry out the same exercise with a primitive tetragonal unit cell.

3.8.3.   Transformation of Miller indices and coordinates in reciprocal space

Both Miller indices and the x^*, y^*, z^* coordinates in reciprocal space transform in the same way as the real space vectors (Section 3.8.2); they are termed *covariant* with the real space vectors.

Self-assessment 3.4. A unit cell transforms as $\mathbf{a}' = \mathbf{a} + \mathbf{2b}$, $\mathbf{b}' = -\mathbf{a} + \mathbf{b}, \mathbf{c}' = \mathbf{c}$. Transform the Miller indices $(\bar{3}1\bar{2})$ to their values in the transformed unit cell.

3.8.4. Transformation of lattice directions

Lattice directions $[uvw]$ and the parameters' reciprocal cell vectors $\mathbf{a}^*$, $\mathbf{b}^*$, $\mathbf{c}^*$ and coordinates x, y, z in real space are *contravariant* with the unit-cell vectors. The transformation of a direction $[uvw]$ to $[u'v'w']$ may be derived by the following argument.

A direction $\mathbf{r}_{[uvw]}$ in a lattice is written with respect to a unit cell $\mathbf{a}$, $\mathbf{b}$, $\mathbf{c}$ and the transformed cell $\mathbf{a}'$, $\mathbf{b}'$, $\mathbf{c}'$ as

$$\mathbf{r}_{[uvw]} = u\mathbf{a} + v\mathbf{b} + w\mathbf{c} = u'\mathbf{a}' + v'\mathbf{b}' + w'\mathbf{c}' \qquad (3.32)$$

Substituting for $\mathbf{a}'$, $\mathbf{b}'$ and $\mathbf{c}'$, using the transformation from Self-assessment 3.4 gives

$$u\mathbf{a} + v\mathbf{b} + w\mathbf{c} = u'(\mathbf{a} + 2\mathbf{b}) + v'(-\mathbf{a} + \mathbf{b}) + \mathbf{w}'\mathbf{c}$$

Collecting terms in $\mathbf{a}$, $\mathbf{b}$ and $\mathbf{c}$ leads to

$$u = u' - v', \quad v = 2u' + b', \quad w = w' \qquad (3.33)$$

which, in matrix form, is

$$\begin{pmatrix} u \\ v \\ w \end{pmatrix} = \begin{pmatrix} 1 & \bar{1} & 0 \\ 2 & 1 & 0 \\ 0 & 0 & 1 \end{pmatrix} \begin{pmatrix} u' \\ v' \\ w' \end{pmatrix} \qquad (3.34)$$

Hence, the required transformation is the inverse, that is,

$$\begin{pmatrix} u' \\ v' \\ w' \end{pmatrix} = \begin{pmatrix} 1/3 & 1/3 & 0 \\ 2/3 & 1/3 & 0 \\ 0 & 0 & 1 \end{pmatrix} \begin{pmatrix} u \\ v \\ w \end{pmatrix} \qquad (3.35)$$

or

$$\begin{pmatrix} u' \\ v' \\ w' \end{pmatrix} = (\mathbf{S}^{\mathrm{T}})^{-1} \begin{pmatrix} u \\ v \\ w \end{pmatrix} \qquad (3.36)$$

Note that $(\mathbf{S}^{\mathrm{T}})^{-1} = (\mathbf{S}^{-1})^{\mathrm{T}}$. Comments on the reciprocity of F and I unit cells can be found in the Web Materials, Appendix WA2.

3.9.   Bilbao Crystallographic Server

The *Bilbao Crystallographic Server* [9] was initiated in 1997 at the Materials Laboratory of the University of the Basque Country, Spain, and offers gratis a very wide range of crystallographic and solid-state programs, databases and utilities. The programs for the variety of interrelated topics offered by the Server are grouped into *shells*, of which the CELLTRAN link in the Structure Utilities shell is of interest in the context of unit-cell transformations. In use, the entries comprise (a) the unit-cell dimensions a, b, c, α, β, γ and (b) the transformation matrix.

3.9.1.   Unit-cell transformation by the Bilbao Crystallographic Server

If the equations of a transformation are $\mathbf{a}' = \mathbf{a} + \mathbf{b}$, $\mathbf{b}' = \mathbf{b}$, $\mathbf{c}' = -\mathbf{a} + \mathbf{c}$, the matrix $\mathbf{S}$ is

$$\begin{pmatrix} 1 & 1 & 0 \\ 0 & 1 & 0 \\ \bar{1} & 0 & 1 \end{pmatrix}$$

and $\mathbf{S}^{\mathrm{T}}$ is

$$\begin{pmatrix} 1 & 0 & \bar{1} \\ 1 & 1 & 0 \\ 0 & 0 & 1 \end{pmatrix}$$

The input to the Server comprises a, b, c, α, β, γ and the matrix $\mathbf{S}^{\mathrm{T}}$. The output is obtained by clicking the tab 'Show'; the conventional transformed unit cell is the final entry of the output results. Further notes on the Bilbao Crystallographic Server are available in the literature [10].

Example 3.4. (a) Consider a unit-cell (I) transformation

$$\mathbf{a}' = \mathbf{a} + 2\mathbf{b}, \ \mathbf{b}' = -\mathbf{a} - \mathbf{c}, \ \mathbf{c}' = \mathbf{b} + \mathbf{c} : \mathbf{S} = \begin{pmatrix} 1 & 2 & 0 \\ \bar{1} & 0 & \bar{1} \\ 0 & 1 & 1 \end{pmatrix}$$

If the unit-cell parameters are $a = 5\,\text{Å}$, $b = 10\,\text{Å}$, $c = 15\,\text{Å}$, $\alpha = \beta = \gamma = 90°$ (vol $= 750\,\text{Å}^3$). Using the Bilbao Crystallographic Server, the transformed cell is given by inputting the unit-cell dimensions and the matrix

$$\mathbf{S}^{\mathrm{T}} = \begin{pmatrix} 1 & \bar{1} & 0 \\ 2 & 0 & 1 \\ 0 & \bar{1} & 1 \end{pmatrix}$$

The results given by the Server for cell (II) are $a = 20.6155$, $b = 15.8114$, $c = 18.0278$, $\alpha = 142.120$, $\beta = 57.440$, $\gamma = 94.400$ (vol $= 2250.0000$). Note that the resulting unit cell has three times the volume of the original unit cell; clearly, the transformed cell would not be a conventional choice.

(b) If a lattice direction for the unit cell (I) is $[1\,0\,1]$, what is $[u'v'w']$ for the transformed unit cell (II)? The transformed lattice direction is given as

$$\begin{pmatrix} u' \\ v' \\ w' \end{pmatrix} = (\mathbf{S}^{\mathrm{T}})^{-1} \begin{pmatrix} u \\ v \\ w \end{pmatrix} = \begin{pmatrix} 1/3 & 1/3 & \overline{1/3} \\ 2/3 & 1/3 & \overline{1/3} \\ 2/3 & 1/3 & 2/3 \end{pmatrix} \begin{pmatrix} 1 \\ 0 \\ 1 \end{pmatrix} = \begin{pmatrix} 0 \\ \bar{1} \\ 0 \end{pmatrix}$$

that is, the direction is $[0\,\bar{1}\,0]$.

3.10. Transformation Mnemonic

Once the transformations are understood, the following mnemonic may be used to perform the calculations directly. The following diagrams show a matrix $\mathbf{S}$ and its inverse $\mathbf{S}^{-1}$, and the arrows show the directions of multiplication. The mnemonic provides a simple way of performing the transformations, and the following examples illustrate the procedure; an arrow ↱ should be interpreted as $\mathbf{a}'$ in terms of $\mathbf{a}$; here, $\mathbf{a}$ implies a, b, c and so on.

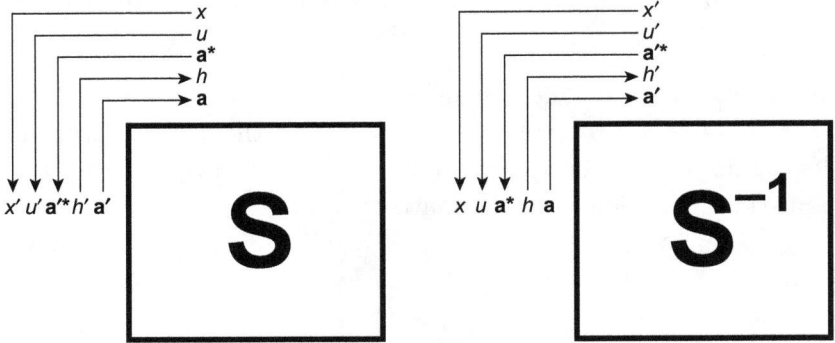

Let the matrix $\mathbf{S}$ transforming $\mathbf{a}$ to $\mathbf{a}'$ be

$$\begin{pmatrix} 2/3 & 1/3 & 1/3 \\ \overline{1/3} & 1/3 & 1/3 \\ \overline{1/3} & \overline{2/3} & 1/3 \end{pmatrix}$$

Since $\mathbf{a}'$ represents the triplet $\mathbf{a}'$, $\mathbf{b}'$, $\mathbf{c}'$, then following the arrow implies that $\mathbf{a}'$, $\mathbf{b}'$ and $\mathbf{c}'$ are given through the mnemonic by

which means

$$\mathbf{a}' = \frac{2}{3}\mathbf{a} + \frac{1}{3}\mathbf{b} + \frac{1}{3}\mathbf{c}$$
$$\mathbf{b}' = -\frac{1}{3}\mathbf{a} + \frac{1}{3}\mathbf{b} + \frac{1}{3}\mathbf{c}$$
$$\mathbf{c}' = -\frac{1}{3}\mathbf{a} - \frac{2}{3}\mathbf{b} + \frac{1}{3}\mathbf{c}$$

The inverse matrix $\mathbf{S}^{-1}$ is $\begin{pmatrix} 1 & \bar{1} & 1 \\ 0 & 1 & \bar{1} \\ 1 & 1 & 1 \end{pmatrix}$. If, for example, $\mathbf{x}'$ represents the triplet x', y, z, then $\mathbf{x}'$ evolves through the mnemonic as

$$x' = x + z$$
$$y' = -x + y + z$$
$$z' = x - y + z$$

Example 3.5. As a final example, consider the unit-cell transformation $\mathbf{a}' = \mathbf{a} + 2\mathbf{b}$, $\mathbf{b}' = -2\mathbf{a} - \mathbf{b}$, $\mathbf{c}' = 2\mathbf{c}$ together with the following diagram:

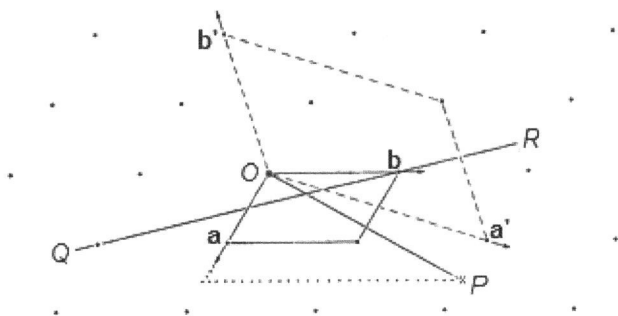

The matrix $\mathbf{S}$ is 1 2 0/-2 -1 0/0 0 2 (concise notation) and its transpose $\mathbf{S}^{\mathrm{T}}$ is 1 -2 0/2 -1 0/0 0 2.

Given that the line QR has the Miller indices (210) with respect to the cell I $(\mathbf{a}, \mathbf{b}, \mathbf{c})$, its indices when referred to cell II $(\mathbf{a}', \mathbf{b}', \mathbf{c}')$ are then

$$(210) \begin{pmatrix} 1 & \bar{2} & 0 \\ 2 & \bar{1} & 0 \\ 0 & 0 & 2 \end{pmatrix} = (4\bar{5}0)$$

Example 3.5. (*Continued*) The lattice direction for the line OP is $[3/2\ 2\ 0]$ in terms of cell I. When referred to cell II, the direction is given by

$$\begin{pmatrix} u' \\ v' \\ w' \end{pmatrix} = (\mathbf{S}^{\mathrm{T}})^{-1} \begin{pmatrix} u \\ v \\ w \end{pmatrix}$$

that is,

$$\begin{pmatrix} \bar{\tfrac{1}{3}} & \tfrac{2}{3} & 0 \\ \tfrac{2}{3} & \tfrac{1}{3} & 0 \\ 0 & 0 & 1/2 \end{pmatrix} \begin{pmatrix} \tfrac{3}{2} \\ 2 \\ 0 \end{pmatrix} = \begin{pmatrix} \tfrac{5}{6} \\ \bar{\tfrac{1}{3}} \\ 0 \end{pmatrix}$$

The diagram is scaled, so that the results can be verified by direct measurement.

Self-assessment 3.5. A direction in a lattice is given as $[10\bar{2}]$ with respect to a unit cell $\mathbf{a}$, $\mathbf{b}$, $\mathbf{c}$. Transform the zone symbol to the value in the unit cell transformed as $\mathbf{a}' = \mathbf{a}+2\mathbf{b}$, $\mathbf{b}' = -\mathbf{a}+\mathbf{c}$, $\mathbf{c}' = \mathbf{a} + 2\mathbf{b} + \mathbf{c}$. *Note*: It is useful to be able to invert a 3×3 matrix manually; alternatively, it can be done with the program MATOPS.

Problems

3.1. Given a unit cell with $a = 5.000$, $b = 6.000$, $c = 7.000\,\text{Å}$, $\alpha = \gamma = 90.00°$ and $\beta = 120°$. (a) What is the most probable crystal system? (b) Calculate the length of $[321]$.

3.2. Do any of the following unit cells represent a lattice? If so, give their conventional names. (a) Monoclinic A; (b) orthorhombic B; (c) tetragonal F; (d) cubic $A + C$; (e) triclinic I.

3.3. Outline a rhombohedron within a cubic F unit cell. What is the ratio of $V_{\text{rhomb}}/V_{\text{cubic}}$?

3.4. A primitive unit cell is given the dimensions $a = b = 3.155$ Å, $c = 8.965$ Å, $\alpha = \beta = 90°$, $\gamma = 120°$. Subsequently, it turns out to be a triply primitive hexagonal unit cell. Study Fig. 3.8 and determine the equations for the transformation $R_{\text{hex}} \rightarrow R_{\text{obv}}$, and calculate the parameters of the rhombohedral unit cell in the obverse setting.

3.5. An orthogonal P unit cell (I) is transformed to cell (II) by the equations $\mathbf{a}' = \mathbf{a} - \mathbf{b}$, $\mathbf{b}' = \mathbf{a} + \mathbf{b}$, $\mathbf{c}' = \mathbf{c}$. (a) Determine the volume V_{II} of cell (II) in terms of V_{I}. (b) Transform the point $(0.123, -0.671, 0.314)$ in cell (I) to its value in cell (II).

3.6. A unit cell of a lattice has *inter alia* a $\bar{4}$ symmetry axis along z. (a) What is the matrix for a $\bar{4}$ symmetry operation? (b) What points are generated from the points x, y, z by a $\bar{4}$ symmetry operation? (c) What are the subgroups of point group $\bar{4}$? (d) Show that these results imply the tetragonal conditions of $a = b \not c$; $\alpha = \beta = \gamma = 90°$.

3.7. A monoclinic F unit cell has the dimensions $a = 6.105$, $b = 6.985$, $c = 8.125$ Å, $\beta = 115°$. A monoclinic C unit cell is equivalent to the F unit cell and as it is of smaller volume it is the conventional choice. (a) Use the Bilbao Crystallographic Server to determine a', b', c' and β' for the C unit cell. (b) What is the ratio of V_C/V_F?

3.8. A unit cell has the dimensions $a = b = 5.1720$, $c = 7.1730$ Å, $\alpha = \gamma = 90°$, $\beta = 105.0°$. Calculate the volume $V_{\mathbf{c}}$ of the unit cell and the volume V^* of the reciprocal unit cell. Afterwards, the results may be checked with the program RECIP.

3.9. Measurements on crystal a unit cell revealed the relationships $a \not b \not c$; $\alpha \not \beta \not 90°$, $\gamma = 90°$. Does this cell represent a diclinic system? Comment.

3.10. An orthorhombic primitive unit cell has the dimensions $a = 5.000$ Å, $b = 10.00$ Å, $c = 15.00$ Å. The cell is transformed according to the equations

$$\mathbf{a}' = 3\mathbf{a} - \mathbf{b}$$
$$\mathbf{b}' = \mathbf{b}$$
$$\mathbf{c}' = -2\mathbf{a} + \mathbf{b} + \mathbf{c}$$

(a) Transform the (conventional) primitive unit cell using the Bilbao Crystallographic Server, and record the new unit-cell dimensions.

(b) Set up matrices to determine transformed values of (i) $(10\bar{2})$, (ii) $[310]$, (iii) $x = 0.3150$, $y = 0.4700$, $z = -0.5175$.

3.11. Calcite $CaCO_3$ is trigonal (hexagonal) with $a = 4.990\,\text{Å}$, $c = 17.06\,\text{Å}$, $\alpha = \beta = 90°$, $\gamma = 120°$. Determine the angle between the lattice directions $[104]$ and $[\bar{1}14]$ in the reciprocal lattice.

3.12. A primitive orthorhombic unit cell has the dimensions $a = 3.000\,\text{Å}$, $b = 4.000\,\text{Å}$, $c = 5.000\,\text{Å}$, $\alpha = \beta = \gamma = 90.00°$. Determine the values for a, b and c under the transformation $\mathbf{a}' = \mathbf{a} + 2\mathbf{b}$, $\mathbf{b}' = \mathbf{a} - \mathbf{c}$, $\mathbf{c}' = \mathbf{b} + 2\mathbf{c}$. Use the program METTENS or the Bilbao Crystallographic Server, or otherwise. *Note that in METTENS, the matrix is entered as **rows** of matrix* $\mathbf{S}$, *whereas in the Bilbao Crystallographic Server, the matrix is entered as **columns** of* $\mathbf{S}$ ($\equiv$ *rows of* $\mathbf{S}^\mathrm{T}$).

3.13. A unit cell has the exact dimensions $a = 6\,\text{Å}$, $b = 9\,\text{Å}$, $c = 12\,\text{Å}$, $\alpha = 105°$, $\beta = 90°$, $\gamma = 110°$. (a) What crystal system is represented by these parameters? (b) Calculate the length of the vector from the origin to the lattice point $[35\bar{4}]$ to four significant figures. (c) Calculate the angle between the vectors $\mathbf{p} = [111]$ and $\mathbf{q} = [35\bar{4}]$ to four significant figures.

Answers to Self-Assessments

3.1. (a) It is not a lattice. For example, a vector $\mathbf{p}$ placed with its tail at a point such as T should terminate on another lattice point. Clearly, this is not so.

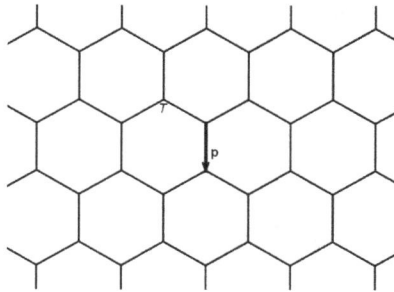

Hexagonal 'honeycomb' array of points

(b) Centring the a-b planes of an array of primitive hexagonal unit cells degrades the symmetry at each point to that of the orthorhombic system. The original point symmetry ($\frac{6}{m}mm$) becomes mmm. The following diagram shows four hexagonal unit cells in projection on to the x-y plane; each cell has

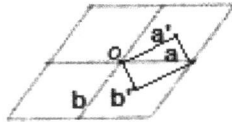

a point at its centre: $\mathbf{a}$ and $\mathbf{b}$ refer to the hexagonal P unit cell, whereas $\mathbf{a}'$ and $\mathbf{b}'$ define a smaller, orthorhombic P unit cell.

3.2. $V_c = abc\sqrt{1 - \cos^2\beta - \cos^2\gamma} = 210\,\text{Å}^3 \times (1 - 0.030154 - 0.116978)^{1/2} = 193.9\,\text{Å}^3$. $a^* = \frac{bc\sin\alpha}{V} = 42\,\text{Å}^2/193.94\,\text{Å}^3 = 0.2166\,\text{Å}^{-1}$. Similarly, $b^* = 0.1777\,\text{Å}^{-1}$, $c^* = 0.1454\,\text{Å}^{-1}$. $\cos\alpha^* = \frac{\cos\beta\cos\gamma - \cos\alpha}{\sin\beta\sin\gamma} = \frac{-0.17365\times(-0.34202)}{0.98481\times0.93969} = 0.064178$, so that $\alpha^* = 86.32°$. Similarly, $\beta* = 79.35°$, $\gamma^* = 69.68°$. $V^* = 1/V_c = 1/193.94\,\text{Å}^3 = 5.156 \times 10^{-3}\,\text{Å}^{-3}$.

3.3. (a) Consider Fig. 3.14, but with orthogonal axes. The new unit cell is a true lattice, and the symmetry is unaltered. However, one interaxial angle in the P cell is no longer 90°. Thus, the B-centred cell represents a true orthorhombic lattice; conventionally, it would be reoriented to C-centred.

(b) Centring the B faces of a tetragonal unit cell reduces its symmetry to that of a centred orthorhombic unit cell. The symmetry is restored apparently by centring the A faces also. But the diagram below shows that the arrangement is no longer a lattice: the two **r** vectors do not have identical environments.

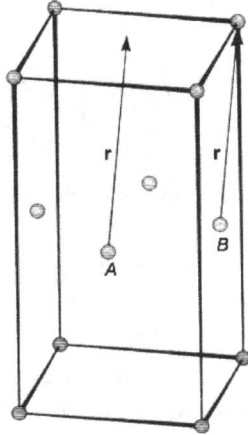

[It could be made a true lattice by centring also the C faces. But the resulting tetragonal F cell would then be transformed to tetragonal I.]

3.4. Following Section 3.7, the matrix equation is given by the row matrix:

$$(h'\ k'\ l') = (\bar{3}\ 1\ \bar{2}) \begin{pmatrix} 1 & \bar{1} & 0 \\ 2 & 1 & 0 \\ 0 & 0 & 1 \end{pmatrix} (\bar{1}\ 4\ \bar{2})$$

3.5.

$$\mathbf{S} = \begin{pmatrix} 1 & 2 & 0 \\ \bar{1} & 0 & 1 \\ 1 & 2 & 1 \end{pmatrix} \quad \text{and} \quad \mathbf{S}^{\mathrm{T}} = \begin{pmatrix} 1 & \bar{1} & 1 \\ 2 & 0 & 2 \\ 0 & 1 & 1 \end{pmatrix}$$

Then,

$$[u'\ v'\ w'] = \begin{pmatrix} \bar{1} & 1 & \bar{1} \\ \bar{1} & 1/2 & 0 \\ 1 & 1/2 & 1 \end{pmatrix} \begin{pmatrix} 1 \\ 0 \\ \bar{2} \end{pmatrix} = [1\ \bar{1}\ \bar{1}]$$

$$(\mathbf{S}^{\mathrm{T}})^{-1}$$

References

[1] Bravais A, *Mem. Acad. Roy. Sci. France* **9**, 255 (1846).

[2] Burns G and Glazer AM, *Space Groups for Solid State Scientists*. Elsevier (2013).

[3] Ladd M, *Symmetry of Crystals and Molecules*. Oxford University Press (2014).

[4] Ladd M and Palmer R, *Structure Determination by X-ray Crystallography*, 5th edn. Springer Science+Business Media (2013).

[5] Streitwolf H-W, *Group Theory in Solid-State Physics*. Macdonald (1971).

[6] Gibbs JW, *The Elements of Vector Analysis*. New Haven (1881), particularly p. 80ff. Reprinted by Forgotten Books (2016). Online at ⟨https://www.for gottenbooks.com/en/books/ElementsofVectorAnalysis_10268178⟩.

[7] Hodgkin DMC, *Biogr. Mems. Fell. R. Soc.* **26**, 28 (1980).

[8] Ewald PP, *Z. Phys.* **144**, 465 (1913).

[9] International Union of Crystallography. *Bilbao Crystallographic Server*. Online at ⟨http://www.cryst.ehu.es/⟩.

[10] Dauter Z and Jaskolski M. <https://mcl1.ncifcrf.gov/dauter_pubs/284.pdf>.

Chapter 4

Space Groups

You cannot make a man by standing a sheep on its
hind legs. But by standing a whole flock of sheep
in that position you can make a crowd of men.
Max Beerbohm

Key Topics

- Space groups in one, two and three dimensions
- Asymmetric unit
- International tables
- Change of origin
- Symmorphic space groups
- Alternative settings of space groups
- Space groups and crystal structures
- Matrix representation of space-group operations

4.1. Introduction

The previous chapters have considered the symmetry of finite bodies, of which atoms and molecules are those bodies of interest here, and also the concept of a lattice in terms of the 14 arrangements deduced by Bravais. The regular repetition of a motif of a given point-group symmetry in a fixed orientation at, or in relation to, each point of a lattice builds up a crystal structure. The process is known mathematically as *convolution*, which is discussed in Appendix A10 and may be illustrated by the following example.

The diagram in Fig. 4.1(a) represents a pair of Na^+ and Cl^- ions of spacing d. It has cylindrical symmetry, point group ∞, and may be considered as a unit of structure represented by the function $p(\mathbf{d})$. The vector $\mathbf{d}$ may be assumed to run from Na^+ to Cl^-; it is the *pattern motif*. Figure 4.1(b) shows a face-centred unit cell

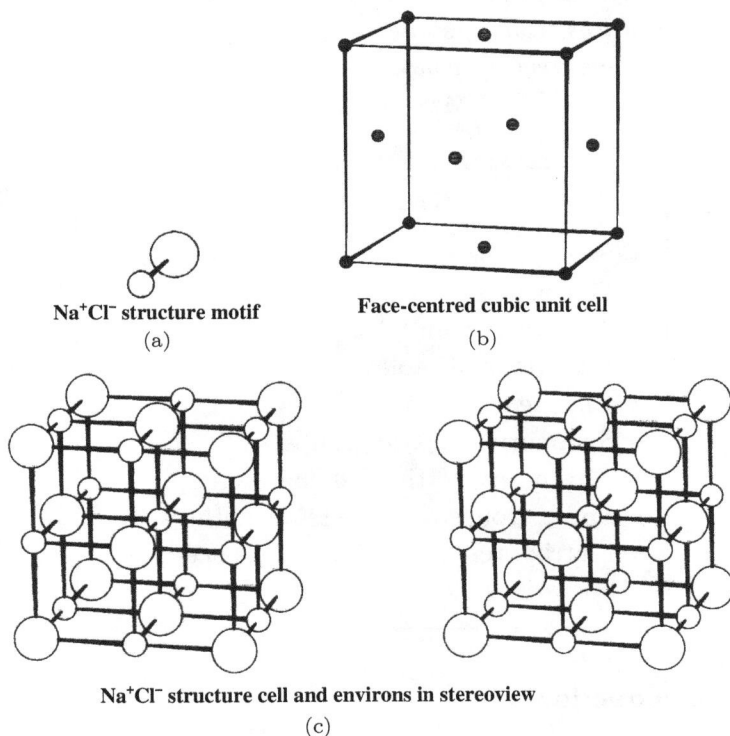

Na^+Cl^- structure motif   **Face-centred cubic unit cell**

(a)						(b)

Na^+Cl^- structure cell and environs in stereoview

(c)

Fig. 4.1.   Building the crystal structure of sodium chloride Na^+Cl^-. (a) An Na^+Cl^- ion-pair forms the structural unit. (b) A face-centred (F) unit cell of a cubic lattice. (c) Stereoview of the sodium chloride structure, obtained by the repetition (convolution) of the Na^+Cl^- structural unit in a fixed orientation at each point of the cubic lattice.

representing a cubic lattice, and may be termed the function $q(\mathbf{d})$. The crystal structure of sodium chloride, $c(\mathbf{d})$, is obtained as the convolution process $c(\mathbf{d}) = p(\mathbf{d}) * q(\mathbf{d})$, which implies the repetition of the Na^+-Cl^- pattern motif in a fixed orientation with respect to the lattice points, as represented in Fig. 4.1(c). A three-dimensional packing of these unit cells forms the macro-structure of sodium chloride.

It should be noted that, although the term 'unit cell' has been defined strictly in Section 3.3 as a region of space [1], it will be found frequently in books and published papers to refer to the *unit cell together with its chemical constituents*. This usage is continued herein; it is obvious in context whether it refers to space alone or to that same space together with its chemical contents.

4.2.   Space Group Defined

The convolution process results in an arrangement of chemical entities, ideally of infinite extent, that are related by symmetry and described by a *space group*. A space group may be defined as an infinite set of symmetry operators, including the unit-cell translations **a**, **b** and **c**, the action of any of which brings the three-dimensional, periodic object to which they refer into self-coincidence. In one and two dimensions, the corresponding terms are *line group* and *plane group*, respectively.

A space group, although an infinite concept, can be applied to crystals in practice because the number of repeating units under examination is very large. For example, the unit cell of sodium chloride is a cube of side 5.640 Å. A crystal of experimental size, say, 0.2, 0.2, 0.2 mm, has a volume of *ca.* 8×10^{-3} mm^3, and the unit-cell volume is *ca.* 1.8×10^{-19} mm^3. Thus, the number of unit cells involved in the diffraction experiment is very large, approximately 4×10^{16}. When bathed in an X-ray beam, a small crystal behaves, from a practical point of view, as an infinite, regular array of scattering centres.

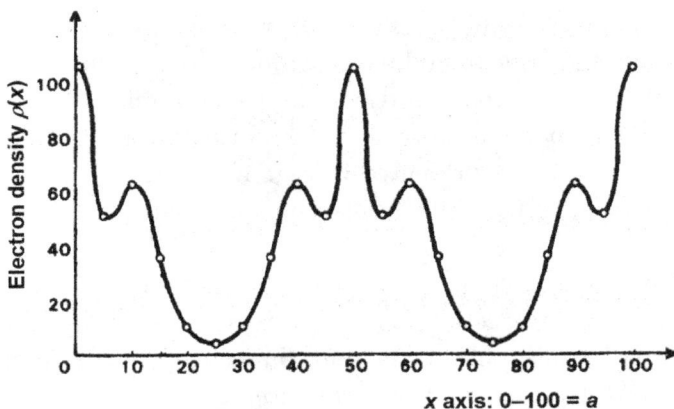

Fig. 4.2. One-dimensional projection of the electron density of pyrite FeS$_2$. Line group *pm*; the asymmetric unit is one-quarter of the unit cell of the line group, 0 to $a/4$.

4.2.1. One-dimensional space groups: Line groups

A combination of the one-dimensional lattice and a group of symmetry 1 or *m* produces a *line group*, *p*1 or *pm*. A simple structure, such as that of pyrite FeS$_2$, can be represented in projection on to the *x*-axis by the line group *pm*. Figure 4.2 is a plot of its electron density $\rho(x)$ projected on to the *x*-axis, and calculated in steps of $a/100$. The iron atoms occupy positions at fractional coordinates $x = 0$ and $1/2$, while the sulphur atoms lie at $x = ca. \pm 0.1$ and $(1/2 \pm 0.1)$. The structure is represented by the sufficient portion $x = 0$ to $x = a/4$, on account of the *m* line group symmetry at the points at $x = 0$ and $1/2$ (see also Section 4.12.4).

4.2.2. Asymmetric unit

An *asymmetric unit* of a space group, in one, two or three dimensions, is a region of that space of the appropriate dimension which, when acted upon by the symmetry of the space group, repeats that unit region so as to fill the space entirely [1]. The asymmetric unit is an integral submultiple of the space group; in the example of pyrite considered in the previous section, the asymmetric unit is one-quarter of the unit cell.

Notwithstanding the above strict definition of asymmetric unit, common usage in books and published papers has extended it to include its chemical contents, as in the case of the unit cell. The common usage is continued herein as the context will reveal always the intended application.

The definition of asymmetric unit means symbolically that

$$\text{Asymmetric unit} \xrightarrow{+ \text{ Space group}} \text{Infinite crystal}$$

In practice, a crystal is truncation of the ideally infinite crystal of its material. One-dimensional groups can form useful illustrations of structural principles but are not of great importance in crystal structure determination.

4.3. Two-Dimensional Space Groups: Plane Groups

Plane groups provide a useful introduction to the three-dimensional groups (space groups), and are important also when considering structures in plane projection. The plane groups are listed by symbol in Table 4.1, and the two-dimensional lattices are given in Table 3.1. Following the discussion in Section 4.1, plane groups may be formed by combining the data from these two tables. A plane pattern motif is repeated by the translations **a** and **b** of a two-dimensional lattice; in this way, a total of seventeen plane groups may be derived. Plane group $p1$ is easily understood, and a more revealing start may be made with plane groups $p2$ and $p1m1$ (pm).

4.3.1. Plane groups in the oblique and rectangular systems

In Fig. 4.3(a), a number of oblique unit cells of a two-dimensional lattice are shown, while Fig. 4.3(b) is the same diagram but populated with a number of entities in accordance with the two-fold symmetry at each lattice point. These symbols in a unit cell, each designated conventionally by a small circle ○, need not necessarily be regarded as chemical contents but rather as markers that reveal the symmetry

Table 4.1. The seventeen plane groups.

System name and unit-cell symbol	Point group	Plane group Full symbol	Short symbol
Oblique, p	1	$p1$	$p1$
	2	$p211$	$p2$
Rectangular, p and c	m	$p1m1$	pm
		$p1g1$	pg
		$c1m1$	cm
	$2mm$	$p2mm$	pmm
		$p2mg$	pmg
		$p2gg$	pgg
		$c2mm$	cmm
Square, p	4	$p4$	$p4$
	$4mm$	$p4mm$	$p4m$
		$p4gm$	$p4g$
Hexagonal, p	3	$p3$	$p3$
	$3m$	$p3m1$	$p3m1$
		$p31m$	$p31m$
	6	$p6$	$p6$
	$6mm$	$p6mm$	$p6m$

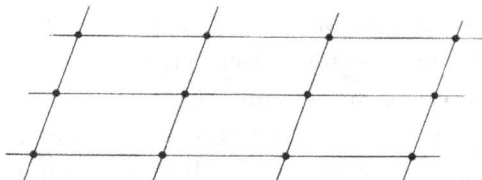

(a)

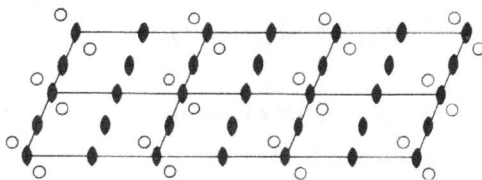

(b)

Fig. 4.3. (a) Two-dimensional oblique net with p unit cells outlined; the symmetry at each lattice point is two-fold rotation (2). (b) The unit cells of (a) together with a motif (O) representing a single structural entity. Note that the lines are no part of the lattice; they merely act as an aid to the appreciation of the lattice geometry.

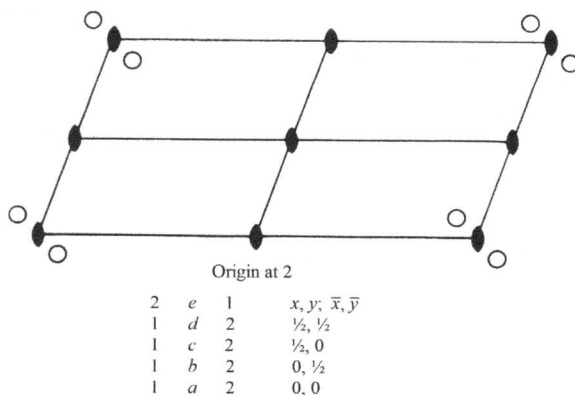

Origin at 2

2	e	1	$x, y;\ \bar{x}, \bar{y}$
1	d	2	$\frac{1}{2}, \frac{1}{2}$
1	c	2	$\frac{1}{2}, 0$
1	b	2	$0, \frac{1}{2}$
1	a	2	$0, 0$

Fig. 4.4. A single unit cell and its environs from Fig. 4.3(b); additional two-fold symmetry points within the unit cell are revealed by the completed diagram.

relationships of the space group. In a crystal unit cell, however, they are the chemical entities occupying the space of that unit.

The pattern motif when completed at the corners of a unit cell by the translations **a** and **b** reveals the total symmetry that is present. In general, a single unit cell is representative of the entire lattice, as shown in Fig. 4.4, provided that the pattern is completed with respect to all four corners of the unit-cell diagram. The asymmetric unit is one-half of the unit cell in this plane group.

The origin point 0, 0 of symmetry 2 is, by convention, the top left-hand corner of the unit cell with **a** running from top to bottom of the drawing and **b** running from left to right. Two-fold rotation points, other than those at the corners of the cell, are revealed at 0, $\frac{1}{2}$; $\frac{1}{2}$, 0; $\frac{1}{2}$, $\frac{1}{2}$ in the completed unit cell. After a little practice, such symmetry elements can be identified by inspection. However, at first, they may be located by taking a marker x, y and considering how every other such point on the diagram may be reached by a single symmetry operation of the group, including the basic translations of $\pm$**a** and/or $\pm$**b** as necessary.

A list of fractional coordinates accompanies the plane group diagram; they are divided into two sets in this group. The *general equivalent positions* of x, y coordinates have site symmetry 1, and together they reveal the plane group symmetry. *Special equivalent positions* are sites of point-group symmetry elements of the plane group. There are four such sets in $p2$, and they lie on the two-fold symmetry elements. In any group, each set of special equivalent positions must conform to the symmetry of the plane group.

The data below the drawing of the plane group in Fig. 4.4 are firstly a statement of the origin position, then the coordinates of the general equivalent positions followed by the coordinates of the special equivalent positions, there being four sets in this example. Each coordinate line lists, in order, the number of coordinates in the set, the Wyckoff notation for the site in the given group, the site symmetry and the coordinates of the set. The general position $\bar{x}$, $\bar{y}$ could have been chosen as $1 - x$, $1 - y$, but it is generally more convenient to work in the neighbourhood of the origin.

The number of sites in a set of special equivalent positions is a sub-multiple of the number of sites in the corresponding set of general equivalent positions. Exceptions to this rule may arise in structures that are disordered with respect to general space group rules; often, fewer sites than expected are occupied on average, and in a random manner. An example of disorder is found in the structure of ice at low temperatures (*ca.* 95 K), which has a tendency for a tetrahedral disposition of bonds around each oxygen atom, as shown in Fig. 4.5. In any one molecule, however, only two of these tetrahedral orientations bonds carry hydrogen atoms at any instant; the small circles on the diagram should be regarded statistically as a distribution of half-hydrogen atoms.

Consider next the plane group pm, illustrated in Fig. 4.6 together with the centred group cm, both of which belong to the rectangular two-dimensional system. The special equivalent positions in pm lie on the mirror lines. The origin in these two groups lies on an m line and is undefined in the y direction. In practice, the origin is fixed on introduction of the first chemical entity into the cell. New features arise from this plane group.

In cm, the coordinate list is headed by the expression $(0, 0; 1/2, 1/2)+$, which is a convenient way of indicating the translations that

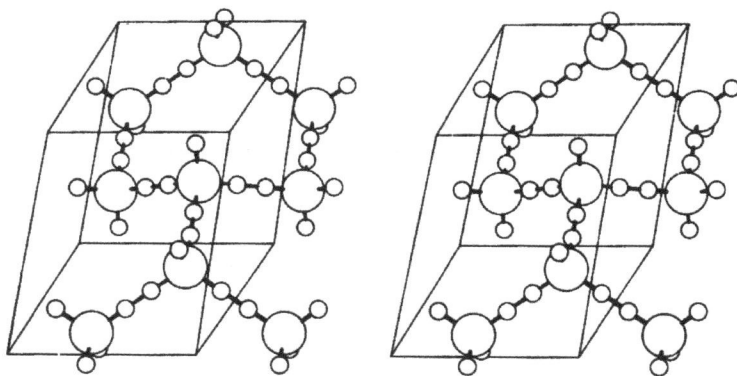

Fig. 4.5. Stereoview of the unit cell and environs of the hydrogen-bonded structure of ice at 90 K. The circles represent, in decreasing order of size, oxygen and statistical half-hydrogen atoms. A tetrahedral array of bonds is latent at each oxygen atom, but from any one oxygen atom only two of these directions carry hydrogen atoms at any instant. [*Crystal Structures: Lattices and Solids in Stereoview*, Mark Ladd (1998). Reproduced by permission of Ellis Horwood Ltd. UK/Woodhead Publishing, UK.]

are to be added to the coordinates listed below to complete the sets for the group. Thus, the general equivalent positions are x, y; $\bar{x}, y$; $1/2 + x, 1/2 + y$ $1/2 - x, 1/2 + y$.

Centring in the presence of mirror lines introduces *glide line* symmetry at $x = \pm 1/4$ which relates points such as x, y and $1/2 - x, 1/2 + y$. Glide symmetry arises when mirror symmetry features in centred unit cells; the symbol cg is an alternative, non-standard notation for cm. There is only one set of special position in cm; the centring condition requires the use of both m lines $[0, y]$ and $[1/2, y]$.[a] If two sets are written, by analogy with pm, there obtains the pairs $0, y$; $1/2, 1/2 + y$ and $1/2, y$; $0, 1/2 + y$. However, these two sets are related by a change of origin to the point $0, 1/2$ and do not constitute different arrangements in cm.

If glide symmetry is introduced into a primitive rectangular unit cell, a new plane group of symbol pg is defined, of which Fig. 4.7 is an example. In illustration (a), one unit cell and its immediate

---

[a]The notation $[p, q]$ is used here to represent the line $x = p, y = q$, by analogy to $[uvw]$ for lattice directions.

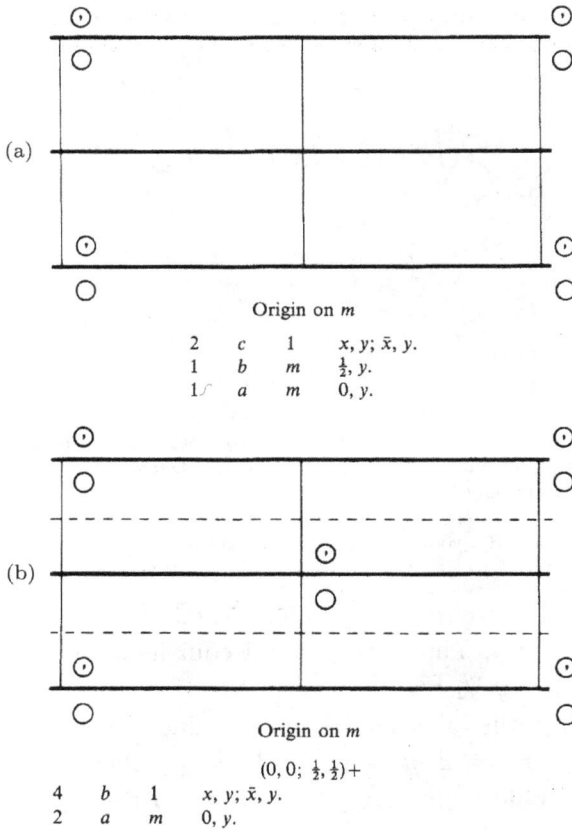

Origin on m

2	c	1	$x, y; \bar{x}, y.$
1	b	m	$\frac{1}{2}, y.$
1	a	m	$0, y.$

Origin on m

$(0, 0; \frac{1}{2}, \frac{1}{2})+$

4	b	1	$x, y; \bar{x}, y.$
2	a	m	$0, y.$

Fig. 4.6. Unit cell and environs of the plane groups (a) *pm* and (b) *cm*. The data lines show, in order, the position of the origin, the site symmetry of the coordinate positions, the Wyckoff notation for the set of coordinates, the symmetry of the sites, and the coordinate list. [*Structure Determination by X-ray Crystallography*, Mark Ladd and Rex Palmer, fifth edn. (2013). Reproduced by permission of Springer Science + Business Media, NY.]

environment are shown; there are two motifs per unit cell. A motif at any corner of the unit cell is related to that at the centre of the cell by a glide line g that is parallel to a and lies at $\pm b/4$; the translational component is $(a + b)/2$.

In relation to Section 4.2.2, Fig. 4.7(b) illustrates the asymmetric unit, which is one-half of the unit cell in this example. According to the strict definition, the asymmetric unit is the region of one-half

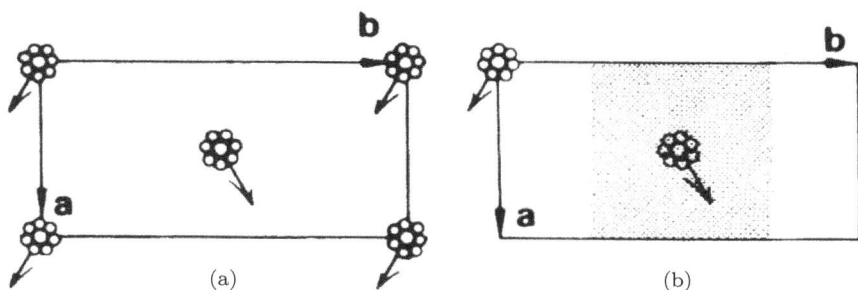

Fig. 4.7. A pattern in plane group *pg*. (a) A unit cell containing two pattern motifs and showing its immediate environment. (b) The asymmetric unit of the plane group (see Section 4.2.2).

of the unit-cell space; in general parlance, the asymmetric unit here would contain a single pattern motif.

Self-assessment 4.1. Make a conventional drawing of plane group *pg* to show the symmetry elements and general equivalent positions. List data appropriate to this group.[b]

The rectangular system includes also plane groups based on plane point group 2*mm*; nominally, there would appear to be eight such groups:

$$p2mm^* \quad p2mg^* \quad p2gm \quad p2gg^*$$
$$c2mm^* \quad c2mg \quad c2gm \quad c2gg$$

Of these, only those marked with an asterisk are unique: *p2mg* and *p2gm*, for example, are equal under interchange of the *x*- and *y*-axes, and three designations for the centred symbols are also non-standard, equivalent descriptors; plane group *p2gg* will be studied as an example of this class.

It is useful to consider first the point group 2*mm* in this case. It is recognized easily by removing the unit-cell symbol and then replacing all translational elements by its non-translational counterpart: $p2gg \rightarrow p2gg \rightarrow 2gm \rightarrow 2mm$. Then, from Table 2.4, the

---

[b] Answers to all Self-assessments are given at the end of the chapter.

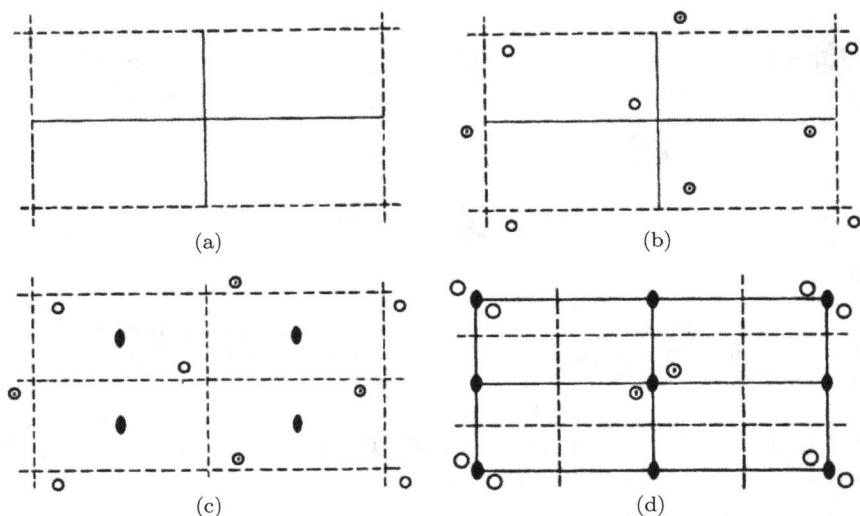

Origin at 2

4	c	1	x, y; $\bar{x}, \bar{y}$; $\frac{1}{2}+x, \frac{1}{2}-y$; $\frac{1}{2}-x, \frac{1}{2}+y$
2	b	2	$\frac{1}{2}, 0$; $0, \frac{1}{2}$
2	a	2	$0, 0$; $\frac{1}{2}, \frac{1}{2}$

Fig. 4.8. Plane group p2gg. (a) Unit-cell framework with the origin on gg. (b) Equivalent positions added. (c) Additional symmetry elements revealed. (d) Standard diagram of the unit cell and environs for plane group p2gg; the origin is at 2, a point of two-fold rotation. [*Structure Determination by X-ray Crystallography*, Mark Ladd and Rex Palmer, fifth edn. (2013). Reproduced by permission of Springer Science+Business Media, NY.]

orientation of the symmetry elements is derived. In point group 2mm, and plane group p2mm, the two m lines meet in the two-fold rotation point; this does not occur when glide symmetry is present.

In Fig. 4.8(a), a working origin is taken at gg; then, in (b), the general equivalent positions are entered. Figure 4.8(c) shows the positions of the two-fold rotation points, and, in (d), the diagram is completed in its standard orientation with the origin on 2. The coordinate data for the plane group are given at the bottom of the diagram. Special equivalent positions occur on Wyckoff sites a and b, and an entity occupying these sites must have a symmetry that is compatible with such a site.

> **Self-assessment 4.2.** In plane group $p2gg$, a special position
> set is chosen with the coordinates 0, 0 and $1/2$, 0. Do they satisfy
> the plane group symmetry? If not, what do they represent?

This discussion has made use of an important feature of symmetry groups, namely, that the interaction of two symmetry operations of a group leads to a third operation in the group, as discussed in Section 2.3. Symbolically, in $p2gg$,

$$gg \equiv 2 \tag{4.1}$$

4.3.2. Plane groups in the square and hexagonal systems

In these plane groups, further new features arise on account of the presence of four-fold or six-fold symmetry; plane group $p4gm$ will be studied as an example.

The symbol implies, in terms of Table 2.4, four-fold rotation at a point, glide lines normal to x and y, and m lines at 45° to x and y. It is not to be expected that the operator 4 will lie at the intersection of the two g lines, and a device can be introduced here that will be useful in many other plane groups and also in the three-dimensional (space) groups; it depends upon the general specification of symbols listed in Table 2.4 and Eq. (4.1). The following positions of the symmetry elements are adopted:

$$4 \text{ is the point } 0, 0$$
$$g \text{ is the line } [p, y]$$
$$m \text{ is the line } [q, q]$$

Then the following scheme can be applied, making use of Eq. (4.1), so that the values for p and q can be deduced:

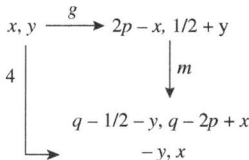

$$x, y \xrightarrow{g} 2p - x, 1/2 + y$$

with 4 and m operations leading to $q - 1/2 - y, q - 2p + x$ and $-y, x$.

Generating the general equivalent positions for $p4gm$

One and the same point is reached by two paths, and, by comparing coefficients, it follows that $q = 1/2$ and $p = q/2$ so that $p = 1/4$, and the diagrams of Fig. 4.9 follow. Then, the eight general equivalent positions may be written as: $\pm(x, y; \bar{y}, x; 1/2 + y, 1/2 + x; 1/2 - x, 1/2 + y)$.

Example 4.1. The argument for the change in coordinates by reflection across the line $[x, r]$, and in similar situations, may be justified through the following diagram:

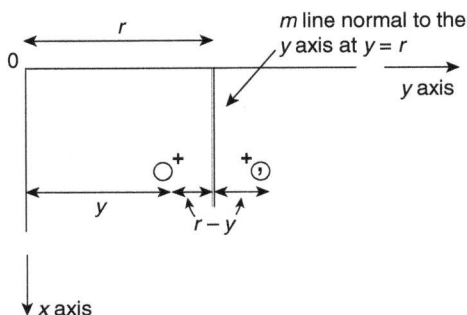

The distances of the two points from the mirror line are each $r - y$. Thus, the y coordinate after reflection is $2r - 2y + y = 2r - y$.

$$x, y \xrightarrow{\ m_{[x, r]}\ } x, 2r - y$$

The same argument is applicable to other symmetry elements and in three dimensions. The shift of a symmetry element is one-half the shift of the related coordinate; for reference, it may be termed the *half-shift rule*.

Now, the plane group drawing can be made, as shown in Fig. 4.9. In the higher symmetry systems, it may be convenient to draw two diagrams, one to show the general equivalent positions and another to show the symmetry elements. By completing the general equivalent positions at all corners of the unit cell, as well as those within the cell, the total array of symmetry elements is revealed (see Section 4.3.1).

The hexagonal system will be studied via a problem; the only new feature will be the evaluation of the coordinates of a point x, y, z

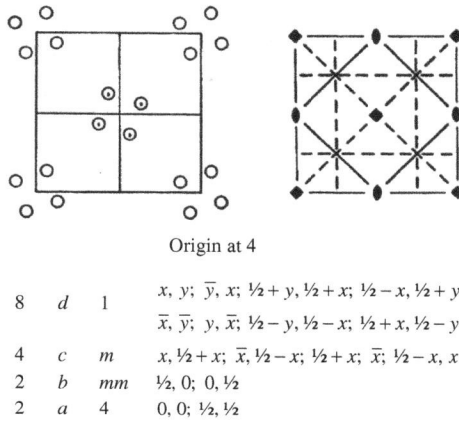

Origin at 4

8	d	1	$x, y;\ \bar{y}, x;\ \frac{1}{2}+y, \frac{1}{2}+x;\ \frac{1}{2}-x, \frac{1}{2}+y$
			$\bar{x}, \bar{y};\ y, \bar{x};\ \frac{1}{2}-y, \frac{1}{2}-x;\ \frac{1}{2}+x, \frac{1}{2}-y$
4	c	m	$x, \frac{1}{2}+x;\ \bar{x}, \frac{1}{2}-x;\ \frac{1}{2}+x;\ \bar{x};\ \frac{1}{2}-x, x$
2	b	mm	$\frac{1}{2}, 0;\ 0, \frac{1}{2}$
2	a	4	$0, 0;\ \frac{1}{2}, \frac{1}{2}$

Fig. 4.9. Unit cell and environs for plane group $p4gm$, origin on 4, showing the general equivalent positions, the symmetry elements and coordinate data.

after rotation by 60° or 120° with respect to hexagonal axes. In this context, Appendix A4 may be helpful.

Diagrams that show the general equivalent positions and the symmetry elements of the seventeen plane groups are given in Fig. 4.10; the asymmetric unit in each group is the region of space containing one scalene triangle.

Self-assessment 4.3. By means of a diagram show that a position x, y reflected across an m line $[q, \bar{q}]$ in the square system produces a point with coordinates $q + y$, $q + x$.

4.4. Three-Dimensional Space Groups

The 230 space groups were described independently by Fedorov [2], Schönflies [3] and Barlow [4] as ways of arranging infinite arrays of points regularly in three-dimensional space, commensurate with the Bravais lattices. The large majority of known crystalline substances occur in the triclinic, monoclinic and orthorhombic crystal systems, and space groups in the low symmetry classes will be studied first. A small number of space groups of higher symmetry will be

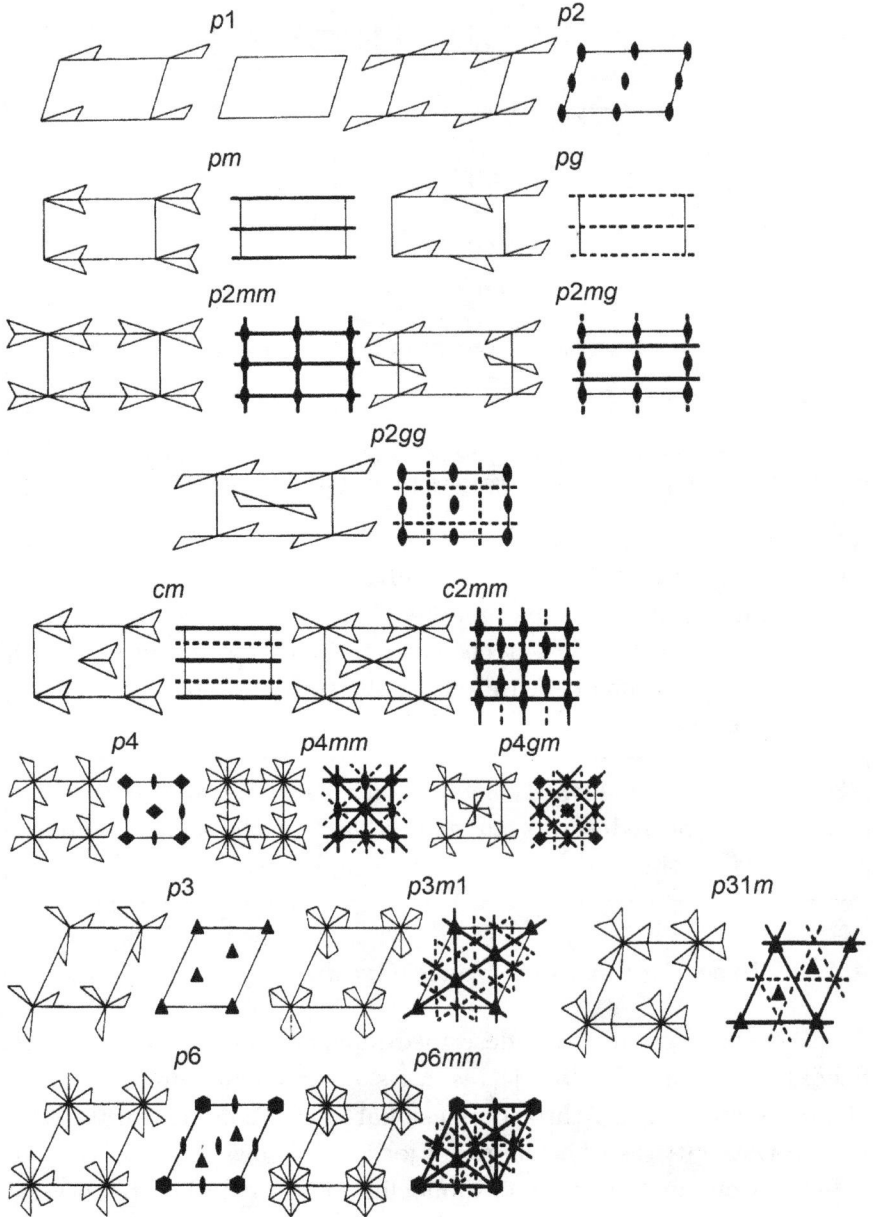

Fig. 4.10. Diagrams to show the general equivalent positions (scalene triangles) and symmetry elements of the seventeen plane groups.

discussed also in order to introduce features that do not arise in the low symmetry groups.

There are different ways of deriving the 230 space groups. It may be done on a mathematical basis or it can be done in more illustrative manner. Group theory concepts are the basis in both cases, particularly those expressed in Section 2.3 and utilized through Eq. (4.1), namely, that the combination of two symmetry operations in a set is equivalent to another symmetry operation in the set. It is the view of the author that a first introduction to the subject benefits from a descriptive procedure, and that method will be used herein. A short description of the matrix representation of space group operations is provided at the end of the chapter.

4.4.1. Triclinic space groups

The triclinic space groups are $P1$ and $P\bar{1}$. In the latter group, a point of inversion (aka centre of symmetry) is taken at the origin of the unit cell and occurs at other related sites in the unit cell.

Self-assessment 4.4. By making a drawing for space group $P\bar{1}$, determine the number of sets of special equivalent positions and their coordinates for this space group.

4.4.2. Monoclinic space groups

The monoclinic system has point groups 2, m and $2/m$, and two lattices described conventionally by P and C unit cells, and space groups $P2$ and $C2$ in class 2 are considered first.

4.4.2.1. *Class 2*

Space group $P2$ is illustrated in Fig. 4.11: space group diagrams are presented here normally with the translation a running from top to bottom, the origin as the top left-hand corner and translation b running from left to right. The $+$ and $-$ signs refer to the z coordinate, $+z$ running upwards towards the observer. The symbol $\rightarrow$ indicates a two-fold

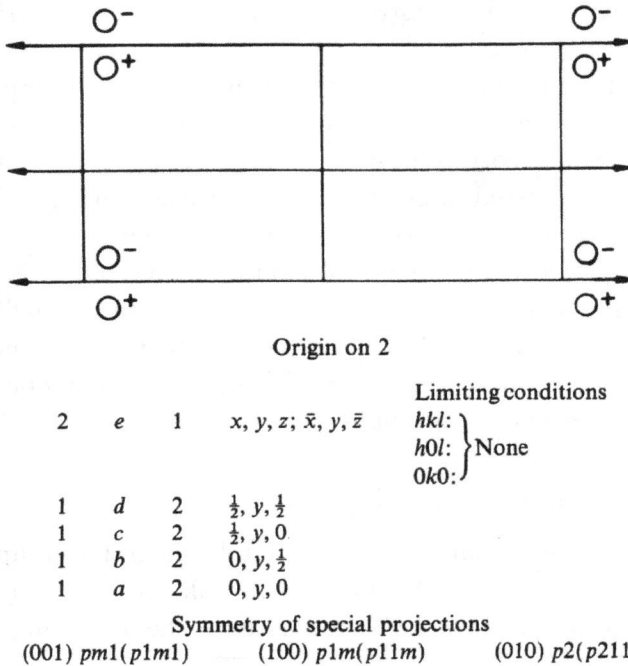

Origin on 2

				Limiting conditions
2	e	1	$x, y, z; \bar{x}, y, \bar{z}$	hkl: ⎫
				$h0l$: ⎬ None
				$0k0$: ⎭
1	d	2	$\frac{1}{2}, y, \frac{1}{2}$	
1	c	2	$\frac{1}{2}, y, 0$	
1	b	2	$0, y, \frac{1}{2}$	
1	a	2	$0, y, 0$	

Symmetry of special projections

(001) $pm1(p1m1)$ (100) $p1m(p11m)$ (010) $p2(p211)$

Fig. 4.11. The unit cell and environs of space group $P2$ in the (001) projection, showing the general equivalent positions, the symmetry elements and selected space-group data.

axis lying in the plane: the space group drawing reveals these axes at $0, y, 0$ and $1/2, y, 0$ as shown on the diagram; however, they occur also at $0, y, 1/2$ and $1/2, y, 1/2$, and relate points such as $x, y\ z$ and $\bar{x}, \bar{y}, 1 - z$. Another feature worthy of mention here are the lists the symmetry of special projections, on to the a-b, b-c and c-a planes. The reader will recognize them as plane group symbols.

The data for $P2$ and $C2$ and, indeed, most plane groups and space groups carry associated *limiting conditions*. They express restrictions of the hkl indices of the spectra obtained by X-ray diffraction from crystals. They may be ignored for now, but their derivation and significance will be discussed after studying X-ray diffraction in subsequent chapters.

When the symmetry motif is combined with a monoclinic C-centred unit cell, or alternatively when Fig. 4.11 is centred,

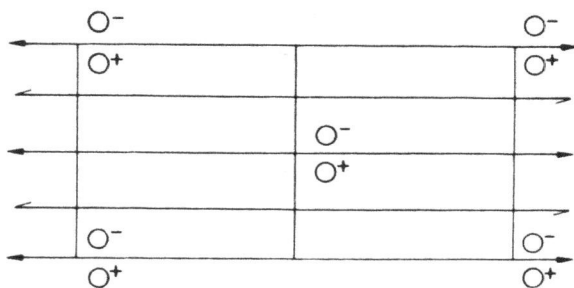

Origin on 2

				Limiting conditions
			$(0, 0, 0; \frac{1}{2}, \frac{1}{2}, 0) +$	
4	c	1	$x, y, z; \bar{x}, y, \bar{z}.$	$hkl: h + k = 2n$
				$h0l: (h = 2n)$
				$0k0: (k = 2n)$
2	b	2	$0, y, \frac{1}{2}.$	As above
2	a	2	$0, y, 0.$	

Symmetry of special projections

(001) $cm1(c1m1)$ (100) $p1m(p11m)$ $b' = b/2$ (010)$p2(p211)$ $a' = a/2$

Fig. 4.12. The unit cell and environs of space group $C2$ in the (001) projection, showing the general equivalent positions, the symmetry elements and selected space-group data. Space group $C2$ may be regarded as formed by an addition of the C-centring translations $1/2$, $1/2$, 0 to the coordinates of space group $P2$.

space group $C2$ is produced, as shown in Fig. 4.12. As the centring of a group with m symmetry introduces glide symmetry, so centring a group with rotational symmetry leads to screw axis symmetry.

A *screw axis* is of the form R_n, where $n = 2, 3, 4$ or 6 ($n < R$). Its operation has the effect of a rotation of $(360/R)^\circ$ about the axis combined with a translation of $1/n$ of the repeat distance in the direction of the axis. Figure 4.13 illustrates the symmetry operation associated with a 6_1 screw axis. The operation is a rotation by 60° combined with a translation of $1/6$ of the repeat distance in the vertical direction. A 2_1 screw axis in $C2$ relates points such as x, y, z to $1/2 - x$, $1/2 + y$, $\bar{z}$. The symbols and translations for all screw axes are listed in Table 2.6.

There are only two sets of special positions in $C2$ compared with four sets in $P2$, and for similar reason as discussed with pm and cm (Section 4.3.1). The meaning of the parenthetical

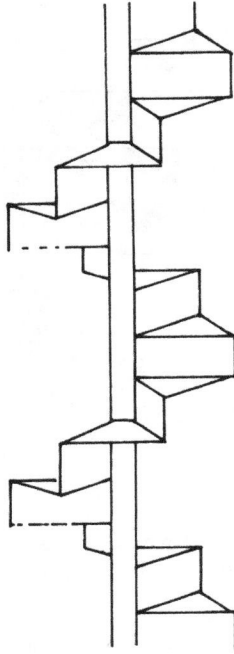

Fig. 4.13. A spiral staircase as an illustration a six-fold screw axis 6_1 symmetry. [Reproduced with permission from Bragg WL, *The Crystalline State*, Vol. 1, Bell and Sons (1939).]

$(0, 0, 0; 1/2, 1/2, 0)+$ has been discussed in relation to cm; the difference here is the added third dimension. On account of the centring, the unit cells of certain special projections appear doubled. Thus, for the (100) projection of $C2$ a new value b', equal to $b/2$, is specified so as to define the smaller unit cell.

4.4.2.2. *Class m*

Monoclinic space groups in class m should present little difficulty. The possible symbols for this class are summarized by the following scheme, with asterisks indicating the four unique groups:

$$Pm^* \quad Pa \quad Pc^* \quad Pn$$
$$Cm^* \quad Ca \quad Cc^* \quad Cn$$

It is not difficult to show that $Pa \equiv Pn \equiv Pc, Ca \equiv Cm$ and $Cn \equiv Cc$.

4.4.2.3.   *Class* $\frac{2}{m}$

Attention is given next to space groups in class $\frac{2}{m}$; space group $\frac{P2_1}{c}$c occurs in practice to the extent of *ca.* 36% of crystals examined. In this group, as well as the screw axes, glide planes are present. The symbols and translations for all glide planes are listed in Table 2.7.

The following scheme links the primitive and centred space groups in this class:

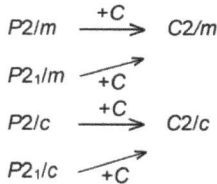

$$
\begin{array}{ll}
P2/m \xrightarrow{\;+C\;} C2/m \\
P2_1/m \;\;\nearrow{+C} \\
P2/c \xrightarrow{\;+C\;} C2/c \\
P2_1/c \;\;\nearrow{+C}
\end{array}
$$

Unnecessary or equivalent symbols have been avoided, and it may be shown that, in this system, with the setting of y unique, $B \equiv P$ and $F \equiv I \equiv A \equiv C$. Additionally, equivalent arrangements differ only by a change of axes; for example, a crystal of space group $\frac{P2_1}{c}$ can be represented equally well, albeit not conventionally, by the symbol $\frac{P2_1}{a}$ or $\frac{P2_1}{n}$.

In space group $P2_1/c$, on account of the translational symmetry, the centre of symmetry does not lie at the intersection of 2_1 and c. It is desirable for the centre of symmetry to be taken at the origin, and from the full meaning of the symbol the specification may be given as

$$
\begin{array}{l}
\bar{1} \text{ at } 0,\, 0,\, 0 \\
2_1 \text{ the line } [p, y, r] \\
c \text{ the plane } (x, q, z)
\end{array}
$$

Then, the following scheme can be written:

---

c $P2_1/c$ is an equivalent way of writing $P\frac{2_1}{c}$.

$$x, y, z \xrightarrow{\;4^{+}_{2[0,0,z]}/n_{(x,y,0)}\;} \tfrac{1}{2}-y, \tfrac{1}{2}+x, \tfrac{1}{2}-z \xrightarrow{\;b_{(s,y,z)}\;} 2s-\tfrac{1}{2}+y, x, \tfrac{1}{2}-z$$

$$\downarrow c_{(t,t,z)}$$

$$t-x, t-2s+\tfrac{1}{2}-y, \bar{z}$$

$$\xrightarrow{\;\bar{1}_{(\frac{1}{4},\frac{1}{4},0)}\;} \tfrac{1}{2}-x, \tfrac{1}{2}-y, \bar{z}$$

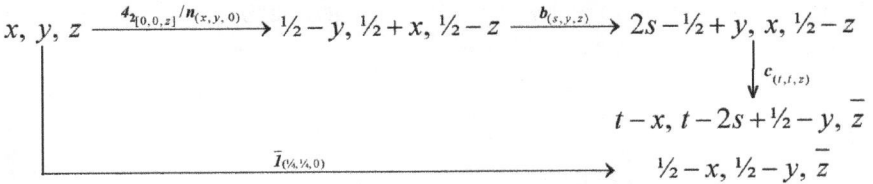

The function of this scheme and others like it depends upon the combination of symmetry operations, of which Eq. (4.1) is a special case, together with the information contained in the Hermann–Mauguin space-group symbol. In point group $2/m$, and using Eq. (2.8), the following group relationships exist:

$G_{2/m}$	1	2	m	$\bar{1}$
1	1	2	m	$\bar{1}$
2	2	1	$\bar{1}$	m
m	m	$\bar{1}$	1	2
$\bar{1}$	$\bar{1}$	m	2	1

Multiplication (Cayley) table for point group *P2/m*

So, in the scheme for $P2_1/c$, the routes $1 \to 2 \to 3$ and $1 \to 4$ arrive at one and the same point, so that by comparing coefficients $p = 0$, $q = -1/4$ and $r = 1/4$. Note that in all space groups, a symmetry operation acting at or through a point will, when the translation vectors are applied, lead to a similar symmetry operations at $a/2$, $b/2$ and $c/2$. This rule is illustrated by the following diagram, using space group $P2$ as an example:

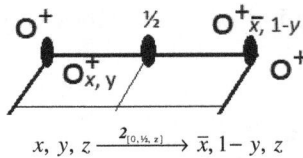

$$x, y, z \xrightarrow{\;2_{[0,\frac{1}{2},z]}\;} \bar{x}, 1-y, z$$

For convenience, in the above scheme for $P2_1/c$, $q = -1/4$ may be replaced by $q = 1/4$, and the usual diagram, with the convention of $\bar{1}$ at the origin, can be drawn now (Fig. 4.14). The completed diagram

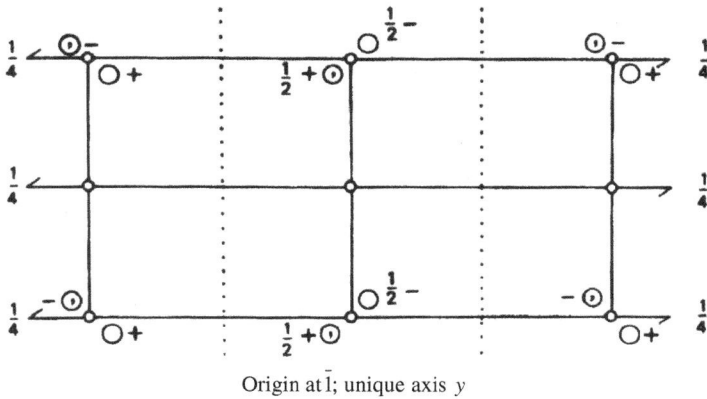

Origin at $\bar{1}$; unique axis y

4	e	1	$x, y, z;\ \bar{x}, \bar{y}, \bar{z};\ \bar{x}, \tfrac{1}{2}+y, \tfrac{1}{2}-z;\ x, \tfrac{1}{2}-y, \tfrac{1}{2}+z$

2	d	$\bar{1}$	$\tfrac{1}{2}, 0, \tfrac{1}{2};\ \tfrac{1}{2}, \tfrac{1}{2}, 0$
2	c	$\bar{1}$	$0, 0, \tfrac{1}{2};\ 0\ \tfrac{1}{2}, 0$
2	b	$\bar{1}$	$\tfrac{1}{2}, 0, 0;\ \tfrac{1}{2}, \tfrac{1}{2}, \tfrac{1}{2}$
2	a	$\bar{1}$	$0, 0, 0;\ 0, \tfrac{1}{2}, \tfrac{1}{2}$

Fig. 4.14. The unit cell and environs of space group $P2_1/c$ in the (001) projection, showing the equivalent positions, the symmetry elements and coordinate data.

confirms that symmetry elements occur at the half-translation spacings along the a, b and c directions. The fractions on the diagram indicate the heights of the motifs in units of c; thus, the 2_1 screw axes lie at $[0, y, \pm^1/_4]$ and $[^1/_2, y, \pm^1/_4]$. Note that it is always permissible in crystallography to change a coordinate value by ± 1 by virtue of translational symmetry.

Self-assessment 4.5. Show by a drawing, or otherwise, that monoclinic space groups Pa, Pc and Pn represent one and the same arrangement of points in three-dimensional space.

The centred space groups in this class follow easily from the foregoing studies. Figure 4.15 is a stereoview showing the zinc and iodine atoms in the structure of diiodo-$(N, N, N', N'$-tetramethy

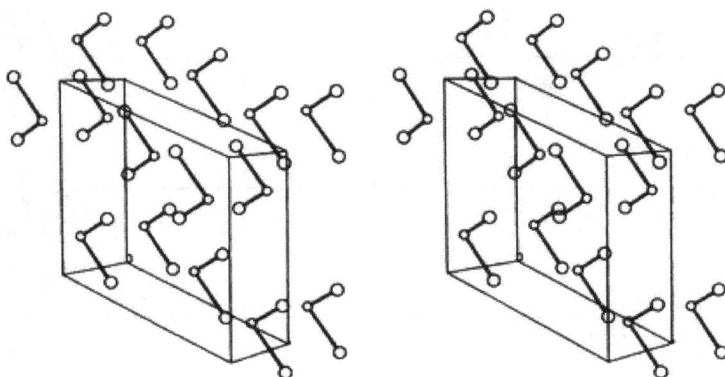

Fig. 4.15. Stereoview of the zinc and iodine atom positions in the crystal structure of diiodo-(N, N, N', N'-tetramethylethylenediamine)zinc(II) which crystallizes in space group $C2/c$. The atoms lie in sets of special equivalent positions of the space group.

lethylenediamine)zinc(II) $I_2[(CH_3)_2NCH_2CH_2N(CH_3)_2]Zn$, which crystallizes in space group $C2/c$. The zinc and iodine atoms occupy special equivalent positions. What are they?

4.5. International Tables for Crystallography

In studying space groups, it is important to refer from time to time to the standard literature on this subject. Illustrated tables of space groups appeared first in 1935 as two volumes entitled *Internationale Tabellen zur Bestimmung von Kristallstrukturen*, edited by WH Bragg *et al.* and published by Bornträger. A major revision of the space group tables was carried out under the editorship of NFM Henry and K Lonsdale and published in 1952 as the *International Tables for X-ray Crystallography*, Vol. I; revisions were made in 1963 and 1969; it is referred to here as Vol. I [5].

The International Union of Crystallography produces and updates now a wide range of crystallographic data. The *International Tables for Crystallography* is in nine volumes *A*, *A*1, *B*−*G* and a

Monoclinic 2/*m* $P \ 1 \ 2_1/c \ 1$ No .14 $P \ 2_1/c$
C_{2h}^5

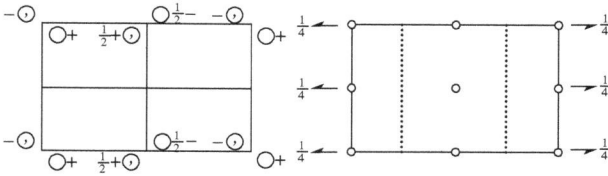

Origin at Ī; unique axis *b* 2ND SETTING

Number of positions, Wyckoff notation, and point symmetry	Coordinates of equivalent positions	Conditions limiting possible reflections
		General:
4 e 1	$x,y,z; \ \bar{x},\bar{y},\bar{z}; \ x,\frac{1}{2}+y,\frac{1}{2}-z; \ x,\frac{1}{2}+y,\frac{1}{2}-z.$	hkl: No conditions $h0l$: $l = 2n$ $0k0$: $k = 2n$
		Special: as above, plus
2 d Ī	$\frac{1}{2},0,\frac{1}{2}; \ \frac{1}{2},\frac{1}{2},0.$	
2 c Ī	$0,0,\frac{1}{2}; \ 0,\frac{1}{2},0.$	hkl: $k + l = 2n$
2 b Ī	$\frac{1}{2},0,0; \ \frac{1}{2},\frac{1}{2},\frac{1}{2}.$	
2 a Ī	$0,0,0; \ 0,\frac{1}{2},\frac{1}{2}.$	

Symmetry of special projections

(001) *pgm*; $a' = a, b' = b$ (100) *pgg*; $b' = b, c' = c$ (010) *p2*; $c' = c/2, a' = a$

Fig. 4.16. Diagrams of the general equivalent positions and symmetry elements for space group $P2_1/c$. [© The International Union of Crystallography. *International Tables for X-ray Crystallography*, Volume I (1969); reproduced by permission of the International Union of Crystallography.]

Database [6]. Volume A, often referred to as ITA, is a much revised version of Vol. I, and discusses all aspects of space group symmetry. Vol. I is a less detailed source than the current tables, but probably forms an easier introduction for those meeting the subject for the first time. Illustrative extracts from these two sources of space-group data will be presented here.

$P2_1/c$ C_{2h}^5 $2/m$ Monoclinic

No. 14 $P12_1/c1$ Patterson symmetry $P12/m1$

UNIQUE AXIS b, CELL CHOICE 1

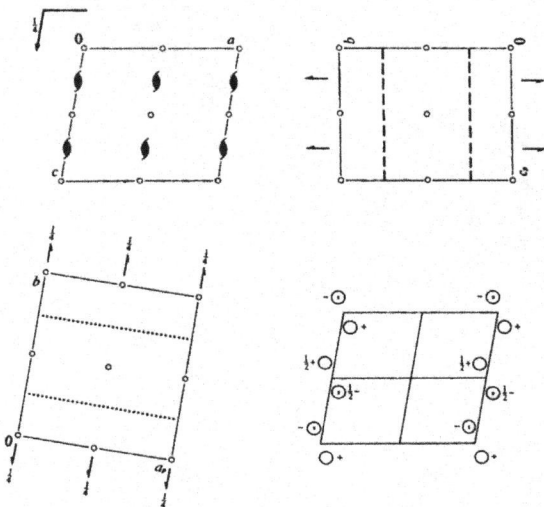

Origin at $\bar{1}$

Asymmetric unit $0 \le x \le 1$; $0 \le y \le \frac{1}{4}$; $0 \le z \le 1$

Symmetry operations

(1) 1   (2) $2(0,\frac{1}{2},0)$ $0,y,\frac{1}{4}$ (3) $\bar{1}$ $0,0,0$ (4) c $x,\frac{1}{4},z$

Fig. 4.17.   Diagrams of the general equivalent positions and symmetry elements for space group $P2_1/c$. [© The International Union of Crystallography. *International Tables for Crystallography*, Volume A (2005); reproduced by permission of the International Union of Crystallography.]

The example space group $P2_1/c$ is shown in Fig. 4.16 as it appears in Vol. I. However, Vol. I uses two diagrams for each space group, one shows the general equivalent positions while the other shows the symmetry elements. This feature is particularly useful in the higher

CONTINUED   **No. 14** $P2_1/c$

Generators selected   (1); $t(1,0,0)$; $t(0,1,0)$; $t(0,0,1)$; (2); (3)

Positions

Multiplicity, Wyckoff letter, Site symmetry		Coordinates				Reflection conditions
						General:
4	e	1	(1) x,y,z (2) $\bar{x},y+\frac{1}{2},\bar{z}+\frac{1}{2}$ (3) $\bar{x},\bar{y},\bar{z}$ (4) $x,\bar{y}+\frac{1}{2},z+\frac{1}{2}$			$h0l$: $l=2n$ $0k0$: $k=2n$ $00l$: $l=2n$
						Special: as above, plus
2	d	$\bar{1}$	$\frac{1}{2},0,\frac{1}{2}$ $\frac{1}{2},\frac{1}{2},0$			hkl : $k+l=2n$
2	c	$\bar{1}$	$0,0,\frac{1}{2}$ $0,\frac{1}{2},0$			hkl : $k+l=2n$
2	b	$\bar{1}$	$\frac{1}{2},0,0$ $\frac{1}{2},\frac{1}{2},\frac{1}{2}$			hkl : $k+l=2n$
2	a	$\bar{1}$	$0,0,0$ $0,\frac{1}{2},\frac{1}{2}$			hkl : $k+l=2n$

Symmetry of special projections

Along [001] $p2gm$ Along [100] $p2gg$ Along [010] $p2$
$\mathbf{a}'=\mathbf{a}$, $\mathbf{b}'=\mathbf{b}$ $\mathbf{a}'=\mathbf{b}$ $\mathbf{b}'=\mathbf{c}$, $\mathbf{a}'=\frac{1}{2}\mathbf{c}$ $\mathbf{b}'=\mathbf{a}$
Origin at $0,0,z$ Origin at $x,0,0$ Origin at $0,y,0$

Maximal non-isomorphic subgroups

I   [2] $P1c1$ $(Pc, 7)$ 1; 4
      [2] $P12_11$ $(P2_1, 4)$ 1; 2
      [2] $P\bar{1}$ (2)   1; 3

IIa   none
IIb   none

Maximal isomorphic subgroups of lowest index

IIc   [2] $P12_1/c1$ ($\mathbf{a}'=2\mathbf{a}$ or $\mathbf{a}'=2\mathbf{a},\mathbf{c}'=2\mathbf{a}+\mathbf{c}$) $(P2_1/c, 14)$; [3] $P12_1/c1$ ($\mathbf{b}'=3\mathbf{b}$) $(P2_1/c, 14)$

Minimal non-isomorphic supergroups

I   [2] $Pnna$ (52); [2] $Pmna$ (53); [2] $Pcca$ (54); [2] $Pbam$ (55); [2] $Pccn$ (56); [2] $Pbcm$ (57); [2] $Pnnm$ (58); [2] $Pbcn$ (60);
      [2] $Pbca$ (61); [2] $Pnma$ (62); [2] $Cmce$ (64)

II   [2] $A12/m1$ $(C2/m, 12)$; [2] $C12/c1$ $(C2/c, 15)$; [2] $I12/c1$ $(C2/c, 15)$; [2] $P12_1/m1$ ($\mathbf{c}'=\frac{1}{2}\mathbf{c}$) $(P2_1/m, 11)$;
      [2] $P12/c1$ ($\mathbf{b}'=\frac{1}{2}\mathbf{b}$) $(P2/c, 13)$

Fig. 4.17.   (*Continued*)

symmetry systems. In addition, the illustrations in Vol. I include the *crystal system, point-group symbol, full space group symbol* in the Hermann–Mauguin (H–M) notation, the *space group number*, and the *short* H–M *symbol* with the equivalent *Schönflies symbol* below it. The final line lists the *special projections* and their plane groups and the conditions limiting X-ray reflections.

The same data can be found in ITA as the next illustration shows, but many more crystallographic and structural features are presented. Two pages are devoted to a space group. Much of the first page will be familiar from what has been discussed already. The space group symbol is given in its full form when there is a possibility of ambiguity. Thus, the unique y-axis in $P2_1/c$ is shown by the full symbol $P1\frac{2_1}{c}1$.

The *Patterson symmetry* of $P2_1/c$ corresponds to the space group symmetry with glide planes and screw axes replaced by their non-translational counterparts, namely, $P\frac{2}{m}$, or $P1\frac{2}{m}1$ in full symbol. Patterson symmetry may be derived by the direct combination of a Bravais lattice and a point group. Patterson symmetry is symmorphic (Section 4.7), and is demonstrated for the simple example of space group Cc by the following tabulated coordinate data:

General equivalent positions	Vectors	Vectors with $x_0 = 2x$, $z_0 = 2z$
x, y, z	$\pm(2x, 0, 2z)$	$\pm(x_0, 0, z_0)$
$\bar{x}, y, \bar{z}$	$\pm(1/2, 1/2, 0)$	$\pm(1/2, 1/2, 0)$
$1/2 + x, \, 1/2 + y, \, z$	$\pm(1/2 + 2x, 1/2, 2z)$	$\pm(1/2 + x_0, 1/2, z_0)$
$1/2 - x, \, 1/2 + y, \, \bar{z}$	$\pm(1/2 - 2x, 1/2, 2\bar{z})$	$\pm(1/2 - x_0, 1/2, \bar{z}_0)$
	$\pm(1/2, 1/2, 0)$	$\pm(1/2, 1/2, 0)$
	$\pm(2x, 0, 2z)$	$\pm(x_0, 0, z_0)$

It is evident that the vector coordinates correspond to the symmetry $C2/m$.

The *symmetry operations* of the group are listed in ITA, and numbered so as to coincide with the coordinates of general equivalent positions given on the second page. The *generators selected* are those operators that produce the complete set of symmetry operators of the group. In the example of $P2_1/c$, they are **1** (identity), the $t(100)$, $t(010)$ and $t(001)$ translations, and symmetry operators numbered (2) and (3) from the list of operators. By applying these operators to a position x, y, z the coordinates of all general equivalent positions are obtainable. For example, operation (2) **2**(0, $1/2$, 0) $1/4$, y, 0 means a two-fold screw axis with a translation of $1/2$ along b acting around the line [$1/4$, y, 0] in the ($1/2$, $1/2$, 0)+, producing the action $x, y, z \xrightarrow{\;2_{1,[1/4,y,0]}\;} 1/2 - x, 1/2 + y, \bar{z}$. The second page contains also the special projections and their symmetries, as was discussed in relation to Vol. I, and other crystallographic data which will not be considered herein. For further elaboration of ITA and other aspects of crystal symmetry, the reader is referred to the literature [5–8].

4.6. General Change of Origin

In crystal structure studies, it may happen that a change of origin is desirable. The procedure is quite straightforward and will be illustrated by the following example.

Example 4.2. A set of coordinates referred to a given origin has the values $x.y, z; 1/2 - x, y, z; 1/2 + x, 1/2 - y, z; \bar{x}, 1/2 + y, z$. A new origin is set at $1/4$, 0, 0. In order to change the coordinates to this origin, the coordinates of the new origin are subtracted from the set of coordinates: $x - 1/4$ $y, z; 1/4 - x, \bar{y}, z; 1/4 + x, 1/2 - y, z; -x - 1/4, 1/2 + y, z$. Then, new variables are assigned as $x_o = x - 1/4, y_o = y, z_o = z$ to give $x_o, y_o, z_o; \bar{x}_o, \bar{y}_o, z_o; 1/2 + x_o, 1/2 - y_o, z_o; 1/2 - x_o, 1/2 + y_o, z_o$. This procedure can be applied generally, and, for convenience, the subscript can be dropped.

4.7. Symmorphic Space Groups, Non-symmorphic Space Groups and Hierarchy of Naming

A space group is termed *symmorphic* if the generating operations, excluding primitive and centring translations, leave at least one common point unmoved. The combination of the Bravais lattices with symmetry operators having no translational components yields the 73 symmorphic space groups. In general, there is at least one special position of point-group symmetry, and the permitted generators of symmorphic space groups are point-group symmetry operations. The Hermann–Mauguin symbols of symmorphic space groups do not contain glide or screw operations.

The remaining 157 space groups that contain elements of non-primitive translational symmetry (screw axes and glide planes) are termed *non-symmorphic*. It is not correct to state that a symmorphic space group contains neither glide planes nor screw axes. For example, $C2$ contains screw axes and Cm has glide planes, as discussed in Section 4.4.2. Furthermore, it is not necessary that glide planes or screw axes should be absent in a symmorphic group even in the absence of centring. For example, the symmorphic space group $P321$ has three 2_1 axes parallel to the three two-fold axes of the space group.

Many space groups, both symmorphic and non-symmorphic, can be named with more than one symbol and a system of preferred naming exists, albeit not without certain exceptions.

4.7.1. Hierarchy of space group names

In the Hermann–Mauguin notation, there is a hierarchy of naming where more than one symmetry element exists in a given orientation; for example, Cm has both mirror (m) planes and glide (a) planes normal to the y-axis. The general order for naming symmetry planes in a space group symbol is $m > a > b > c > n > d$, unless a particular reason exists for a departure from this rule. For example, $P2/c$ is used instead of $P2/a$ because $P2/c + C$ gives $C2/c$, whereas $P2/a + C$ gives $C2/m$. Again, the orthorhombic space group *Ibca* would be *Ibaa* according to the rule, but the standard name shows the relationship to *Pbca*.

4.8. Orthorhombic Space Groups

In the orthorhombic system, the point groups that exist are 222, $mm2$ and mmm, and must be considered together with the lattices that are designated by P, C, I and F unit-cell descriptors. Additionally, in the $mm2$ class, A-centring also is required, leading to a total of 59 space groups in this system.

4.8.1. Class 222

Recall that the three symmetry axes are parallel to the x-, y- and z-axes. The simplest space group in this class is $P222$. Also, with 2 replaced by 2_1 an apparently large number of space groups would appear to arise. However, they are reduced to nine unique types on account of equivalence under interchange of axes. For example, $P2_122$ and $P22_12$, which are also examples of non-symmorphic space groups, are equivalent under interchange of the x- and y-axes. A second 2_1 axis gives rise to the equivalence relation $P2_12_12 \equiv P22_12_1 \equiv P2_122_1$; it is straightforward to demonstrate that the combination of two 2_1 axes introduces a mutually perpendicular two-fold axis. The placing of three two-fold screw axes must be arranged such that no two of them intersect; otherwise, space group $P2_12_12$, or an equivalent, is formed.

When C centring is added, translations parallel to the x- and y-axes are introduced, so that while $C222 \equiv C2_122 \equiv C22_12, C222_1$ is a new group, as it introduces translation parallel to the z-axis. Arguments such as this lead to the deduction of nine unique space groups in this class, according to the scheme outlined hereunder:

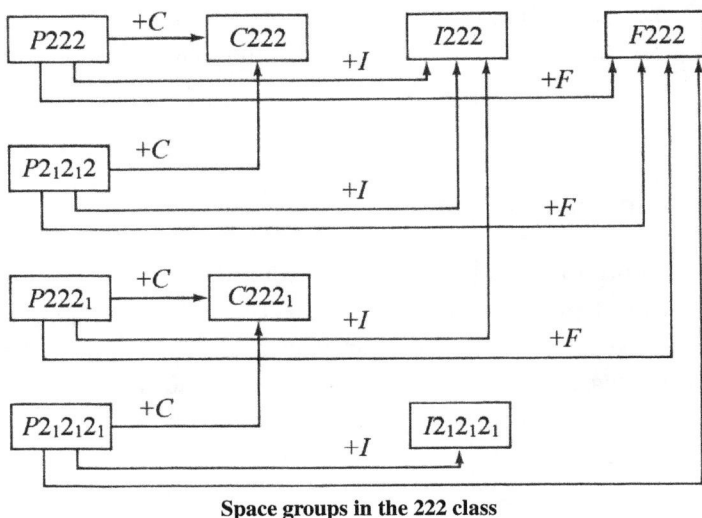

Space groups in the 222 class

In space group $P2_12_12_1$, the three mutually perpendicular 2_1 screw axes do not intersect, and the space group diagram is shown in Fig. 4.18. There are more ways than one of setting this space group such that no two 2_1 axes intersect, but the figure given demonstrates the standard setting. Although this space group is non-centrosymmetric, the special projections are centric ($p2gg$); projection along a screw axis removes the sense of screw, that is, $2_1 \xrightarrow{\text{projection}} 2$.

Self-assessment 4.6. In the diagrammatic scheme of space groups in class 222, which space groups, if any, are symmorphic?

4.8.2. Class $mm2$

In class $mm2$, A-centring arises in addition to C-centring as a unique space group descriptor. Thus, $Amm2$ is different from $Cmm2$: in $Cmm2$, the unique, two-fold axis is normal to the x-y plane, whereas

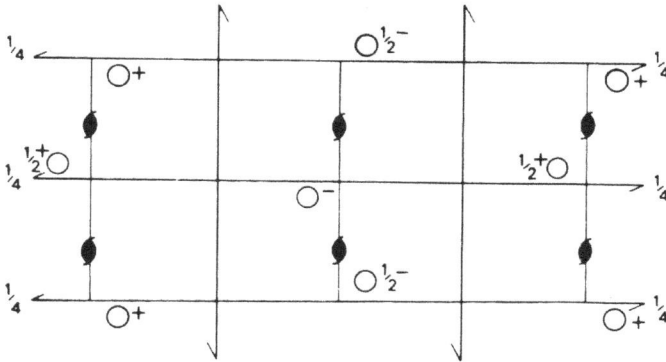

Origin halfway between three pairs of non-intersecting screw axes

				Limiting conditions
4	a	1	$x, y, z; \frac{1}{2} - x, \bar{y}, \frac{1}{2} + z; \frac{1}{2} + x, \frac{1}{2} - y, \bar{z}; \bar{x}, \frac{1}{2} + y, \frac{1}{2} - z.$	hkl: $\left.\begin{array}{l} 0kl: \\ h0l: \\ hk0: \end{array}\right\}$ None $h00: h = 2n$ $0k0: k = 2n$ $00l: l = 2n$

Symmetry of special projections
(001) p2gg   (100) p2gg   (010) p2gg

Fig. 4.18.   The unit cell and environs of space group $P2_12_12_1$ in the (001) projection, showing the general equivalent positions, the symmetry elements and selected space-group data. Note that the three 2_1-axes are non-intersecting. What arises if any two of these axes intersect?

in $Amm2$ ($\equiv Bmm2$) it is normal to the y-z ($\equiv x$-z) plane, so that it constitutes a different spatial arrangement. Space groups such as these are termed *polar space groups* and the corresponding point groups are polar point groups.

The essence of a polar point group is that it contains a line (axis) known as a *polar direction* joining points that are unmoved by the point-group operations. In polar space groups, the origin is not fixed by the symmetry: compare point groups 2, m and $2/m$.; Which group/s is/are polar? In a polar group, there is no symmetry operator $\boldsymbol{R}$ for which $\boldsymbol{R}r = \bar{r}$, and a crystal with a polar axis may show different physical properties at each end of the polar axis. Such crystals can exhibit pyroelectric and piezoelectric effects; they develop an electric charge of their surfaces when subjected to heat or other stress:

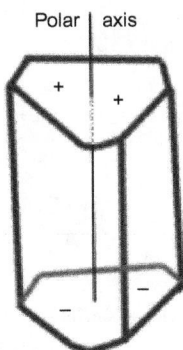

Pyroelectric crystal under applied heat

One form of potassium dihydrogen phosphate, a crystal that shows a pyroelectric effect, has been reported in space $Fdd2$, and is an example which allows a discussion of the diamond (d) glide plane. From Table 2.7, the symmetry translations for the d glide are of the type $(a \pm b)/4$, and, in space group $Fdd2$, the d-glide planes are normal to the x- and y-axes.

Diamond glide planes occur only in conjunction with I and F unit cells in the orthorhombic, tetragonal and cubic crystal systems. Although $Pdd2$ does not exist, it can be used as an approach to $Fdd2$. As before, a specification can be made, again using just the meaning of the Hermann–Mauguin space group symbol together with knowledge of the symmetry operations, placing the origin at 2, by convention:

$$2 \text{ is the line } [0, 0, \; z]$$
$$d \text{ is the plane } (p, y, z)$$
$$d \text{ is the plane } (x, q, z)$$

As with the earlier examples, a scheme can be set up as follows, and is illustrated with the diagram below. The d-glide has a directional quality, as will appear shortly, so the second d-glide operation is symbolized as $-d$:

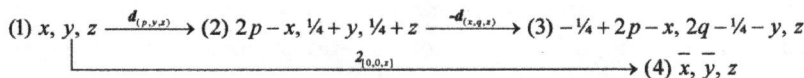

(1) $x, y, z \xrightarrow{\;d_{(p,y,z)}\;}$ (2) $2p - x, \frac{1}{4} + y, \frac{1}{4} + z \xrightarrow{\;-d_{(x,q,z)}\;}$ (3) $-\frac{1}{4} + 2p - x, 2q - \frac{1}{4} - y, z$

$\xrightarrow{\;\;\;\;\;\;\;\;\;2_{(0,0,z)}\;\;\;\;\;\;\;\;\;}$ (4) $\bar{x}, \bar{y}, z$

Scheme for space group $Fdd2$

Since points (3) and (4) are one and the same ($\boldsymbol{dd} \equiv \boldsymbol{2}$), $p = q = 1/8$ and the coordinate list can be started: x, y, z; $\bar{x}, \bar{y}, z$; $1/4 - x, 1/4 + y$; $1/4 + z$; $1/4 + x, 1/4 - y, 1/4 + z$; these coordinates are plotted on the partial diagram hereunder, which shows the upper, right octant of the unit cell:

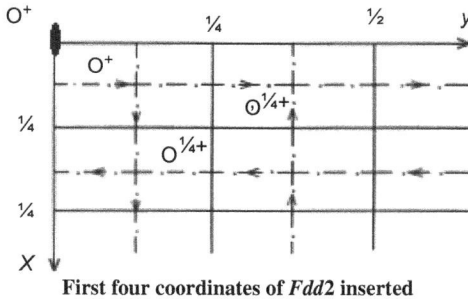

First four coordinates of *Fdd2* inserted

If these four coordinate positions are repeated with the F translations $(0, 1/2, 1/2; 1/2, 0, 1/2; 1/2, 1/2, 0)$, the complete diagram for space group $Fdd2$ is obtained. Figure 4.19 shows diagrams of the general equivalent positions and symmetry elements for this space group. In fact, if the m planes in $Fmm2$ are replaced by d-glides, $Fdd2$ results.

In the d-glide operation, the amount of the glide translation is one-half of the resultant of the two possible glide translations in the x- and y-directions. The arrows show the direction of the horizontal component of the translation when the z-component is positive. In $Fdd2$, the d-glides are separated by $1/4$; there are also separations of $1/4$ in space group $Fddd$. Similar remarks apply to the d-glides in the tetragonal and cubic systems. A further useful discussion on d-glide symmetry, particularly in relation to space group $Fddd$, may be found in the literature [8].

Self-assessment 4.7. (a) What are the coordinates of the fourth general equivalent position lying in the upper left quadrant of the diagram for $Fdd2$? (b) What are coordinates of the two-fold axis lying within the upper left octant of this space group diagram?

(a)

(b)

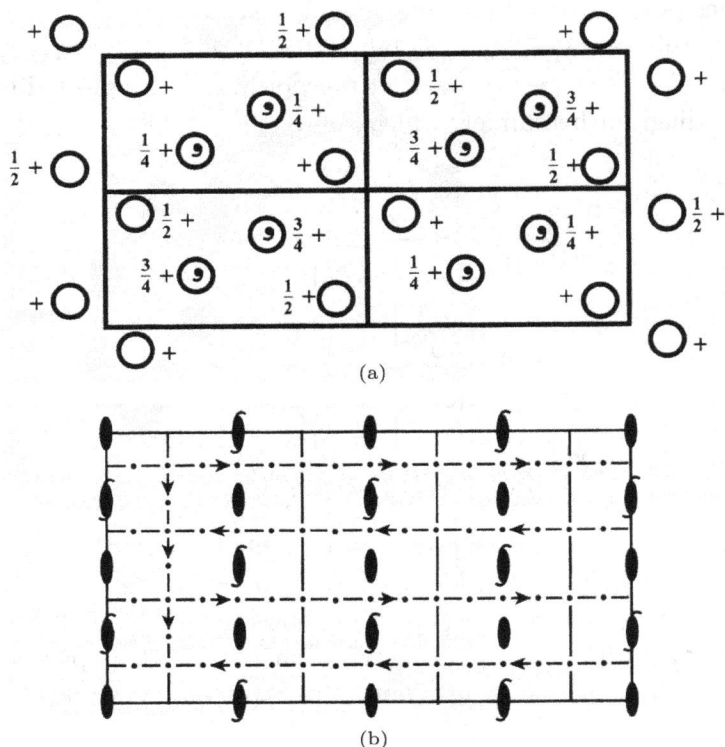

Fig. 4.19. Diagrams showing (a) the general equivalent positions, (b) the symmetry elements, for the space group *Fdd2* in the (001) projection. Note the translations of $1/4$ and their directional character.

4.8.3. Class *mmm*

The orthorhombic class *mmm* is centrosymmetric; its full symbol is $\frac{2}{m}\frac{2}{m}\frac{2}{m}$. The centre of symmetry may be defined by the intersection of three symmetry planes. However, in the presence of screw axes or glide planes, the centre of symmetry will be set off from the intersection of the symmetry planes by an amount dependent upon the nature of the symmetry planes and their relative orientation. The half-shift rule, already described, may be applied to locate the correct orientation of the symmetry elements in order that the centre of symmetry is sited at the origin.

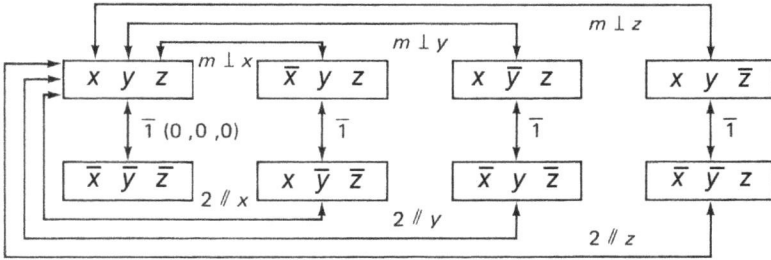

Fig. 4.20. Sign changes of a general point x, y, z under the symmetry operations of point group mmm; the arrow symbols indicate the symmetry changes. What symmetry operations link the coordinates in the other boxes in the diagram?

First, it is useful to consider how a point x, y, z changes in signs according to the symmetry, and it is illustrated in Fig. 4.20. It may be seen that reflection produces a single sign change, rotation produces two sign changes, and inversion changes the signs of x, y and z. These relationships apply in the monoclinic and orthorhombic crystal systems, and in other situations where the symmetry operators are either parallel to or contain the reference axes.

It has been shown that $\boldsymbol{m}_{[010]}\,\boldsymbol{m}_{[100]} \equiv \boldsymbol{2}_{[00z]}$. The addition of a centre of symmetry enhances the relationship to $\boldsymbol{m}_{[001]}\,\boldsymbol{m}_{[010]}\,\boldsymbol{m}_{[100]}$, or $\dfrac{2}{\boldsymbol{m}_{[001]}}\,\dfrac{2}{\boldsymbol{m}_{[010]}}\,\dfrac{2}{\boldsymbol{m}_{[100]}}$ in full, $\equiv \overline{\boldsymbol{1}}_{[000]}$.[d] However, from the foregoing, it would be expected that in the presence of translational symmetry, the site of the $\overline{1}$ point will not lie at the intersection of the three symmetry planes in all cases; its position can be determined by the half-shift rule.

Consider the space group $Pbcn$, for example. The full symbol notation implies the following setting, taking the origin at $\overline{1}$:

$$\overline{1} \ at \ 0, \ 0, \ 0$$
$$b \ \| \ (p, \ y, \ z)$$
$$c \ \| \ (x, \ q, \ z)$$
$$n \ \| \ (x, \ y, \ r)$$
$$2_P \ [x, \ B, \ C]$$
$$2_Q \ [A, \ y, \ C']$$
$$2_R \ [A', \ B', \ z]$$

[d] In all, eight such points per unit cell.

and the nine unknown parameters will be determined now based on the fact that $ncb \equiv \bar{1}$. The following scheme is set up:

$$(1)\ x,\ y,\ z \xrightarrow{\ b_{(p,y,z)}\ } (2)\ 2p-x,\ \tfrac{1}{2}+y,\ z \xrightarrow{\ c_{(x,q,z)}\ } (3)\ 2p-x,\ 2q-\tfrac{1}{2}-y,\ \tfrac{1}{2}+z$$

$$\tfrac{1}{2}+2p-x,\ 2q-y,\ 2r-\tfrac{1}{2}-z\ (4) \xleftarrow{\ n_{(x,y,r)}\ }$$

$$\xrightarrow{\ \bar{1}_{0,0,0}\ } (5)\ \bar{x},\ \bar{y},\ \bar{z}$$

Scheme for space group *Pbcn*

Since points (4) and (5) are one and the same, $p = -\tfrac{1}{4}, q = 0$, $r = \tfrac{1}{4}$, and it has been shown that $p = \tfrac{1}{4}$ is equally correct. Now, the complete set of coordinates of general equivalent positions and other relevant data can be determined and added to the space group

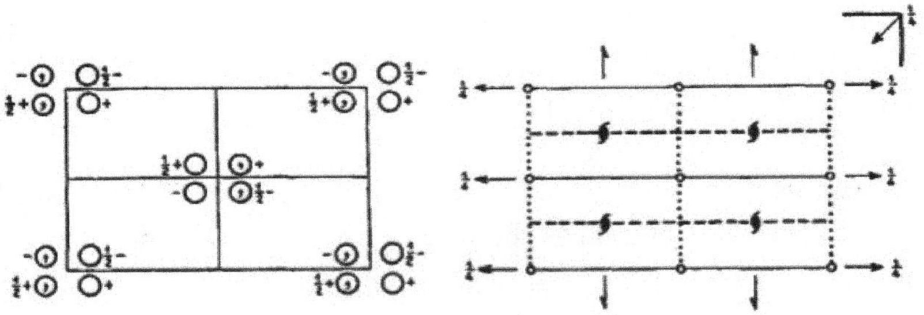

Origin at $\bar{1}$

Number of positions, Wyckoff notation, and site symmetry			Coordinates of equivalent positions
8	*d*	1	(1) x, y, z; (2) $\tfrac{1}{2}-x, \tfrac{1}{2}+y, z$; (3) $x, \bar{y}, \tfrac{1}{2}+z$; (4) $\tfrac{1}{2}+x, \tfrac{1}{2}+y, \tfrac{1}{2}-z$ (5) $\bar{x}, \bar{y}, \bar{z}$; (6) $\tfrac{1}{2}+x, \tfrac{1}{2}-y, \bar{z}$; (7) $\bar{x}, y, \tfrac{1}{2}-z$; (8) $\tfrac{1}{2}-x, \tfrac{1}{2}-y, \tfrac{1}{2}+z$
4	*c*	2	$0, y, \tfrac{1}{4}$; $0, \bar{y}, \tfrac{3}{4}$; $\tfrac{1}{2}, \tfrac{1}{2}+y, \tfrac{1}{4}$; $\tfrac{1}{2}, \tfrac{1}{2}-y, \tfrac{3}{4}$;
4	*b*	$\bar{1}$	$0, \tfrac{1}{2}, 0$; $0, \tfrac{1}{2}, \tfrac{1}{2}$; $\tfrac{1}{2}, 0, 0$; $\tfrac{1}{2}, 0, \tfrac{1}{2}$
4	*a*	$\bar{1}$	$0, 0, 0$; $0, 0, \tfrac{1}{2}$; $\tfrac{1}{2}, \tfrac{1}{2}, 0$; $\tfrac{1}{2}, \tfrac{1}{2}, \tfrac{1}{2}$

Symmetry of special projections

(001) *cmm*; $a' = a, b' = b$ (100) *pgm*; $b' = b/2, c' = c$ (010) *pgm*; $a' = a, c' = c/2$

Fig. 4.21. The general equivalent positions, symmetry elements and space-group data for the unit cell and environs of space group *Pbcn* in the (001) projection.

diagrams, as shown in Fig. 4.21; the symbol in the top right-hand corner of the figure indicates an n-glide plane at $c = \pm 1/4$.

This type of scheme, together with the information on sign changes, allows the nature and position of the two-fold axes to be determined:

(1) x, y, z; (2) $1/2 - x, 1/2 + y, z$; (3) $x, \bar{y}, 1/2 + z$; (4) $1/2 + x, 1/2 + y, 1/2 - z$;

(5) $\bar{x}, \bar{y}, \bar{z}$; (6) $1/2 + x, 1/2 - y, \bar{z}$; (7) $\bar{x}, y, 1/2 - z$; (8) $1/2 - x, 1/2 - y, 1/2 + z$

From the pair of coordinates 1, 6 on the list of equivalent positions, a 2_1 axis lies on the line $[x, 1/4, 0]$; similarly, points 1 and 7 give symmetry 2 along $[0, y, 1/4]$ and points 1 and 8 give 2_1 symmetry along $[1/4, 1/4, z]$. Thus, the full symbol is $P\frac{2_1}{b}\frac{2}{c}\frac{2_1}{n}$.

Centred space groups in this class can be treated usually as $P + a$ centring condition. It will not always lead directly to the standard coordinates of equivalent positions, but they may be obtained by an appropriate change of origin.

Self-assessment 4.8. The coordinates in the sequence $(1) \rightarrow (2)$ $\rightarrow (3) \rightarrow (4)$ listed below represent successive reflections across planes normal, respectively, to x, y and z for a space group in the mmm class; the symmetry at the origin is $\bar{1}$. Determine the positions of all symmetry elements and the full space group symbol.

(1) $x, y, z \rightarrow$ (2) $2p - x, 1/2 + y, 1/2 + z$

(3) $2p - x, 2q - 1/2 - y, 1/2 + z \rightarrow$ (4) $1/2 + 2p - x, 1/2 - y, 2r - z$

4.9. Standard and Alternative Setting of Space Groups

Alternative settings exist for the space group symbols while maintaining right-handed reference axes.

Let $C\frac{2}{c}$ be an example in the monoclinic system with y as the unique axis and the standard (**abc**) setting. If a change is made from **abc** to a setting of **bca**, it is not difficult to show that the symbol is revised to $A\frac{2}{a}$. The orthorhombic system is a little more complicated, since the settings could be **abc** (standard), **cab, bca, a$\overline{c}$b, ba$\overline{c}$** and $\overline{c}$**ba**, and examples will be drawn from the *mmm* class.

Certain space groups in the *mmm* class do not take the standard origin on $\overline{1}$. An example is space group *Pban* for which the origin is selected at a 222 point of symmetry; the International Tables lists also the alternative setting on $\overline{1}$. One reason for the choice of 222 is that it leads to a neater set of general equivalent positions, easier to manipulate manually if required. The change from one choice to the other involves a simple change of origin from 0, 0, 0 (222) to $\overline{1/4, 1/4}, 0$ ($\overline{1}$), as exemplified in Section 4.6.

Consider next the transformation of space group *Pbcn* from the **abc** (standard) setting to the **bca** setting, which is an example of working *from the standard symbol to a non-standard symbol*. Translation symbols are indicated in bold font and glide symbols are in italics in Fig. 4.22, which shows *Pbcn* in the standard setting; the transformed symbols are indicated by primes.

The transformation implies that the first position symbol **a′** (new) was **b** (old), and the associated glide which was c is now b, as it is in the **b′** (new) direction. Likewise, the second position symbol **b′** was **c** become **b′** and the n-glide name remains as it as. The third position symbol **c′** was **a** and the associated glide which was b is now a. Thus,

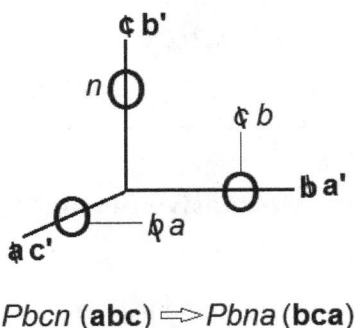

$$\text{Pbcn}\ (\textbf{abc}) \Rightarrow \text{Pbna}\ (\textbf{bca})$$

Fig. 4.22. Illustration of the transformation of a space group symbol from the **abc** (standard) setting to the **bca** (a′b′c′) setting in the *mmm* class; the symbol **O** represents a symmetry plane and a thin line indicates the direction of glide translation.

the new symbol is *Pbna*. When working from a non-standard symbol to the standard symbol, the procedure is reversed, as the following exercise will demonstrate. A complete set of transformations is tabulated in the literature [5, 6].

Self-assessment 4.9. Transform the orthorhombic space group *Bbmb* in the **bca** setting to its form in the standard (**abc**) setting.

4.10. Tetragonal Space Groups

A total of 68 tetragonal space groups emerge from the two lattices and seven classes in this system; three tetragonal groups will be considered here. Space groups in class 4 are straightforward. In space group $\bar{4}$, the origin is best chosen on the inversion axis; this space group can be treated then as $(P\bar{4} + I)$, which leads to the following partial space group diagram, enabling the complete diagram to be built up.

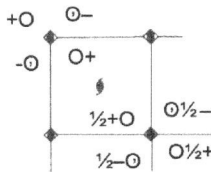

Partial diagram for space group $I\bar{4}$

4.10.1. Class $\frac{4}{m}$

In this centrosymmetric class, space group $P\frac{4_2}{n}$ will be examined. The symbol indicates the presence of a 4_2-axis along z and an n-glide plane normal to it. The space group diagram can be obtained in the manner already described. First, from Section 4.2.1, it may be expected that the combination of a four-fold rotation and a centre of symmetry will lead to a four-fold inversion axis:

$$x, y, z \xrightarrow{\ \bar{1}\ } \bar{x}, \bar{y}, \bar{z} \xrightarrow{\ 4\ } y, \bar{x}, \bar{z}$$

$$x, y, z \xrightarrow{\qquad \bar{4} \qquad} \bar{y}, x, \bar{z}$$

Operations: $4_2\, \bar{1} \equiv \bar{4}$

Space group $P\frac{4_2}{n}$ will be set with $\bar{4}$ at the origin, with its inversion point at $c/2$. Hence, the n-glide planes must lie at $\pm c/4$ in the diagram since they themselves are separated by $c/2$ in the cell. Then, the space group may be set as

$$\bar{4} \text{ at } 0, 0, 0$$
$$n \parallel (x, y, \tfrac{1}{4})$$
$$\bar{1} \text{ at } (p, q, r)$$

Then, the following scheme can be derived ($\bar{4}^2$ means two successive $\bar{4}$ operations):

$$x, y, z \xrightarrow{\ \bar{4}^2_{(0,0,0)}\ } \bar{x}, \bar{y}, z \xrightarrow{\ n(x,\,y,\,1/4)\ } \tfrac{1}{2} - x, \tfrac{1}{2} - y, \tfrac{1}{2} - z$$

$$x, y, z \xrightarrow{\qquad \bar{1}_{(p,q,r)} \qquad} 2p - x, 2p - y, 2r - z$$

Scheme for space group $P4_2/n$

The points $\tfrac{1}{2} - x, \tfrac{1}{2} - y, \tfrac{1}{2} - z$ and $2p - x, 2q - y, 2r - z$ are one and the same, so that $p = q = r = \tfrac{1}{4}$; the space-group diagram is shown in Fig. 4.23.

The diagram shows that 2 is a subgroup of both $\bar{4}$ and 4_2. An alternative setting of this space group takes $\bar{1}$ at the origin, and a simple change of origin can be applied to achieve this setting.

Self-assessment 4.10. What are the coordinates, Wyckoff notation and site symmetry for the sets of special equivalent positions in space group $P\frac{4_2}{n}$, origin on $\bar{1}$?

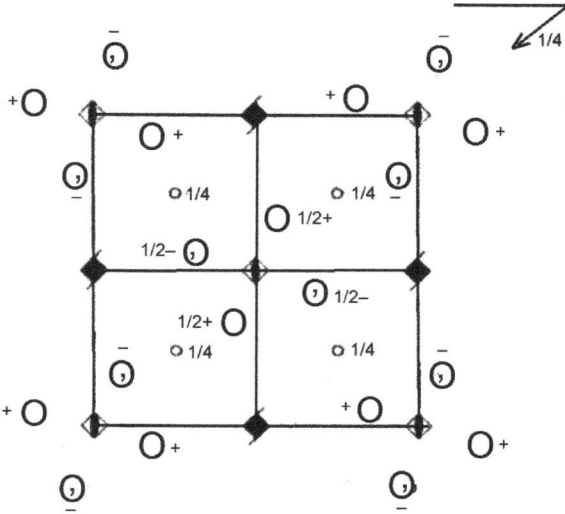

Fig. 4.23. Unit cell and environs of the tetragonal space group $P4_2/n$. The origin is on $\bar{4}$ but may be reset to $\bar{1}$ by the change of origin procedure.

4.10.2. Class $4mm$

Space groups in this class can be derived readily by setting the origin on the four-fold axis; the origin is not defined with respect to the z-direction. The symmetry action with a diagonal plane was considered under Section 4.3.2 and in Self-assessment 4.3. Consider space group $P4_2bc$. This group can be set as follows:

$$4_2 \text{ at } [0, 0, z]$$
$$b \parallel (p, y, z)$$
$$c \parallel (q, q, z)$$

Then, applying the symmetry operations leads to the following scheme:

$$x, y, z \xrightarrow{\ b_{(p,y,z)}\ } 2p - x, \tfrac{1}{2} + y, z \xrightarrow{\ c_{(q,q,z)}\ } q - \tfrac{1}{2} - y, q - 2p + x, \tfrac{1}{2} + z$$
$$\xdownarrow{} \xrightarrow{\ 4_{2[0,0,z]}\ } \bar{y}, x, \tfrac{1}{2} + z$$

Scheme for space group $P4_2bc$

It shows that $q = 1/2$ and $p = q/2 = 1/4$, from which the coordinates of the general equivalent positions may be deduced. After the first four coordinates have been derived from the scheme, the fact that 2 is a sub-group of 4_2 can be used to deduce the remaining four coordinates:

$$x, y, z; \ 1/2 - x, 1/2 + y, z; \ \bar{y}, x, 1/2 + z; \ 1/2 - y, 1/2 - x, 1/2 + z$$
$$\bar{x}, \bar{y}, z; \ 1/2 + x, 1/2 - y, z; \ y, \bar{x}, 1/2 + z; \ 1/2 + y, 1/2 + x, 1/2 + z$$

The reader is invited to derive the sets of special equivalent positions for this space group; they may be checked from references [5] or [6].

4.10.3.   Class $\frac{4}{m}mm$

As a final example in the tetragonal system, space group $P\frac{4_2}{n}bc$ will be studied. The space groups in class $\frac{4}{m}mm$ are a little more complex than those studied so far, but they can be solved by applying the same procedure as before. Obvious choices for the origin are $\bar{4}, \frac{4_2}{n}$ or $\bar{1}$. The second of these is used here, the others being available in the literature [5,6]; the space group can be set then as follows:

$$4_2 \text{ at } 0, 0, z$$
$$n \parallel (x, y, r)$$
$$\bar{1} \text{ at } p, q, 0$$
$$b \parallel (s, y, z)$$
$$c \parallel (t, t, z)$$

It is permissible to choose $z = 0$ for the origin on either 4_2 or $\bar{1}$, element, since it is not defined with respect to z. More than one scheme is required in order to solve completely this space group for the positions of all symmetry elements, as will be shown. Scheme 1 involves the symmetry elements $4_2, n$ and $\bar{1}$; recalling that $4_2{}^2 \equiv 2$, the following scheme (1) can be applied:

$$x, y, z \xrightarrow{\ 4_{2[0,0,z]}^{\ 2}\ } \bar{x}, \bar{y}, z \xrightarrow{\ n_{(x,y,r)}\ } \tfrac{1}{2}-x, \tfrac{1}{2}-y, 2r-z$$
$$\xrightarrow{\qquad \bar{1}_{p,q,0}\qquad} 2p-x, 2q-y, \bar{z}$$

Scheme (1) for space group $P\frac{4_2}{n}bc$

Scheme (1) leads to two expressions for one and the same position; hence, $p = q = \tfrac{1}{4}$ and $r = 0$. Scheme (2) involves the symmetry elements $\frac{4_2}{n}$, b and c, together with $\bar{1}$ as just determined:

$$x, y, z \xrightarrow{\ 4_{2[0,0,z]}/n_{(x,y,0)}\ } \tfrac{1}{2}-y, \tfrac{1}{2}+x, \tfrac{1}{2}-z \xrightarrow{\ b_{(s,y,z)}\ } 2s-\tfrac{1}{2}+y, x, \tfrac{1}{2}-z$$

$$\Big\downarrow c_{(t,t,z)}$$

$$t-x, t-2s+\tfrac{1}{2}-y, \bar{z}$$

$$x, y, z \xrightarrow{\qquad\qquad \bar{1}_{(\frac{1}{4},\frac{1}{4},0)}\qquad\qquad} \tfrac{1}{2}-x, \tfrac{1}{2}-y, \bar{z}$$

Scheme (2) for space group $P\frac{4_2}{n}bc$

Similarly, from this scheme, $t = \tfrac{1}{2}$ and $s = t/2 = \tfrac{1}{4}$. These results lead to the following arrangement of the symmetry elements $4_2, n, \bar{1}, b(a), c$; many more symmetry elements feature on a fully completed diagram:

Portion of space group $P\frac{4_2}{n}bc$

The standard setting of this space group places the origin on $\bar{4}$, and Fig. 4.24 shows the space group in this setting as presented in ITA. The reader is referred to Section 4.5 for a description of the items in this figure; note particularly the usefulness of the numbering that relates the individual symmetry elements to the coordinates of the general equivalent position. The topics on subgroups are not treated in this book, but a discussion of them is given in the literature [5–8].

$P4_2/nbc$ **D_{4h}^{11}** **$4/mmm$** **Tetragonal**

No. 133 $P\ 4_2/n\ 2/b\ 2/c$ Patterson symmetry $P4/mmm$

ORIGIN CHOICE 1

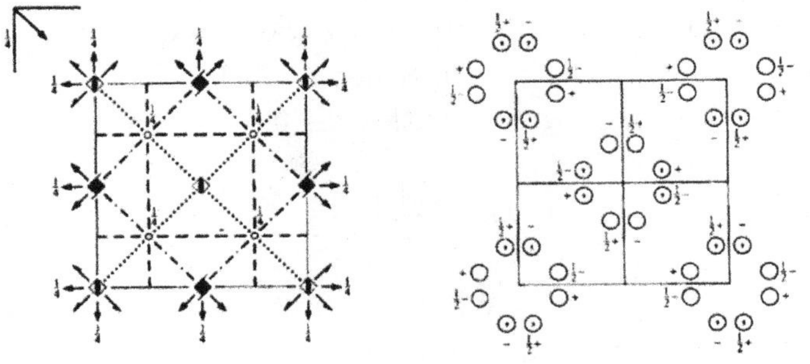

Origin at $\bar{4}12_1/c$, at $-\frac{1}{4},\frac{1}{4},-\frac{1}{4}$ from $\bar{1}$

Asymmetric unit $0 \leq x \leq \frac{1}{2};\ \ 0 \leq y \leq \frac{1}{2};\ \ 0 \leq z \leq \frac{1}{4}$

Symmetry operations

(1) 1	(2) 2 0,0,z	(3) $4^+(0,0,\frac{1}{2})$ $0,\frac{1}{2},z$	(4) $4^-(0,0,\frac{1}{2})$ $\frac{1}{2},0,z$
(5) 2 $0,y,\frac{1}{4}$	(6) 2 $x,0,\frac{1}{4}$	(7) $2(\frac{1}{2},\frac{1}{2},0)$ $x,x,0$	(8) 2 $x,\bar{x}+\frac{1}{2},0$
(9) $\bar{1}$ $\frac{1}{4},\frac{1}{4},\frac{1}{4}$	(10) $n(\frac{1}{2},\frac{1}{2},0)$ $x,y,\frac{1}{4}$	(11) $\bar{4}^+$ $0,0,z;$ $0,0,0$	(12) $\bar{4}^-$ $0,0,z;$ $0,0,0$
(13) a $x,\frac{1}{4},z$	(14) b $\frac{1}{4},y,z$	(15) c $x,\bar{x},z$	(16) c x,x,z

Fig. 4.24. Unit cell and environs for space group $P\frac{4_2}{n}bc$ showing the equivalent positions, symmetry elements and space-group data (*Continued on p.* 145). [© The International Union of Crystallography. *International Tables for Crystallography*, Volume A (2005); reproduced by permission of the International Union of Crystallography.]

4.11. Space Groups in the Trigonal, Hexagonal and Cubic Systems

The space groups in these three systems have, in general, fewer representatives among known crystal structures, except in the case of elements and alloys; a small number of these space groups will be considered here.

4.11.1. Trigonal space groups

As discussed in Chapter 3, three-fold symmetry is compatible with both the rhombohedral and the hexagonal lattices that are used in the trigonal system. As an example, the data for space group $R3m$ in

Generators selected   (1); $t(1,0,0)$; $t(0,1,0)$; $t(0,0,1)$; (2); (3); (5); (9)

Positions

Multiplicity, Wyckoff letter, Site symmetry	Coordinates				Reflection conditions
					General:
16 k 1	(1) x,y,z	(2) $\bar{x},\bar{y},z$	(3) $\bar{y}+\frac{1}{2},x+\frac{1}{2},z+\frac{1}{2}$	(4) $y+\frac{1}{2},\bar{x}+\frac{1}{2},z+\frac{1}{2}$	$hk0: h+k=2n$
	(5) $\bar{x},y,\bar{z}+\frac{1}{2}$	(6) $x,\bar{y},\bar{z}+\frac{1}{2}$	(7) $y+\frac{1}{2},x+\frac{1}{2},\bar{z}$	(8) $\bar{y}+\frac{1}{2},\bar{x}+\frac{1}{2},\bar{z}$	$0kl: k=2n$
	(9) $\bar{x}+\frac{1}{2},\bar{y}+\frac{1}{2},\bar{z}+\frac{1}{2}$	(10) $x+\frac{1}{2},y+\frac{1}{2},\bar{z}+\frac{1}{2}$	(11) $y,\bar{x},\bar{z}$	(12) $\bar{y},x,\bar{z}$	$hhl: l=2n$
	(13) $x+\frac{1}{2},\bar{y}+\frac{1}{2},z$	(14) $\bar{x}+\frac{1}{2},y+\frac{1}{2},z$	(15) $\bar{y},\bar{x},z+\frac{1}{2}$	(16) $y,x,z+\frac{1}{2}$	$00l: l=2n$
					$h00: h=2n$
					Special: as above, plus
8 j ..2	$x,x+\frac{1}{2},0$	$\bar{x},\bar{x}+\frac{1}{2},0$	$\bar{x},x+\frac{1}{2},\frac{1}{2}$	$x,\bar{x}+\frac{1}{2},\frac{1}{2}$	$hkl: h+k+l=2n$
	$\bar{x}+\frac{1}{2},\bar{x},\frac{1}{2}$	$x+\frac{1}{2},x,\frac{1}{2}$	$x+\frac{1}{2},\bar{x},0$	$\bar{x}+\frac{1}{2},x,0$	
8 i .2.	$x,0,\frac{1}{4}$	$\bar{x},0,\frac{1}{4}$	$\frac{1}{4},x+\frac{1}{2},\frac{1}{4}$	$\frac{1}{4},\bar{x}+\frac{1}{2},\frac{1}{4}$	$hkl: h+k=2n$
	$\bar{x}+\frac{1}{2},\frac{1}{2},\frac{1}{4}$	$x+\frac{1}{2},\frac{1}{2},\frac{1}{4}$	$0,\bar{x},\frac{1}{4}$	$0,x,\frac{1}{4}$	
8 h .2.	$x,0,\frac{1}{4}$	$\bar{x},0,\frac{1}{4}$	$\frac{1}{4},x+\frac{1}{2},\frac{3}{4}$	$\frac{1}{4},\bar{x}+\frac{1}{2},\frac{3}{4}$	$hkl: h+k=2n$
	$\bar{x}+\frac{1}{2},\frac{1}{2},\frac{3}{4}$	$x+\frac{1}{2},\frac{1}{2},\frac{3}{4}$	$0,\bar{x},\frac{3}{4}$	$0,x,\frac{3}{4}$	
8 g 2..	$0,0,z$	$\frac{1}{2},\frac{1}{2},z+\frac{1}{2}$	$0,0,\bar{z}+\frac{1}{2}$	$\frac{1}{2},\frac{1}{2},\bar{z}$	$hkl: h+k,l=2n$
	$\frac{1}{2},\frac{1}{2},\bar{z}+\frac{1}{2}$	$0,0,\bar{z}$	$\frac{1}{2},\frac{1}{2},z$	$0,0,z+\frac{1}{2}$	
8 f 2..	$0,\frac{1}{2},z$	$0,\frac{1}{2},z+\frac{1}{2}$	$0,\frac{1}{2},\bar{z}+\frac{1}{2}$	$0,\frac{1}{2},\bar{z}$	$hkl: h+k,l=2n$
	$\frac{1}{2},0,\bar{z}+\frac{1}{2}$	$\frac{1}{2},0,\bar{z}$	$\frac{1}{2},0,z$	$\frac{1}{2},0,z+\frac{1}{2}$	
8 e $\bar{1}$	$\frac{1}{4},\frac{1}{4},\frac{1}{4}$	$\frac{3}{4},\frac{3}{4},\frac{1}{4}$	$\frac{1}{4},\frac{3}{4},\frac{3}{4}$	$\frac{3}{4},\frac{1}{4},\frac{3}{4}$ $\frac{3}{4},\frac{1}{4},\frac{1}{4}$ $\frac{1}{4},\frac{3}{4},\frac{1}{4}$ $\frac{3}{4},\frac{3}{4},\frac{3}{4}$ $\frac{1}{4},\frac{1}{4},\frac{3}{4}$	$hkl: h,k,l=2n$
4 d $\bar{4}$..	$0,0,0$	$\frac{1}{2},\frac{1}{2},\frac{1}{2}$	$0,0,\frac{1}{2}$	$\frac{1}{2},\frac{1}{2},0$	$hkl: h+k,l=2n$
4 c 2.22	$0,\frac{1}{2},0$	$0,\frac{1}{2},\frac{1}{2}$	$\frac{1}{2},0,\frac{1}{2}$ $\frac{1}{2},0,0$		$hkl: h+k,l=2n$
4 b 222.	$0,0,\frac{1}{2}$	$\frac{1}{2},\frac{1}{2},\frac{3}{4}$	$\frac{1}{2},\frac{1}{2},\frac{1}{4}$	$0,0,\frac{3}{4}$	$hkl: h+k,l=2n$
4 a 222.	$0,\frac{1}{2},\frac{1}{4}$	$0,\frac{1}{2},\frac{3}{4}$	$\frac{1}{2},0,\frac{1}{4}$	$\frac{1}{2},0,\frac{3}{4}$	$hkl: h+k,l=2n$

Symmetry of special projections

Along [001] $p4mm$
$a'=\frac{1}{2}(a-b)$ $b'=\frac{1}{2}(a+b)$
Origin at $0,0,z$

Along [100] $p2mm$
$a'=\frac{1}{2}b$ $b'=c$
Origin at $x,0,\frac{1}{4}$

Along [110] $p2mm$
$a'=\frac{1}{2}(-a+b)$ $b'=\frac{1}{2}c$
Origin at $x,x,0$

Maximal non-isomorphic subgroups

I
    [2] $P\bar{4}b2$ (117)   1; 2; 7; 8; 11; 12; 13; 14
    [2] $P\bar{4}2c$ (112)   1; 2; 5; 6; 11; 12; 15; 16
    [2] $P4_2bc$ (106)   1; 2; 3; 4; 13; 14; 15; 16
    [2] $P4_222$ (93)   1; 2; 3; 4; 5; 6; 7; 8
    [2] $P4_2/n11$ ($P4_2/n$, 86)   1; 2; 3; 4; 9; 10; 11; 12
    [2] $P2/n12/c$ ($Ccce$, 68)   1; 2; 7; 8; 9; 10; 15; 16
    [2] $P2/n2/b1$ ($Pban$, 50)   1; 2; 5; 6; 9; 10; 13; 14

IIa   none
IIb   none

Maximal isomorphic subgroups of lowest index
IIc   [3] $P4_2/nbc$ ($c'=3c$) (133); [9] $P4_2/nbc$ ($a'=3a,b'=3b$) (133)

Minimal non-isomorphic supergroups
I   none
II   [2] $C4_2/mmc$ ($P4_2/mcm$, 132); [2] $I4/mcm$ (140); [2] $P4/nbm$ ($c'=\frac{1}{2}c$) (125)

Fig. 4.24.   (*Continued*)

Fig. 4.25 illustrate the general equivalent positions and the symmetry elements for this space group with both obverse (standard setting) rhombohedral axes and hexagonal axes, the latter showing the triply primitive hexagonal H unit cell. The lattice points at $\pm(1/3, 2/3, 2/3)$

$R\,3\,m$
C_{3v}^{5}

No. 160 $R\,3\,m$ $3\,m$ Trigonal

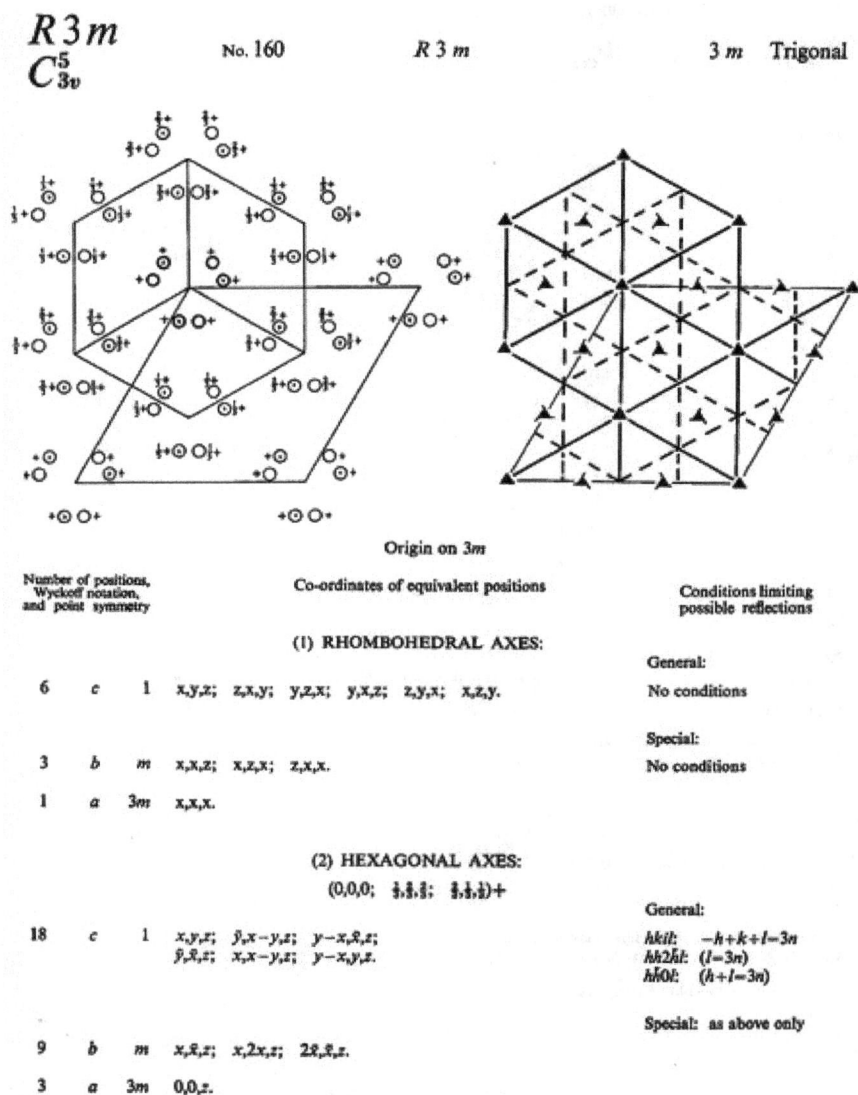

Origin on $3m$

Number of positions, Wyckoff notation, and point symmetry			Co-ordinates of equivalent positions	Conditions limiting possible reflections

(1) RHOMBOHEDRAL AXES:

General:

6 c 1 x,y,z; z,x,y; y,z,x; y,x,z; z,y,x; x,z,y.   No conditions

Special:

3 b m x,x,z; x,z,x; z,x,x.   No conditions

1 a 3m x,x,x.

(2) HEXAGONAL AXES:
$(0,0,0;\ \tfrac{2}{3},\tfrac{1}{3},\tfrac{1}{3};\ \tfrac{1}{3},\tfrac{2}{3},\tfrac{2}{3})+$

General:

18 c 1 $x,y,z;\ \bar y,x-y,z;\ y-x,\bar x,z;$
$\bar y,\bar x,z;\ x,x-y,z;\ y-x,y,z.$

hkil: $-h+k+l=3n$
hh2hl: $(l=3n)$
hh0l: $(h+l=3n)$

Special: as above only

9 b m $x,\bar x,z;\ x,2x,z;\ 2\bar x,\bar x,z.$

3 a 3m $0,0,z.$

Fig. 4.25. Diagrams of the unit cell and environs for space group $R3m$, showing the general equivalent positions and symmetry elements. [© The International Union of Crystallography. *International Tables for X-ray Crystallography*, Volume I (1969); reproduced by permission of the International Union of Crystallography.]

in the latter cell act as a form of centring, so that the mirror planes are interleaved by glide planes, and special conditions arise which will be noted shortly (see also Fig. 3.8ff).

A rhombohedron may be thought of as a cube that is extended along [111] (see Solution 3.3). Not surprisingly, the coordinates of the general equivalent positions follow the cyclic sequence in two sets:

$$x, y, z; \ z, x, y; \ y, z, x \quad \text{and} \quad y, x, z; \ z, y, x; \ x, z, y$$

In terms of the setting on hexagonal axes, the coordinates of three general equivalent positions may be obtained from Eq. (A4.6) with $\phi = \gamma = 120°$, so that

$$x, y, z \xrightarrow{\ 3_{[0,\,0,\,z]}\ } \bar{y}, x - y, z \xrightarrow{\ 3_{[0,\,0,\,z]}\ } y - x, y, z$$

The other three coordinates may be obtained by reflection of x, y, z across the plane normal to the y-axis, followed by three-fold rotation. Thus, the totality of general equivalent positions is

$$\begin{pmatrix} x, y, z; \ \bar{y}, x - y, z; \ y - x, y, z; \\ y, x, z; \ x - y, \bar{y}, z; \ \bar{x}, y - x, z. \end{pmatrix} (0, 0, 0; {}^2/_3, {}^1/_3, {}^1/_3; {}^1/_3, {}^2/_3, {}^2/_3) +$$

In space groups related to point group $3m$, while there is no crystallographic difference in the point groups that could be labelled $3m1$ and $31m$, the two space groups $P3m1$ and $P31m$ correspond to different spatial arrangements, as shown in Figs. 4.26 and 4.27. Considering first $P3m1$, a three-fold axis is placed at the origin, with the mirror planes set normal to the reference axes: Table 2.4 relates to the positions of m in $P3m1$; recall, $\bar{2}$ along x or $y \equiv m \perp x$ or y. When

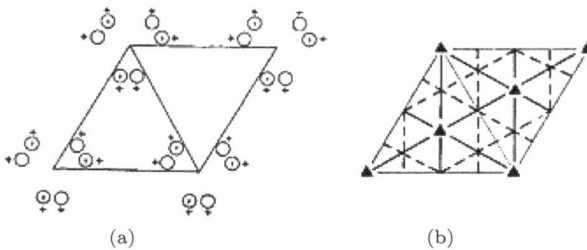

(a)   (b)

Fig. 4.26.   Unit cell and environs for space group $P3m1$, origin on 3, and showing (a) the general equivalent positions and (b) the symmetry elements.

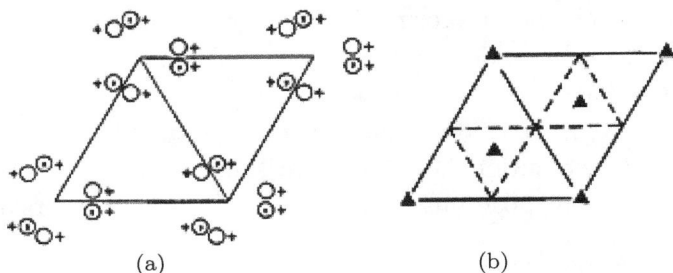

Fig. 4.27. Unit cell and environs for space group $P31m$, origin on 3, and showing (a) the general equivalent positions and (b) the symmetry elements. The mirror planes are rotated by 30° in relation to their position in $P3m1$; they contain the unit-cell boundaries.

the equivalent points are completed around the corners of the cell, three-fold axes are revealed along the lines $\pm[2/3, 1/3, z]$. The third position in the space group symbol is important as it distinguishes between the two space groups. Similar pairs exist for space groups related to 32 and $\bar{3}m$.

The general equivalent positions can be generated from Eq. (A4.3) which shows that a three-fold rotation about an axis normal to the $x - y$ plane generates $\bar{y}, x - y, z$ from x, y, z. Then, the mirror plane normal to the y-axis generates $y - x, y, z$ from x, y, z. Thus, the general and special equivalent positions in $P3m1$ evolve as

$$6 \quad e \quad 1 \quad x, y, z; \ \bar{y}, x - y, z; \ y - x, \bar{x}, z$$
$$x, x - y, z; \ y - x, y, z; \bar{y}, \bar{x}, z$$

$$3 \quad d \quad m \quad x, \bar{x}, z; \ x, 2x, z; \ 2\bar{x}, \bar{x}, z$$
$$1 \quad c \quad 3m \quad 2/3, 1/3, z$$
$$1 \quad b \quad 3m \quad 1/3, 2/3, z$$
$$1 \quad a \quad 3m \quad 0, 0, z$$

In the case of $P31m$, the first three general equivalent positions are the same as in $P3m1$. From the diagram in Fig. 4.27, it is evident that reflection across a mirror plane, specifically that, which is the diagonal of the unit cell shown, changes x, y, z to y, x, z; then, the remaining points are obtained by three-fold rotations from that reflected point. Thus, the general and special equivalent positions for this group are as follows:

$$
\begin{array}{llll}
6 & d & 1 & x,y,z;\ \bar{y},x-y,z;\ y-x,\bar{x},z \\
  &   &   & y,x,z;\ x-y,\bar{y},z;\ \bar{x},y-x,z \\[4pt]
3 & c & m & x,0,z;\ 0,x,z;\ \bar{x},\bar{x},z \\
2 & b & 3 & 1/3,2/3,z;\ 2/3,1/3,z \\
1 & a & 3m & 0,0,z
\end{array}
$$

4.11.2. Hexagonal space groups

A consideration of Table 2.4 shows a similarity between the hexagonal and tetragonal systems, the principal changes being six-fold rotation in place of four-fold rotation and a change in the relative orientation of the symmetry elements m and 2 in point groups $\bar{4}2m$ and $\bar{6}m2$. As an example from the hexagonal system, the space group $P6_3cm$ will be examined.

The space group is non-centrosymmetric. The Hermann–Mauguin symbol shows that the 6_3-axis is the line $[0, 0, z]$ and, again, the origin is not defined with respect to the z-direction. In this space group, the mirror planes are also the unit-cell outlines, and the general equivalent positions around these symmetry elements are shown in the partial space-group diagram below:

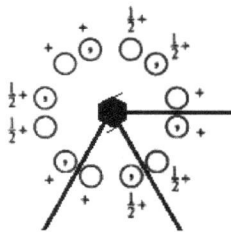

The 6_3 and m symmetry elements in space group $P6_3cm$

It is evident now that the c-glide planes lie mid-way between the mirror planes. If the positions of the m planes and c-glide planes are interchanged, then a space group with an arrangement of points different from that of $P6_3cm$ is obtained, as will be examined in Self-assessment 4.11 [5,6].

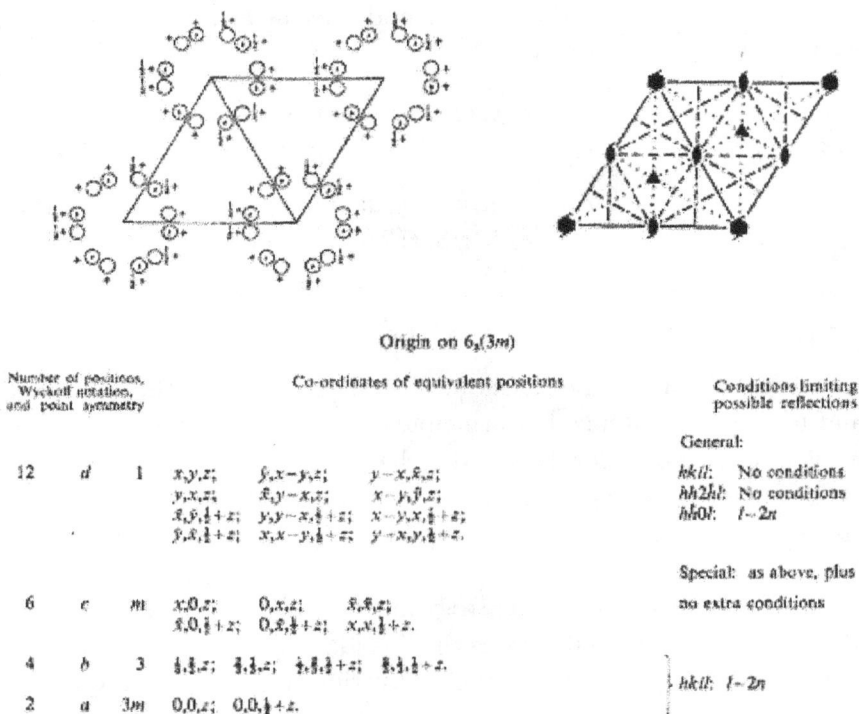

Origin on $6_3(3m)$

Number of positions, Wyckoff notation, and point symmetry			Co-ordinates of equivalent positions			Conditions limiting possible reflections
						General:
12	d	1	$x,y,z;$ $\bar{y},x-y,z;$ $y-x,\bar{x},z;$			$hkil:$ No conditions
			$y,x,z;$ $\bar{x},y-x,z;$ $x-y,\bar{y},z;$			$hh2hl:$ No conditions
			$\bar{x},\bar{y},\tfrac{1}{2}+z;$ $y,y-x,\tfrac{1}{2}+z;$ $x-y,x,\tfrac{1}{2}+z;$			$hh0l:$ $l=2n$
			$\bar{y},\bar{x},\tfrac{1}{2}+z;$ $x,x-y,\tfrac{1}{2}+z;$ $y-x,y,\tfrac{1}{2}+z.$			
						Special: as above, plus
6	c	m	$x,0,z;$ $0,x,z;$ $\bar{x},\bar{x},z;$			no extra conditions
			$\bar{x},0,\tfrac{1}{2}+z;$ $0,\bar{x},\tfrac{1}{2}+z;$ $x,x,\tfrac{1}{2}+z.$			
4	b	3	$\tfrac{1}{3},\tfrac{2}{3},z;$ $\tfrac{2}{3},\tfrac{1}{3},z;$ $\tfrac{1}{3},\tfrac{2}{3},\tfrac{1}{2}+z;$ $\tfrac{2}{3},\tfrac{1}{3},\tfrac{1}{2}+z.$			$hkil:$ $l=2n$
2	a	$3m$	$0,0,z;$ $0,0,\tfrac{1}{2}+z.$			

Fig. 4.28. Unit cell and environs of space group $P6_3cm$, showing (left) the general equivalent positions, (right) the symmetry elements, and (below) the coordinates of general and special equivalent positions and the conditions limiting X-ray reflections from crystals with this space group. [© The International Union of Crystallography. *International Tables for X-ray Crystallography*, Volume I (1969); reproduced by permission of the International Union of Crystallography.]

In the higher symmetry groups, such as that illustrated in Fig. 4.28, the use of two diagrams clarifies the display the space group, one showing the general equivalent positions and the other the symmetry elements.

Self-assessment 4.11. Draw a partial diagram, similar to that above for $P6_3cm$, but with its m planes and c-glide planes interchanged in position. What is the symbol for this space group?

Table 4.2. The 230 space groups; under each system is the class, ITA space group number and symbol.

Triclinic system:

Class 1:	**1** $P1$
Class $\bar{1}$:	**2** $P\bar{1}$

Monoclinic system:

Class 2:	**3** $P2$, **4** $P2_1$, **5** $C2$
Class m:	**6** Pm, **7** Pc, **8** Cm, **9** Cc
Class $\frac{2}{m}$:	**10** $P\frac{2}{m}$, **11** $P\frac{2_1}{m}$, **12** $C\frac{2}{m}$, **13** $P\frac{2}{c}$, **14** $P\frac{2_1}{c}$, **15** $C\frac{2}{c}$

Orthorhombic system:

Class 222:	**16** $P222$, **17** $P222_1$, **18** $P2_12_12$, **19** $P2_12_12_1$, **20** $C222_1$, **21** $C222$, **22** $F222$, **23** $I222$, **24** $I2_12_12_1$
Class mm2:	**25** $Pmm2$, **26** $Pmc2_1$, **27** $Pcc2$, **28** $Pma2$, **29** $Pca2_1$, **30** $Pnc2$, **31** $Pmn2_1$, **32** $Pba2$, **33** $Pna2_1$, **34** $Pnn2$, **35** $Cmm2$, **36** $Cmc2_1$, **37** $Ccc2$, **38** $Amm2$, **39** $Aem2(Abm2)$, **40** $Ama2$, **41** $Aea2(Aba2)$, **42** $Fmm2$, **43** $Fdd2$, **44** $Imm2$, **45** $Iba2$, **46** $Ima2$
Class mmm:	**47** $Pmmm$, **48** $Pnnn$, **49** $Pccm$, **50** $Pban$, **51** $Pmma$, **52** $Pnna$, **53** $Pmna$, **54** $Pcca$, **55** $Pbam$ **56** $Pccn$, **57** $Pbcm$, **58** $Pnnm$, **59** $Pmmn$, **60** $Pbcn$, **61** $Pbca$, **62** $Pnma$, **63** $Cmcm$, **64** $Cmce(Cmca)$, **65** $Cmmm$, **66** $Cccm$, **67** $Cmme(Cmma)$, **68** $Ccce(Cmma)$, **69** $Fmmm$, **70** $Fddd$, **71** $Immm$, **72** $Ibam$, **73** $Ibca$, **74** $Imma$

Tetragonal system:

Class 4:	**75** $P4$, **76** $P4_1$, **77** $P4_2$, **78** $P4_3$, **79** $I4$, **80** $I4_1$
Class $\bar{4}$:	**81** $P\bar{4}$, **82** $I\bar{4}$
Class $\frac{4}{m}$:	**83** $P\frac{4}{m}$, **84** $P\frac{4_2}{m}$, **85** $P\frac{4}{n}$, **86** $P\frac{4_2}{n}$, **87** $I\frac{4}{m}$, **88** $I\frac{4_1}{a}$
Class 422:	**89** $P422$, **90** $P42_12$, **91** $P4_122$, **92** $P4_12_12$, **93** $P4_222$, **94** $P4_22_12$, **95** $P4_322$, **96** $P4_32_12$, **97** $I422$, **98** $I4_122$,
Class 4mm:	**99** $P4mm$, **100** $P4bm$, **101** $P4_2cm$, **102** $P4_2nm$, **103** $P4cc$, **104** $P4nc$, **105** $P4_2mc$, **106** $P4_2bc$, **107** $I4mm$, **108** $I4cm$, **109** $I4_1md$, **110** $I4_1cd$
Class $\bar{4}2m$:	**111** $P\bar{4}2m$, **112** $P\bar{4}2c$, **113** $P\bar{4}2_1m$, **114** $P\bar{4}2_1c$, **115** $P\bar{4}m2$, **116** $P\bar{4}c2$, **117** $P\bar{4}b2$, **118** $P\bar{4}n2$, **119** $I\bar{4}m2$, **120** $I\bar{4}c2$, **121** $I\bar{4}2m$, **122** $I\bar{4}2d$
Class $\frac{4}{m}mm$:	**123** $P\frac{4}{m}mm$, **124** $P\frac{4}{m}cc$, **125** $P\frac{4}{n}bm$, **126** $P\frac{4}{n}nc$, **127** $P\frac{4}{m}bm$, **128** $P\frac{4}{m}nc$, **129** $P\frac{4}{n}mm$, **130** $P\frac{4}{n}cc$, **131** $P\frac{4_2}{m}mc$, **132** $P\frac{4_2}{m}cm$, **133** $P\frac{4_2}{n}bc$, **134** $P\frac{4_2}{n}nm$, **135** $P\frac{4_2}{m}bc$, **136** $P\frac{4_2}{m}nm$, **137** $P\frac{4_2}{n}mc$, **138** $P\frac{4_2}{n}cm$, **139** $I\frac{4}{m}mm$, **140** $I\frac{4}{m}cm$, **141** $I\frac{4_1}{a}md$, **142** $I\frac{4_1}{a}cd$

Trigonal system:

Class 3:	**143** $P3$, **144** $P3_1$, **145** $P3_2$, **146** $R3$
Class $\bar{3}$:	**147** $P\bar{3}$, **148** $R\bar{3}$

(*Continued*)

Table 4.2.　(Continued)

Class 32:	**149** $P312$, **150** $P321$, **151** $P3_112$, **152** $P3_121$, **153** $P3_212$, **154** $P3_221$, **155** $R32$
Class 3m:	**156** $P3m1$, **157** $P31m$, **158** $P3c1$, **159** $P31c$, **160** $R3m$, **161** $R3c$
Class $\bar{3}m$:	**162** $P\bar{3}1m$, **163** $P31c$, **164** $P\bar{3}m1$, **165** $P\bar{3}c1$, **166** $R\bar{3}m$, **167** $R\bar{3}c$

Hexagonal system:

Class 6:	**168** $P6$, **169** $P6_1$, **170** $P6_5$, **171** $P6_2$, **172** $P6_4$, **173** $P6_3$
Class $\bar{6}$:	**174** $P\bar{6}$
Class $\frac{6}{m}$:	**175** $P\frac{6}{m}$, **176** $P\frac{6_3}{m}$
Class 622:	**177** $P622$, **178** $P6_122$, **179** $P6_522$, **180** $P6_222$, **181** $P6_422$, **182** $P6_322$
Class 6mm:	**183** $P6mm$, **184** $P6cc$, **185** $P6_3cm$, **186** $P6_3mc$
Class $\bar{6}m2$:	**187** $P\bar{6}m2$, **188** $P\bar{6}c2$, **189** $P\bar{6}2m$, **190** $P\bar{6}2c$
Class $\frac{6}{m}mm$:	**191** $P\frac{6}{m}mm$, **192** $P\frac{6}{m}cc$, **193** $P\frac{6_3}{m}cm$, **194** $P\frac{6_3}{m}mc$

Cubic system:

Class 23:	**195** $P23$, **196** $F23$, **197** $I23$, **198** $P2_13$, **199** $I2_13$
Class $m\bar{3}$:	**200** $Pm\bar{3}$, **201** $Pn\bar{3}$, **202** $Fm\bar{3}$, **203** $Fd\bar{3}$, **204** $Im\bar{3}$, **205** $Pa\bar{3}$, **206** $Ia\bar{3}$
Class 432:	**207** $P432$, **208** $P4_232$, **209** $F432$, **210** $F4_132$, **211** $I432$, **212** $P4_332$, **213** $P4_132$, **214** $I4_132$
Class $\bar{4}3m$:	**215** $P\bar{4}3m$, **216** $F\bar{4}3m$, **217** $I\bar{4}3m$, **218** $P\bar{4}3n$, **219** $F\bar{4}3c$. **220** $I\bar{4}3d$
Class $m\bar{3}m$:	**221** $Pm\bar{3}m$, **222** $Pn\bar{3}n$, **223** $Pm\bar{3}n$, **224** $Pn\bar{3}m$, **225** $Fm\bar{3}m$, **226** $Fm\bar{3}c$, **227** $Fd\bar{3}m$, **228** $Fd\bar{3}c$, **229** $Im\bar{3}m$, **230** $Ia\bar{3}d$

4.11.3.　Cubic space groups

In studying the space groups of higher symmetry, it can be useful often to refer to their corresponding stereograms. Recall also from the appropriate section of Table 2.4 that the Hermann–Mauguin symbol in the cubic system has the following specification in terms of lattice directions: first position, $\langle 100 \rangle$; second position, $\langle 111 \rangle$; third position, $\langle 110 \rangle$.

Cubic space groups are most common among simple structures. Most metals and certain alloys exhibit the holosymmetric class of the cubic ($m\bar{3}m$) and hexagonal ($\frac{6}{m}mm$) systems. This section will consider space groups in the cubic classes 23 and 432.

4.11.3.1. *Laue class 23*

The simplest cubic space group is $P23$. The rotation of a point x, y, z about [111] generates a set of three points by cyclic transformation: $x, y, z \xrightarrow{3_{[111]}} z, x, y \xrightarrow{3_{[111]}} y, z, x$. The remaining nine general equivalent positions of $P23$ may be generated in more than one way. Starting from point (1) as x, y, z, Table 4.3 shows interrelationships of the general equivalent positions in this space group.

Another fairly straightforward cubic space group is $Pn\bar{3}$, full symbol $P\frac{2}{n}\bar{3}$. The settings of this space group are either at the symmetry point 23 or at a centre of symmetry, actually $\bar{3}$, at $^1/_4$, $^1/_4$, $^1/_4$ from the 23 origin. The first choice gives a simple expression of the coordinates of the general equivalent positions. However, in view of the development of crystallographic computer software, the manual manipulation of coordinates is a rarity. With the second choice of origin, the following settings follow from the full space-group symbol:

$$\bar{1} \text{ at } 0, 0, 0$$
$$n \parallel (x, y, r)$$
$$2 \text{ along } [p, q, z]$$

An analysis based on the half-shift rule, as applied in previous examples, shows that $p = q = {}^1/_4$ and $r = 0$; a space group diagram is shown in Fig.4.29. The reader is invited to deduce the results for p and q in the next exercise.

Self-assessment 4.12. With the aid of the scheme above, deduce the coordinates of the general equivalent positions for space group $Pn\bar{3}$. The stereogram for point group $m\bar{3}$ may be helpful:

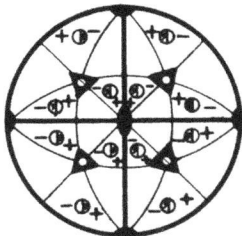

Stereogram for point group $m\bar{3}$

Table 4.3.   General equivalent positions for space group $P23$.

$$(3)\ \bar{x}, y, \bar{z} \quad \xleftrightarrow{\ 3^2_{[1\bar{1}1]}\ } \quad (8)\ \bar{z}, x, \bar{y} \quad \xleftrightarrow{\ 3^2_{[1\bar{1}1]}\ } \quad (10)\ \bar{y}, z, \bar{x}$$

$$\uparrow \qquad\qquad\qquad\qquad \uparrow \qquad\qquad\qquad\qquad \uparrow$$
$$2_y \qquad\qquad\qquad\qquad 2_y \qquad\qquad\qquad\qquad 2_y$$
$$\downarrow \qquad\qquad\qquad\qquad \downarrow \qquad\qquad\qquad\qquad \downarrow$$

$$(1)\ x, y, z \quad \xleftrightarrow{\ 3_{[111]}\ } \quad (5)\ z, x, y \quad \xleftrightarrow{\ 3_{[111]}\ } \quad (9)\ y, z, x$$

$$\uparrow \qquad\qquad\qquad\qquad \uparrow \qquad\qquad\qquad\qquad \uparrow$$
$$2_x \qquad\qquad\qquad\qquad 2_x \qquad\qquad\qquad\qquad 2_x$$
$$\downarrow \qquad\qquad\qquad\qquad \downarrow \qquad\qquad\qquad\qquad \downarrow$$

$$(4)\ x, \bar{y}, \bar{z} \quad \xleftrightarrow{\ 3_{[\bar{1}11]^2}\ } \quad (6)\ z, \bar{x}, \bar{y} \quad \xleftrightarrow{\ 3_{[\bar{1}11]^2}\ } \quad (11)\ y, \bar{z}, \bar{x}$$

$$\uparrow \qquad\qquad\qquad\qquad \uparrow \qquad\qquad\qquad\qquad \uparrow$$
$$2_z \qquad\qquad\qquad\qquad 2_z \qquad\qquad\qquad\qquad 2_z$$
$$\downarrow \qquad\qquad\qquad\qquad \downarrow \qquad\qquad\qquad\qquad \downarrow$$

$$(2)\ \bar{x}, \bar{y}, z \quad \xleftrightarrow{\ 3^2_{[11\bar{1}]}\ } \quad (7)\ \bar{z}, \bar{x}, y \quad \xleftrightarrow{\ 3^2_{[11\bar{1}]}\ } \quad (12)\ \bar{y}, \bar{z}, x$$

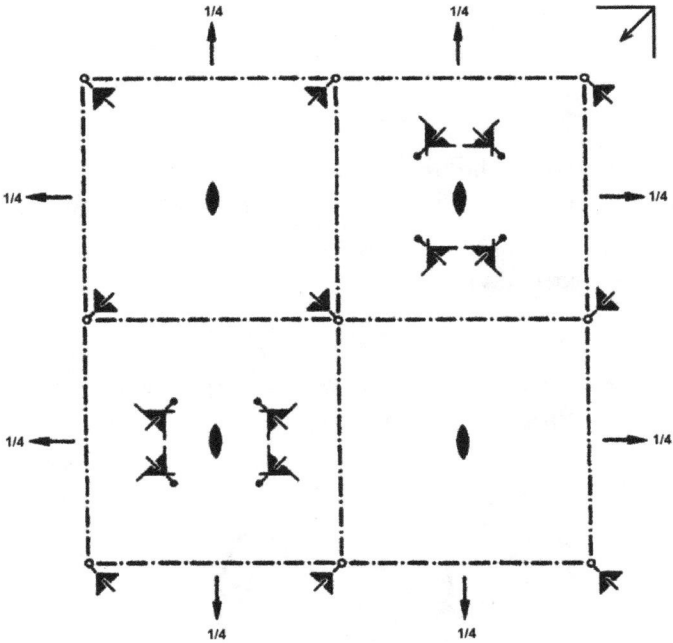

Fig. 4.29.   Diagram for the symmetry elements of space group $Pn\bar{3}$; the origin is on $\bar{1}$ and the n-glide plane is at $c = 0$.

Space group $P2_1 3$, to which reference has been made in Section 2.9 in relation to crystals of sodium chlorate, is another group in the class 23; it is the screw nature which this material adopts in the solid state that gives rise to the chiral property of the crystals of that substance [5, 7, 8].

4.11.3.2. *Laue class 432*

Crystals of viruses and proteins have been reported in this class with space group $F432$, and crystals of strontium disilicide have been said to exhibit space group $P4_3 32$ [9]. Space group $P4_1 32$ would have been an alternative choice for this compound: these two space groups are chiral or enantiomorphic.

A chiral species is non-superposable on its mirror image, as discussed in Section 2.9; the mirror images of a chiral species are enantiomers, with right-handed and left-handed screw arrangements of structure. Protein structures in the 432 class are helical about the principal screw axis; $P4_3 32$ exhibits a left-handed helix, or clockwise screw, along the $+z$ direction.

The stereogram for point group 432 is shown in Fig. 4.30. It is evident that introducing one four-fold axis, say, along [001], in combination with the characteristic three-fold symmetry of this crystal system, leads to four-fold axes also along [100] and [010], as will be

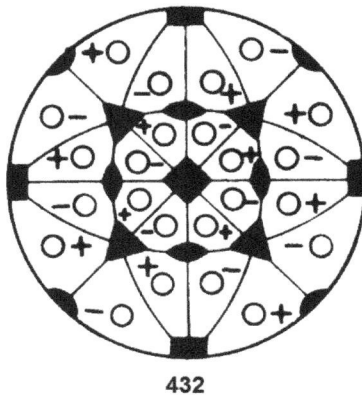

432

Fig. 4.30. Stereogram of the cubic point group 432.

shown in Section 4.13. Consequently, the coordinates of the general equivalent positions in space group $P432$ may be obtained from those of $P23$ by operating on each coordinate in Table 4.3 by any one of the four-fold axes to give 24 general equivalent positions:

General equivalent positions for space group $P432$

$$
\begin{array}{llllll}
x,y,z; & z,x,y; & y,z,x; & \bar{x},\bar{z},\bar{y}; & \bar{y},\bar{x},\bar{z}; & \bar{z},\bar{y},\bar{x}; \\
x,\bar{y},\bar{z}; & z,\bar{x},\bar{y}; & y,\bar{z},\bar{x}; & \bar{x},z,y; & \bar{y},x,z; & \bar{z},y,x; \\
\bar{x},y,\bar{z}; & \bar{z},x,\bar{y}; & \bar{y},z,\bar{x}; & x,\bar{z},y; & y,\bar{x},z; & z,\bar{y},x; \\
\bar{x},\bar{y},z; & \bar{z},\bar{x},y; & \bar{y},\bar{z},x; & x,z,y; & y,x,\bar{z}; & z,y,\bar{x}.
\end{array}
$$

Then, the space group $F432$ referred to above may be given as $P432 + (0, 0, 0; 0, 1/2, 1/2; 1/2, 0\ 1/2; 1/2, 1/2, 0)$, thus generating the total of 96 general equivalent positions.

4.12.   Space Groups and Crystal Structures

At this stage, the features of a crystal structure that may be deduced from knowledge of the space-group symmetry can be considered; four structures will be examined.

4.12.1.   Sodium chloride

A diagram of the sodium chloride structure has been given in Fig. 4.1(c); this compound crystallizes in space group $Fm\bar{3}m$. Its density D_x measures $2.165\ \mathrm{g\,cm^{-3}}$, its relative molar mass M_r is $58.443\ \mathrm{g\,mol^{-1}}$ and the cubic unit cell has a side length a of 5.6402 Å. Then, the number Z_c of formula-entities per unit cell is given by the usual formula, density = mass/volume, that is,

$$
D_x = \frac{Z_c M_r m_u}{a^3} \tag{4.2}
$$

where m_u is the atomic mass unit, equal to 1.66054×10^{-24} g and a^3 is the volume V_c of the unit cell. Thus,

$$
Z_c = \frac{(5.6402 \times 10^{-8}\mathrm{cm})^3 \times 2.165\ \mathrm{g\ cm^{-3}}}{58.443 \times 1.66054 \times 10^{-24}\mathrm{g}} = 4.0027
$$

or 4 as the nearest integer. Reference to the International Tables shows that the most probable arrangement of the species is $4\,\mathrm{Na^+}$

(or Cl$^-$) at 0, 0, 0 (Wyckoff a, point symmetry $m\bar{3}m$) and 4 Cl$^-$ (or Na$^+$) at ¹/₂, ¹/₂, ¹/₂ (Wyckoff b, point symmetry $m\bar{3}m$). Hence, the structure is determined completely from the space-group data.

4.12.2. Potassium aluminium sulphate dodecahydrate

Potassium aluminium sulphate is one of a series of 'alums' with the general formula $MN(SO_4)_2.12H_2O$, where M includes K$^+$, Rb$^+$, Tl$^+$ or NH$_4^+$, and N may be Al^{3+}, Fe or Cr^{3+}; additionally, sulphur may be replaced by selenium. The alums form an isomorphous series of crystalline compounds, that is, they have the same space group and closely similar unit-cell dimensions; the atoms occupying the different alums are similar in position, differing only in their chemical nature.

The alum structures crystallize in space group $Pa\bar{3}$ with the unit-cell dimensions a of 12.2–12.3 Å; there are four formula-entities in the unit cell. From the space-group data, the following positions can be allocated:

$$
\begin{array}{llllll}
4M & a & 0,0,0; & 0,¹/₂,¹/₂; & ¹/₂,0,¹/₂; & ¹/₂,¹/₂,0 \\
4N & b & ¹/₂,¹/₂,¹/₂; & ¹/₂,0,0; & 0,¹/₂,0; & 0,0,¹/₂ \\
8S & c & \pm(x,x,x; & ¹/₂+x,¹/₂-x,\bar{x}; & \bar{x},¹/₂+x, & ¹/₂-x; \\
& & ¹/₂-x,\ \bar{x},¹/₂+x) & & &
\end{array}
$$

The asymmetric unit is $0 - ¹/₂$ along each side of the unit cell, or one-eighth of the cell. The eight sulphur atoms lie along the form of directions $\langle 111 \rangle$, and from the International Tables the probable sites for the 32 oxygen atoms of the sulphate anions are eight in the Wyckoff c positions of symmetry 3 and 24 in general equivalent positions d. The 48 oxygen atoms of the water molecules could occupy two sets of d positions. These allocations were confirmed by a full crystal structure analysis. It is clear that a symmetry analysis reduces significantly the total number of parameters that need to be determined in order to specify the complete crystal structure.

4.12.3. Copper (I) oxide Cu₂O

Copper(I) oxide crystallizes in space group $Pn\bar{3}$ also, with two formula-entities in the cubic unit cell of side 4.267 Å. It is evident

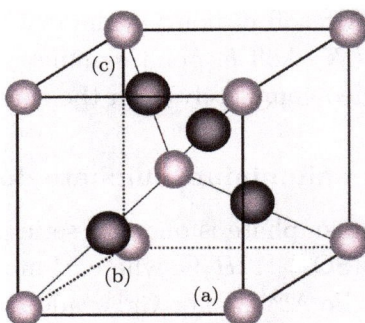

Fig. 4.31.   Unit cell and environs for copper(I) oxide, with the oxygen atoms in Wyckoff *a* positions and the copper atoms in Wyckoff *b*; the spheres in decreasing order of size represent O and Cu species. An alternative placement of copper atoms in Wyckoff *c* positions corresponds to a rotation of the $[Cu_4O]$ structural unit about its vertical axis by $90°$.

that the atoms will occupy special positions in the space group. Reference to the International Tables reveals that, with the origin at the point 23, the relevant special position sets are as follows:

4	*c*	$\bar{3}$	$^3/_4, ^3/_4, ^3/_4; \ ^3/_4, ^1/_4, ^1/_4; \ ^1/_4, ^3/_4, ^1/_4; \ ^1/_4, ^1/_4, ^3/_4$
4	*b*	$\bar{3}$	$^1/_4, ^1/_4, ^1/_4; \ ^1/_4, ^3/_4, ^3/_4; \ ^3/_4, ^1/_4, ^3/_4; \ ^3/_4, ^3/_4, ^1/_4$
2	*a*	23	$0, 0, 0; \ ^1/_2, ^1/_2, ^1/_2$

The oxygen atoms occupy Wyckoff *a*, and the copper atoms are in either *b* or *c*. The alternatives correspond to the two ways of placing the tetrahedral arrangement of copper atoms in the unit cell around the oxygen atom at the centre of the cell; Fig. 4.31 shows them in Wyckoff *b*.

Self-assessment 4.13.   Using the data above, calculate the shortest Cu–O, Cu–Cu and O–O distances.

4.12.4.   Iron sulphide (pyrite) FeS_2

This crystalline material is another cubic crystal exhibiting space group $Pa\bar{3}$; the unit cell contains four formula-entities. The atomic positions in the unit cell are as follows:

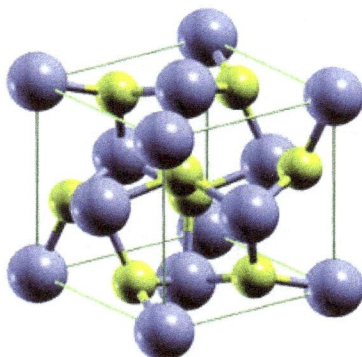

Fig. 4.32.   The unit cell and environs of the crystal structure of pyrite FeS$_2$; the spheres in decreasing order of size represent S and Fe species. The iron atoms are in Wyckoff position a and the sulphur atoms are in Wyckoff c of space group $Pa\bar{3}$.

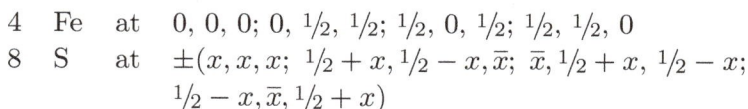

4	Fe	at	0, 0, 0; 0, ½, ½; ½, 0, ½; ½, ½, 0
8	S	at	$\pm(x, x, x;\ \frac{1}{2}+x, \frac{1}{2}-x, \bar{x};\ \bar{x}, \frac{1}{2}+x, \frac{1}{2}-x;$
			$\frac{1}{2}-x, \bar{x}, \frac{1}{2}+x)$

This structure was introduced at the beginning of the chapter as an example of an electron density map. It is evident that the crystal structure is determined by the one parameter x, which has a (fractional) value of *ca.* 0.1. A diagram of the unit cell is shown in Fig. 4.32.

4.13.   Representing Space Group Operations by Matrices

As noted above, a symmetry operation may be written generally by the equation

$$\mathbf{x} = \boldsymbol{R}\mathbf{x} + \mathbf{t} \qquad (4.3)$$

where the vector $\mathbf{x}$ represents the triplet x', y', z' that is obtained by the action of the rotation operator $\boldsymbol{R}$ on $\mathbf{x}(\boldsymbol{x}, \boldsymbol{y}, \boldsymbol{z})$ together with a translational component $\mathbf{t}$; in a point group, $\mathbf{t}$ is zero by definition. Illustrations will be considered now in reference to space groups already studied.

In space group $P2_1/c$, the orientations of the symmetry operations of the space group are as discussed in Section 4.4.2.3. The corresponding matrix operations for this group may be set out in stages. The combination (product) of the stages (1) and (2), shown below,

$$2_{[p,y,r]}$$

$$(1) \quad \underbrace{\begin{pmatrix} \bar{1} & 0 & 0 \\ 0 & 1 & 0 \\ 0 & 0 & \bar{1} \end{pmatrix}}_{\mathbf{R_1}} \underbrace{\begin{pmatrix} x \\ y \\ z \end{pmatrix}}_{\mathbf{x}} + \underbrace{\begin{pmatrix} 2p \\ 1/2 \\ 2r \end{pmatrix}}_{\mathbf{t_1}} = \underbrace{\begin{pmatrix} \bar{x} \\ y \\ \bar{z} \end{pmatrix}}_{\mathbf{x'}} + \underbrace{\begin{pmatrix} 2p \\ 1/2 \\ 2r \end{pmatrix}}_{\mathbf{t_1}}$$

$$c_{(x,q,z)}$$

$$(2) \quad \underbrace{\begin{pmatrix} 1 & 0 & 0 \\ 0 & \bar{1} & 0 \\ 0 & 0 & 1 \end{pmatrix}}_{\mathbf{R_2}} \underbrace{\begin{pmatrix} \bar{x} \\ y \\ \bar{z} \end{pmatrix}}_{\mathbf{x'}} + \underbrace{\begin{pmatrix} 0 \\ 2q \\ 1/2 \end{pmatrix}}_{\mathbf{t_1}} = \underbrace{\begin{pmatrix} \bar{x} \\ \bar{y} \\ \bar{z} \end{pmatrix}}_{\mathbf{x''}} + \underbrace{\begin{pmatrix} 0 \\ 2q \\ 1/2 \end{pmatrix}}_{\mathbf{t_2}} + \underbrace{\begin{pmatrix} 2p \\ 1/2 \\ 2r \end{pmatrix}}_{\mathbf{t_1}}$$

is equivalent to the third stage (3):

$$\bar{1}_{0,0,0}$$

$$(3) \quad \underbrace{\begin{pmatrix} \bar{1} & 0 & 0 \\ 0 & \bar{1} & 0 \\ 0 & 0 & \bar{1} \end{pmatrix}}_{\mathbf{R_3}} \underbrace{\begin{pmatrix} x \\ y \\ z \end{pmatrix}}_{\mathbf{x}} + \underbrace{\begin{pmatrix} 0 \\ 2q \\ 1/2 \end{pmatrix}}_{\mathbf{x''}} + \underbrace{\begin{pmatrix} 2p \\ 1/2 \\ 2r \end{pmatrix}}_{\mathbf{t_3}} = \begin{pmatrix} \bar{1} & 0 & 0 \\ 0 & \bar{1} & 0 \\ 0 & 0 & \bar{1} \end{pmatrix} \begin{pmatrix} \bar{x} \\ \bar{y} \\ \bar{z} \end{pmatrix} + \begin{pmatrix} 0 \\ 0 \\ 0 \end{pmatrix}$$

since $\mathbf{R_2 R_1} = \mathbf{R_3}$ and $\sum \mathbf{t} = 0$. It follows that $p = 0$ and $q = 1/4$, as determined earlier.

This type of procedure is, in essence, the basis of all the schemes discussed for the derivation of the space group under examination from the full Herman–Mauguin symbol. The values of p, q and r that are derived must agree with the equality $\mathbf{t_1} + \mathbf{t_2} = \mathbf{t_3}$.

The process written generally as $\mathbf{R_2 R_1} = \mathbf{R_3}$ and $\sum \mathbf{t} = 0$ means that operation $\mathbf{R_1}$ followed by $\mathbf{R_2}$ is equivalent to the operation of $\mathbf{R_3}$, and the sum of $\mathbf{t_1}$ and $\mathbf{t_2}$ equates to $\mathbf{t_3}$, which is zero in this example. The process may be carried out also by forming first the product $\mathbf{R_2R_1}$ and then operating with the result on $\mathbf{X}$, bearing in mind that $\sum \mathbf{t} = 0$.

A second example studied in this chapter was space group $P4_2bc$, and the following procedure treats the matrices first:

$$
\overset{b_{(p,y,z)}}{\begin{pmatrix} \bar{1} & 0 & 0 \\ 0 & 1 & 0 \\ 0 & 0 & 1 \end{pmatrix}} \overset{c_{(q,q,z)}}{\begin{pmatrix} 0 & \bar{1} & 0 \\ 1 & 0 & 0 \\ 0 & 0 & 1 \end{pmatrix}} + \overset{t_b}{\begin{pmatrix} 2p \\ 1/2 \\ 0 \end{pmatrix}} + \overset{t_c}{\begin{pmatrix} q \\ q \\ 0 \end{pmatrix}}
$$

$$
= \overset{bc}{\begin{pmatrix} 0 & \bar{1} & 0 \\ 1 & 0 & 0 \\ 0 & 0 & 1 \end{pmatrix}} + \overset{t_b+t_c}{\begin{pmatrix} 2p+q \\ 1/2+q \\ 0 \end{pmatrix}}
$$

Also, for the 4_2 symmetry operation,

$$
\overset{4_{2[0,0,0]}}{\begin{pmatrix} 0 & \bar{1} & 0 \\ 1 & 0 & 0 \\ 0 & 0 & 1 \end{pmatrix}} \overset{\mathbf{x}}{\begin{pmatrix} x \\ y \\ z \end{pmatrix}} + \overset{t_{4_2}}{\begin{pmatrix} 0 \\ 0 \\ 0 \end{pmatrix}}
$$

But since $\boldsymbol{b}\,\boldsymbol{c} \equiv 4_2$, it follows that

$$
\begin{pmatrix} 2p+q \\ 1/2+q \\ 0 \end{pmatrix} = \begin{pmatrix} 0 \\ 0 \\ 0 \end{pmatrix}
$$

hence, $q = 1/2$ and $p = q/2 = 1/4$, as was found in Section 4.10.2.

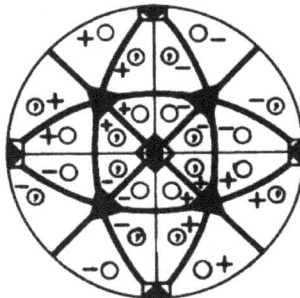

Stereogram for point group $\bar{4}3m$

As a final example in this section, the cubic space group $P\bar{4}3m$ will be studied. A study of the stereogram for point group $\bar{4}3m$ reveals that the combination $3_{[11\bar{1}]}$ followed by $m_{[01\bar{1}]}$ is equivalent to the action of $4_{[00z]}$, as may be confirmed by the following matrix operations. For the combination $3_{[11\bar{1}]}$ followed by $2_{[01\bar{1}]}$:

$$\begin{pmatrix} 1 & 0 & 0 \\ 0 & 0 & 1 \\ 0 & 1 & 0 \end{pmatrix}\begin{pmatrix} 0 & 1 & 0 \\ 0 & 0 & \bar{1} \\ \bar{1} & 0 & 0 \end{pmatrix} + \begin{pmatrix} 0 \\ 0 \\ 0 \end{pmatrix} + \begin{pmatrix} 0 \\ 0 \\ 0 \end{pmatrix} = \begin{pmatrix} 0 & 1 & 0 \\ \bar{1} & 0 & 0 \\ 0 & 0 & \bar{1} \end{pmatrix} + \begin{pmatrix} 0 \\ 0 \\ 0 \end{pmatrix}$$

and $$\begin{pmatrix} 0 & 1 & 0 \\ \bar{1} & 0 & 0 \\ 0 & 0 & \bar{1} \end{pmatrix}\begin{pmatrix} x \\ y \\ z \end{pmatrix} + \begin{pmatrix} 0 \\ 0 \\ 0 \end{pmatrix} = \begin{pmatrix} y \\ \bar{x} \\ \bar{z} \end{pmatrix} + \begin{pmatrix} 0 \\ 0 \\ 0 \end{pmatrix}$$

and this result is equivalent to the operation $\bar{4}_{[00z]}$:

$$\begin{pmatrix} 0 & 1 & 0 \\ \bar{1} & 0 & 0 \\ 0 & 0 & \bar{1} \end{pmatrix}\begin{pmatrix} x \\ y \\ z \end{pmatrix} + \begin{pmatrix} 0 \\ 0 \\ 0 \end{pmatrix} = \begin{pmatrix} y \\ \bar{x} \\ \bar{z} \end{pmatrix}$$

It is evident that the symmetry elements in $P\bar{4}3m$ pass through the origin, as would be expected for a symmorphic space group. For continued discussions on space groups, the reader is referred to the literature [5–8].

Self-assessment 4.14. In an orthorhombic space group, the matrices and translation vectors for a c-glide normal to x and an a-glide normal to y are, respectively,

$$\begin{pmatrix} \bar{1} & 0 & 0 \\ 0 & 1 & 0 \\ 0 & 0 & 1 \end{pmatrix} + \begin{pmatrix} 0 \\ 0 \\ 1/2 \end{pmatrix} \text{ and } \begin{pmatrix} 1 & 0 & 0 \\ 0 & \bar{1} & 0 \\ 0 & 0 & 1 \end{pmatrix} + \begin{pmatrix} 1/2 \\ 0 \\ 0 \end{pmatrix}$$

(a) What symmetry operator arises from the combination $\boldsymbol{ac}$?
(b) What space group is represented? (c) Do the operators $\boldsymbol{a}$ and $\boldsymbol{c}$ commute in this space group?

4.14. The Seitz Symmetry Operator

Space group operations can be defined concisely by the *Seitz operator* $\{R|t\}$, where R is a point-group operator and t is a translation vector, both of which act on a point represented by the vector r; thus,

$$\{R|t\}r = Rr + t \tag{4.4}$$

which is equivalent to the equation with which Section 4.13 began. In Eq. (4.4), R represents *rotational symmetry* (R or $\overline{R}$) and the vector t is a *translation* which may arise through any or all of unit-cell centring, glide plane and screw axis. That Eq. (4.4) represents a group can be determined by applying the group rules discussed in Section 2.3.

4.14.1. Product rule

Consider another operation R' in the group containing R acting together on the vector r. Then, the product of the two Seitz operators is given by

$$\{R|t\}\{R'|t'\}r = \{R|t\}(R'r + t')$$
$$= RR'r + Rt' + t = \{RR'|Rt' + t\}r \tag{4.5}$$

But since Rt' is just another translation, it follows that the product of the two symmetry operators $\{R|t\}\{R'|t'\}$ is also a member of the set, and the product rule holds.

4.14.2. Associativity rule

From Eq. (4.5), it is evident that $\{R'|t'\}\{R|t\}$ is a member of the group, so that the associative rule is obeyed.

4.14.3. Identity rule

The operator $\{R|t\}$ in the form $\{1|0\}$ is the identity operator required by group rules.

4.14.4.   Inverse rule

If the inverse operator $\{R|t\}^{-1}$ be defined as $\{R^{-1}| - R^{-1}t\}$, then

$$\{R|t\}^{-1}\{R^{-1}| - R^{-1}t\} = RR^{-1} + t - t = \{1|0\} \qquad (4.6)$$

which satisfies the requirement of the presence of an inverse.

4.14.5.   Space group $P2_1 2_1 2_1$

The use of the Seitz operator will be considered in relation to space group $P2_1 2_1 2_1$; the three screw axes must be non-intersecting. This space group contains four operators, namely, $1, 2_{1[100]}, 2_{1[010]}$ and $2_{1[001]}$. In the standard setting, the $2_{1[100]}$-axis is set off along b by an amount t. The only sensible values for t are and $1/4$ and $1/2$, but the value of $1/2$ is equivalent to $t = 0$; thus, the value of $1/4$ is chosen. Since $2_{1[010]}$ must not intersect the first symmetry axis, it is displaced by $1/4$ along c. The product of the two 2_1 operations acting on a point $\mathbf{r}(x,\, y,\, z)$ may be written as $\{2_{1[100]}\}\{2_{1[010]}\}\mathbf{r}$, which is equal to

$$\{2_{[100]}|^1\!/_2, \,^1\!/_2, 0\}\{2_{[010]}|(^1\!/_2, 0, \,^1\!/_2)\}\mathbf{r}$$

$$= \{2_{[100]}\}\{2_{[010]}\}\mathbf{r} + \{2_{[100]}\}(^1\!/_2, \,^1\!/_2, 0) + (^1\!/_2, 0, \,^1\!/_2) \qquad (4.7)$$

Expanding Eq. (4.7) concisely in matrix form gives

$$
\overset{2_{[100]}}{\begin{pmatrix}1&0&0\\0&\bar{1}&0\\0&0&\bar{1}\end{pmatrix}}
\overset{2_{[010]}}{\begin{pmatrix}\bar{1}&0&0\\0&1&0\\0&0&\bar{1}\end{pmatrix}}
\overset{\mathbf{r}}{\begin{pmatrix}x\\y\\z\end{pmatrix}} +
\overset{2_{[100]}}{\begin{pmatrix}1&0&0\\0&\bar{1}&0\\0&0&\bar{1}\end{pmatrix}}
\overset{t_a}{\begin{pmatrix}^1\!/_2\\^1\!/_2\\0\end{pmatrix}} +
\overset{t_b}{\begin{pmatrix}0\\^1\!/_2\\^1\!/_2\end{pmatrix}}
$$

$$
= \begin{pmatrix}\bar{1}&0&0\\0&\bar{1}&0\\0&0&1\end{pmatrix}\begin{pmatrix}x\\y\\z\end{pmatrix} + \begin{pmatrix}^1\!/_2\\^1\!/_2\\0\end{pmatrix} + \begin{pmatrix}0\\^1\!/_2\\^1\!/_2\end{pmatrix} = \begin{pmatrix}\bar{x}\\\bar{y}\\z\end{pmatrix} + \begin{pmatrix}^1\!/_2\\0\\^1\!/_2\end{pmatrix}
$$

$$\qquad (4.8)$$

which corresponds to the expected two-fold screw axis parallel to z, but sited at $(^1\!/_4,\, 0,\, z)$ whereby it does not intersect either of the other two 2_1 axes. Hence, the coordinates of the general equivalent positions are

$$x, y, z; \ \frac{1}{2} + x, \frac{1}{2} - y, \bar{z}; \ \bar{x}, \frac{1}{2} + y, \frac{1}{2} - z; \ \frac{1}{2} - x, \bar{y}, \frac{1}{2} + z \quad (4.9)$$

as was illustrated in Fig. 4.18.

Self-assessment 4.15. (a) What is the result if any two 2_1-axes in the above example of $P2_12_12_1$ intersect? (b) What space group results from combining a two-fold axis acting along $[x, 0, 0]$ with space group $P2_12_12_1$?

The application of the Seitz operator in point groups and space groups is discussed in more detail in the literature [6, 8].

Problems

4.1. Draw a diagram showing the general equivalent positions and symmetry elements in plane group *c2mm*; take the origin at *2mm*. List the sets of general and special equivalent positions.

4.2. The diagram contains thick and thin lines and also the letter X superimposed on a black oval. This is intended.

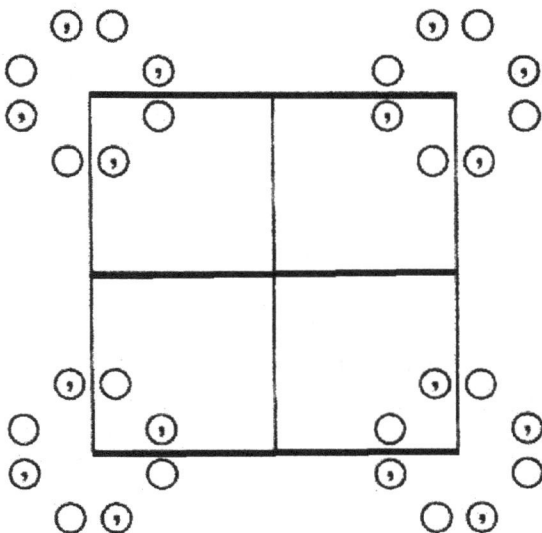

Complete the diagram of *p4m* indicating all symmetry elements. What is the full symbol for this group? List the coordinates of the general and special equivalent positions in their sets, together with the number in each set, its Wyckoff notation and site symmetry.

4.3. Hargreaves and Ritzvi (*Acta Crystallogr.* **15**, 365 (1962)) showed that biphenyl crystallizes in space group $P2_1/c$ with unit-cell dimensions $a = 8.124\,\text{Å}$, $b = 5.635\,\text{Å}$, $c = 9.513\,\text{Å}$, $\beta = 95.1°$; the experimental density D_x is $1.19\,\text{g cm}^{-3}$.

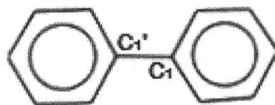

Biphenyl $C_{12}H_{10}$

(a) What can be said about the number and position of the molecules in the unit cell and their conformation?

(b) Calculate the density from the given data.

(c) For this space group, list the coordinates of all possible vectors between pairs of sites together with the multiplicity of each vector.

4.4. Refer to Fig. 4.17. (a) Show how the space-group generators $t_{[100]}, t_{[010]}, t_{[001]}, 2_{[0,y,1/4]} + (0, y, 1/4)$ and $\bar{1}_{[0,0,0]}$ given in ITA lead to the coordinates for space group $P2_1/c$. (b) What additional generator is required to produce space group $C2/c$? (c) The C centring introduces n-glide planes. What is their orientation in the unit cell?

4.5. (a) Determine the coordinates of the general equivalent positions for space group *Pcma*; set the centre of symmetry at the origin. (b) What is the full symbol for this space group? (c) Given that this data refers to the **bca** setting of the space group, what is its standard (**abc**) symbol?

4.6. Continuing from Self-assessment 4.12, determine the coordinates of the special equivalent positions. There are six sets.

4.7. Deduce coordinates for the 24 general equivalent positions of space group $P432$ shown as poles on a stereogram of point group 432 below; Label 1 should be taken as the coordinates x, y, z.

Stereogram of point group 432

4.8. The spinels form a series of compounds with the general formula $M^{II}N^{III}O_4.12H_2O$; spinel itself is $MgAl_2O_4 \cdot 12H_2O$ and crystallizes in space group $Fd\bar{3}m$ with eight (Z_c) formula-entities in the unit cell. Refer to the International Tables [2,3] and suggest how the atoms may be placed in this structure.

4.9. In space group $Pmcb$, a setting is $m||(p, y, z)$, $c||(x, q, r)$ and $b||(x, y, r)$. (a) Take the origin on $\bar{1}$ and use matrices to determine values for p, q and r. (b) What is the full symbol for $Pmcb$? (c) Given that the space group is in the **cab** setting what are the short and full symbols in the standard (**abc**) setting?

4.10. What is the type and position of the symmetry operator for each of the following symmetry operations? (a) $x, y, z \rightarrow \bar{x}, 1/2 + y, 1/2 - z$; (b) $x, y, z \rightarrow 1/2 - y, x, 3/4 + z$; (c) $x, y, z \rightarrow y, x, 1/2 + z$.

4.11. A point $A(x, y)$ is rotated anticlockwise by $90°$ to a point $B(-y, x)$ and about the point Q distant $X = p$ and $Y = q$ from the true origin O. (a) What are the values $-y, x$ in terms in terms of p and q? (b) What are the coordinates of a point x, y, z after a single anticlockwise rotation about a 4_3-axis along $[1/4, 3/4, z]$?

4.12. With respect to hexagonal axes, write the matrices for (a) a $\bar{6}_3$ symmetry operation about the line $[0, 0, z]$ and (b) a c-glide plane operation normal to the y-axis and passing through the origin? (c) Using matrices or an appropriate stereogram from Fig. 2.11, deduce the result of the combination $c\,\bar{6}_3$? (d) What space group is represented by the above combination in (c)?

Answers to Self-Assessments

4.1. Plane group $p1g1$ (pg):

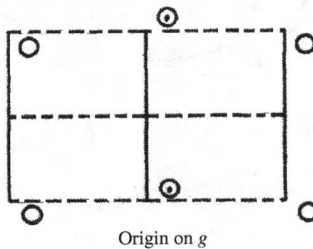

Origin on g

2   a 1 $x, y;\ \bar{x}, {}^1\!/_2 + y$

There are no special equivalent positions because there is only lattice translational symmetry in this group. (Special positions are the sites of physical entities and must, therefore, correspond in point-group symmetry.)

4.2. In the diagram below, the positions 0, 0 and $^1\!/_2$, 0 have been entered with the symbol $\times$ and repeated by the **a** and **b** translations:

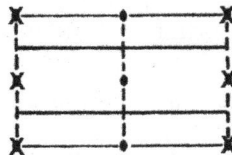

The resulting diagram is not consistent with $p2gg$; it illustrates the plane group $p2mg$ with occupancy of the special position set 0, 0; $^1\!/_2$, 0.

4.3. $P(X, Y)$ is reflected across the m-line to $P'(X', Y')$ to intersect the x- and y-axes at q and $-q$, respectively. From symmetry, the m-line lies at $45°$ to the x- and y-axes. The geometry of the diagram shows that the coordinates after reflection are $X' = q - (-Y) = q + Y$ and $Y' = -q - X = q + X$.

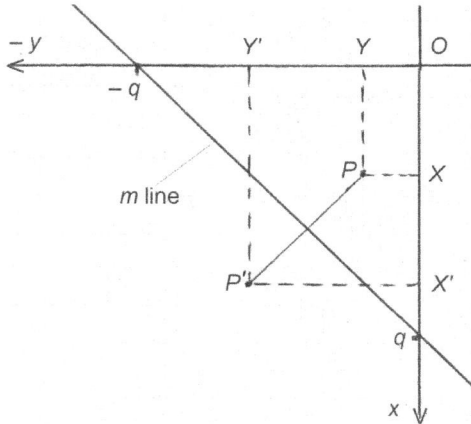

4.4.

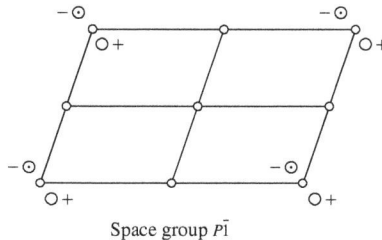

Space group $P\bar{1}$

Eight sets of special equivalent positions: $0, 0, 0$; $0, 0, 1/2$; $0, 1/2, 0$; $1/2, 0, 0$ $1/2, 1/2, 0$; $1/2, 0, 1/2$; $0, 1/2, 1/2$; $1/2, 1/2, 1/2$.

4.5.

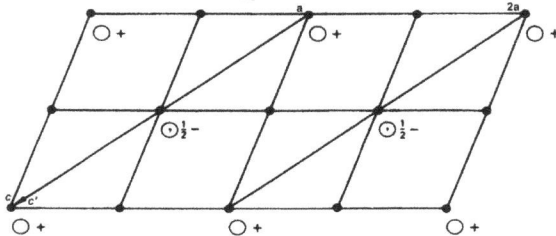

Two unit cells in space group $P2_1/n$

The diagram shows two unit cells for space group Pn drawn on the x-z plane: $x, y, z \xrightarrow{n_{(010)}} 1/2 + x, 1/2 - y, 1/2 + z$. In the transformation $Pn \to Pc$, **a** and **b** are unchanged, but the new value of $\mathbf{c}$ is given by $\mathbf{a}+\mathbf{c}$; now, $x, y, z \xrightarrow{c_{(010)}} x, 1/2 - y, 1/2 + z$. The transformation $Pn \to Pa$ is achieved by interchange of $\mathbf{a}$ and $\mathbf{c}$. Then, the direction of $\mathbf{b}$ is reversed in order to preserve right-handed axes.

4.6. The symmorphic space groups in class 222 are $P222$, $C222$, $I222$ and $F222$. A symmorphic space is compounded from a Bravais lattice, P, C, I or F, and a point group.

4.7. (a)$1/2 - x, 1/2 - y, z$. (b) $1/4, 1/4, 1/4 + z$.

4.8. Since the space group is in class mmm, the sequence $(1) \to (2) \to (3) \to (4)$ shows that the planes are, in order, n, m and a; also, $p = q = r = 1/4$ (recall $-1/4 \equiv 1/4$). The coordinate pairs 1, 2 and 1, 3 and 1, 4 show the positions for the planes are: $n(1/4, y, z)$, $m(x, 1/4, z)$ and $a(x, y, 1/4)$, together with $\bar{1}$ at 0, 0, 0. Now, the coordinate pairs (1), (6) show 2_1 at $[x, 1/4, 1/4]$, (1), (7) show 2_1 at $[0, y, 0]$ and (1), (8) show 2_1 at $[1/4, 0, z]$. Thus, the full symbol for the space group is $P\frac{2_1}{n}\frac{2_1}{m}\frac{2_1}{a}$.

4.9. The transformation is $\mathbf{a} \to \mathbf{b}'$, $\mathbf{b} \to \mathbf{c}'$, $\mathbf{c} \to \mathbf{a}'$. Normal to $\mathbf{a}'$ is a glide that was b glide but is now a c. Normal to $\mathbf{b}'$ was a glide that was b but is now a c. Normal to $\mathbf{c}'$ was an m plane which remains as such. Finally, B-centring (a_-c) becomes C-centring (a'$_-$b'). Assembling this information shows that $Bbmb$ (**bca**) transforms to $Cccm$ in the standard (**abc**) setting. Pictorially, it is shown below, where the primes indicate the transformed symbols:

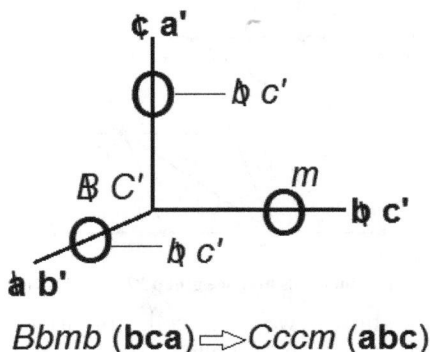

$Bbmb$ (**bca**)$\Rightarrow Cccm$ (**abc**)

4.10. Following the procedures in Section 4.10.1, the coordinates of general equivalent positions are as follows:

$$8 \quad g \quad 1 \quad x, y, z; \; {}^1\!/_2 - x, {}^1\!/_2 - y, z; \; \bar{y}, {}^1\!/_2 + x, {}^1\!/_2 + z;$$
$$ {}^1\!/_2 + y, \bar{x}, {}^1\!/_2 + z$$
$$\bar{x}, \bar{y}, \bar{z}; \; {}^1\!/_2 + x, {}^1\!/_2 + y, \bar{z}; \; y, {}^1\!/_2 - x, {}^1\!/_2 - z;$$
$$ {}^1\!/_2 - y, x, {}^1\!/_2 - z$$

Hence, the following sets of special equivalent positions arise:

$$4 \quad f \quad 2 \quad {}^1\!/_4, {}^1\!/_4, z; \; {}^3\!/_4, {}^3\!/_4, \bar{z}; \; {}^1\!/_4, {}^1\!/_4, {}^1\!/_2 - z;$$
$$ {}^3\!/_4, {}^3\!/_4, {}^1\!/_2 + z$$
$$4 \quad e \quad 2 \quad {}^3\!/_4, {}^1\!/_4, z; \; {}^1\!/_4, {}^3\!/_4, \bar{z}; \; {}^1\!/_4, {}^3\!/_4, {}^1\!/_2 - z;$$
$$ {}^3\!/_4, {}^3\!/_4, {}^1\!/_2 + z$$
$$4 \quad d \quad \bar{1} \quad 0, 0, {}^1\!/_2; \; {}^1\!/_2, {}^1\!/_2, {}^1\!/_2; \; {}^1\!/_2, 0, 0; \; 0, {}^1\!/_2\, 0$$
$$4 \quad c \quad \bar{1} \quad 0, 0, 0; \; {}^1\!/_2, {}^1\!/_2, 0; \; {}^1\!/_2, 0, {}^1\!/_2; \; 0, {}^1\!/_2, {}^1\!/_2$$
$$2 \quad b \quad \bar{4} \quad {}^1\!/_4, {}^1\!/_4, {}^3\!/_4; \; {}^3\!/_4, {}^3\!/_4, {}^1\!/_4$$
$$2 \quad a \quad \bar{4} \quad {}^1\!/_4, {}^1\!/_4, {}^1\!/_4; \; {}^3\!/_4, {}^3\!/_4, {}^3\!/_4$$

4.11. After interchanging the m planes and c glide planes in $P6_3cm$, the following arrangement is obtained:

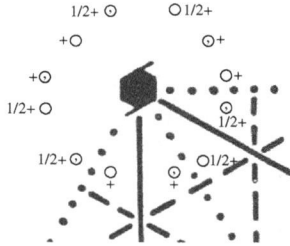

The space group symbol is $P6_3mc$.

4.12. The stereogram indicates that the following scheme gives the desired result:

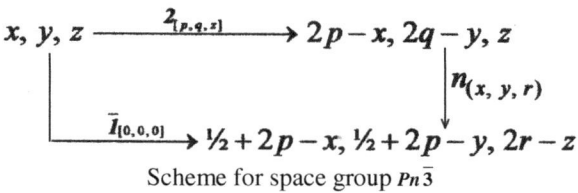

Scheme for space group $Pn\bar{3}$

Hence, $p = q = 1/4$, $r = 0$; the coordinates of the 24 general equivalent positions in space group $Pn3$ are as follows:

(1)	x, y, z	(2)	$\bar{x} + 1/2, \bar{y} + 1/2, z$
(3)	$\bar{x} + 1/2, y, \bar{z} + 1/2$	(4)	$x, \bar{y} + 1/2, \bar{z} + 1/2$
(5)	z, x, y	(6)	$z, \bar{x} + 1/2, \bar{y} + 1/2$
(7)	$\bar{z} + 1/2, \bar{x} + 1/2, y$	(8)	$\bar{z} + 1/2, x, \bar{y} + 1/2$
(9)	y, z, x	(10)	$\bar{y} + 1/2, z, \bar{x} + 1/2$
(11)	$y, \bar{z} + 1/2,$	(12)	$\bar{y} + 1/2, \bar{z} + 1/2, x$
(13)	$\bar{x}, \bar{y}, \bar{z}$	(14)	$x + 1/2, y + 1/2, \bar{z}$
(15)	$x + 1/2, \bar{y}, z + 1/2$	(16)	$\bar{x}, y + 1/2, z + 1/2$
(17)	$\bar{z}, \bar{x}, \bar{y}$	(18)	$\bar{z}, x + 1/2, y + 1/2$
(19)	$z + 1/2, x + 1/2, \bar{y}$	(20)	$z + 1/2, \bar{x}, y + 1/2$
(21)	$\bar{y}, \bar{z}, \bar{x}$	(22)	$y + 1/2, \bar{z}, x + 1/2$
(23)	$\bar{y}, z + 1/2, x + 1/2$	(24)	$y + 1/2, z + 1/2, \bar{x}$

It is interesting to note the symmetry operators that generate the coordinates from a point x, y, z; they correspond in numbering to the listing below:

(1)	1	(2)	$2 \ 1/4, 1/4, z$
(3)	$2 \ 1/4, y, 1/4$	(4)	$2 \ x, 1/4, 1/4$
(5)	$3^+ \ x, x, x$	(6)	$3^+ \ \bar{x}, x + 1/2, \bar{x}$
(7)	$3^+ \ x + 1/2, \bar{x}, \bar{x}$	(8)	$3^+ \ \bar{x} + 1/2, \bar{x} + 1/2, x$
(9)	$3^- \ x, x, x$	(10)	$3^- \ x + 1/2, \bar{x}, \bar{x}$
(11)	$3^- \ \bar{x} + 1/2, \bar{x} + 1/2, x$	(12)	$3^- \ \bar{x}, x + 1/2, \bar{x}$
(13)	$\bar{1} \ 0,0,0$	(14)	$n(1/2, 1/2, 0) \ x, y, 0$
(15)	$n(1/2, 0, 1/2) \ x, 0, z$	(16)	$n(0, 1/2, 1/2) \ 0, y, z$
(17)	$\bar{3}^+ \ x, x, x;$	(18)	$\bar{3}^+ \ \bar{x} - 1, x + 1/2, \bar{x};$
	$0,0,0$		$-1/2, 0, 1/2$
(19)	$\bar{3}^+ \ x - 1/2, \bar{x} + 1, \bar{x};$	(20)	$\bar{3}^+ \ \bar{x} + 1/2, \bar{x} - 1/2, x;$
	$0, 1/2 - 1/2$		$1/2, -1/2, 0$
(21)	$\bar{3}^- \ x, x, x;$	(22)	$\bar{3}^- \ \bar{x} + 1/2, \bar{x} - 1, \bar{x};$
	$0,0,0$		$0, -1/2, \ 1/2$
(23)	$\bar{3}^- \ \bar{x} - 1/2, x + 1/2, x;$	(24)	$\bar{3}^- \ \bar{x} + 1, x - 1/2, \bar{x};$
	$-1/2, 1/2, 0$		$1/2, 0, -1/2$

4.13. Cu–O: $d^2_{\text{Cu}-\text{O}} = (a/4 - 0)^2 + (a/4 - 0)^2 + (a/4 - 0)^2$; then $d_{\text{Cu}-\text{O}} = a\sqrt{3}/4 = 1.848\,\text{Å}$. Cu–Cu: $d^2_{\text{Cu}-\text{Cu}} = (3a/4 - a/4)^2 + (3a/4 - a/4)^2 + (a/4 - a/4)^2$; then $d_{\text{Cu}-\text{Cu}} = a\sqrt{2}/2 = 3.017\,\text{Å}$. O–O: $d^2_{\text{O}-\text{O}} = (a/2)^2$; then $d_{\text{O}-\text{O}} = 2.134\,\text{Å}$.

4.14. (a) Forming the product

$$
\overset{c}{\begin{pmatrix} 1 & 0 & 0 \\ 0 & \bar{1} & 0 \\ 0 & 0 & 1 \end{pmatrix}} \overset{a}{\begin{pmatrix} \bar{1} & 0 & 0 \\ 0 & 1 & 0 \\ 0 & 0 & 1 \end{pmatrix}} + \overset{t_a}{\begin{pmatrix} 1/2 \\ 0 \\ 0 \end{pmatrix}} + \overset{t_c}{\begin{pmatrix} 0 \\ 0 \\ 1/2 \end{pmatrix}} = \overset{2_z}{\begin{pmatrix} \bar{1} & 0 & 0 \\ 0 & \bar{1} & 0 \\ 0 & 0 & 1 \end{pmatrix}} + \overset{t_a+t_c}{\begin{pmatrix} 1/2 \\ 0 \\ 1/2 \end{pmatrix}}
$$

shows that the result of the combination **ac** is a 2_1-symmetry operator along $[1/2, 0, 0]$. (b) The space-group symbol is $Pca2_1$. (c) The operators **a** and **c** commute in this space group, as is shown readily by interchanging the order of the product **ca** to **ac**.

4.15. (a) If any two of the 2_1-axes are allowed to intersect, the resulting space group is $P2_12_12$:

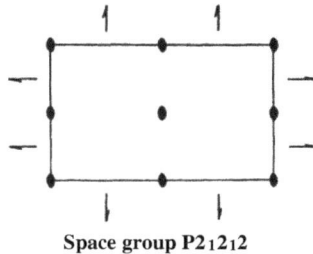

Space group P2₁2₁2

(b) If a two-fold axis along $[x, 0, 0]$ is combined with space group $P2_12_12_1$, the resulting space group is $C222_1$:

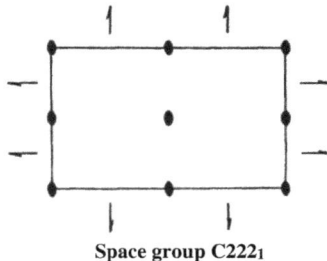

Space group C222₁

References

[1] International Union of Crystallography, *Online Dictionary of Crystallography.* <http://reference.iucr.org/dictionary/Main_Page>.

[2] Federov ES, *Zap. Mineral. Obch.* **28**, 1 (1891); Eng. trans. *Amer. Cryst. Assoc. Mono.* 7 (1971).

[3] Schönflies A, *Kristallsysteme und Kristallstruktur.* Leipzig (1891).

[4] Barlow WW, *Z. Kristallogr.* **23**, 1 (1894).

[5] Hentry NFM and Lonsdale K (eds), *International Tables for X-ray Crystallography*, Vol. I, 3rd edn., International Union of Crystallography, Kynoch Press (1969). Online at <https://archive.org/stream/International TablesForX-rayCrystallographyVol1/HenryLonsdaleEds-internationalTables ForX-rayCrystallographyVol1#page/n71/mode/2up>. *Note*: This source contains the earlier notation of $m3$ and $m3m$ in place of $m\bar{3}$ and $m\bar{3}m$ for those point groups, and for the space groups that derive from them.

[6] Aroyo MI (ed), *International Tables for Crystallography* — Volume A, *Space-Group Symmetry*, 6th edn., International Union of Crystallography, Wiley (2016); 2nd edn. Wiley (2014). Online at ⟨http://www.ITC-Vol.A.HahnT.ed. InternationalTablesforCrystallographyVol.A.SpaceGroupSymmetry5ed.Sprin ger2005ISBN0792365909910s.pdf⟩; also in Program Suite as ITA.pdf.

[7] Ladd M, *Symmetry of Crystals and Molecules.* Oxford University Press (2014).

[8] Burns GG and Glazer AM, *Space Groups for Solid State Scientists*, 3rd edn. Elsevier (2013).

[9] Pringle GE, *Acta Cryst.* **B28**, 2326 (1972).

Chapter 5

X-rays and X-ray Diffraction

If the hand be held between the discharge-tube and the screen,
the darker shadow of the bones is seen within the slightly dark
shadow-image of the hand itself. For brevity's sake I shall use the
expression 'rays'; and to distinguish them from others of this
name I shall call them 'X-rays'.
Wilhelm Röntgen

Key Topics

- Generation and properties of X-rays
- X-ray scattering by atoms and crystals
- Laue and Bragg equations
- Structure factor and its applications
- Reflection conditions
- Geometry of data collection
- Data collection: film and diffractometer
- Intensity corrections
- Statistics of diffraction data

5.1. Introduction

X-radiation (aka X-rays) is an electromagnetic radiation with a wavelength range from 0.1–100 Å (0.01–10 nm); it is also known as Röntgen radiation, after the discoverer [1]. Electromagnetic radiation has a wide range of wavelengths with corresponding designations, as the scheme below shows:

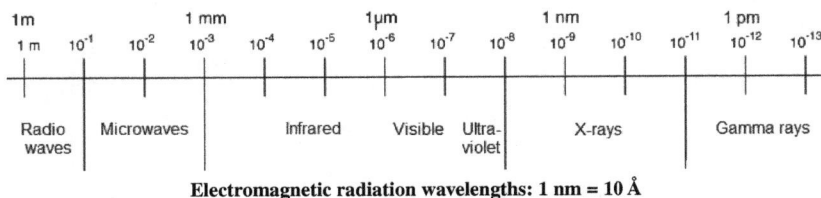

1m			1 mm			1μm			1 nm			1 pm	
1 m	10^{-1}	10^{-2}	10^{-3}	10^{-4}	10^{-5}	10^{-6}	10^{-7}	10^{-8}	10^{-9}	10^{-10}	10^{-11}	10^{-12}	10^{-13}

Radio waves	Microwaves	Infrared	Visible	Ultra-violet	X-rays	Gamma rays

Electromagnetic radiation wavelengths: 1 nm = 10 Å

X-rays have many uses, including radiography, medical diagnosis, airport security, astronomy and diffraction studies. The X-rays used in crystallography lie in the 'soft' X-ray region, with wavelengths between 0.5 Å and 2.5 Å.

5.2. Generation of X-rays

X-rays are produced by the sudden deceleration of rapidly moving electrons impinging on a metal target. An electron falling through a potential difference of V volt acquires an energy of eV electron volt, where e is the electronic charge and V is the applied voltage. The electron-volt (eV) is a measure of energy: 1 eV $= (e/\text{C})$ J $= 1.60218 \times 10^{-19}$ J. In terms of wavelength λ, an energy quantum is expressed by the equation

$$\lambda = hc/(eV) \tag{5.1}$$

where h is Planck's constant and c is the speed of light *in vacuo*. Inserting values of the constants into Eq. (5.1) gives

$$\lambda = 12.398/V \tag{5.2}$$

where V is measured in kV and λ in Å.

5.3. X-rays and 'White' Radiation

One source of X-rays for crystallographic studies is the sealed, hot-cathode tube with a rotating anode, as illustrated in Fig. 5.1. Electrons are emitted from a tungsten filament under excitation by a

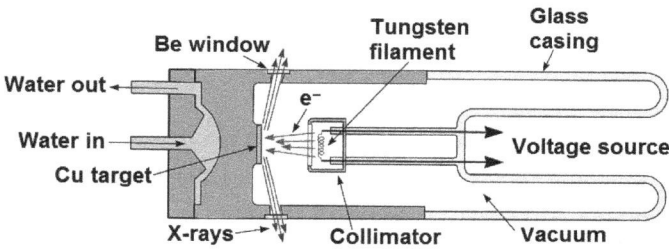

Fig. 5.1. Schematic diagram of a crystallographic X-ray tube. The target material is usually molybdenum or copper and its rotation distributes the dissipation of heat energy. [*Minerals: Their Constitution and Origin*, H-R Wenk and A Bulakh (2004). Reproduced by permission of Cambridge University Press.]

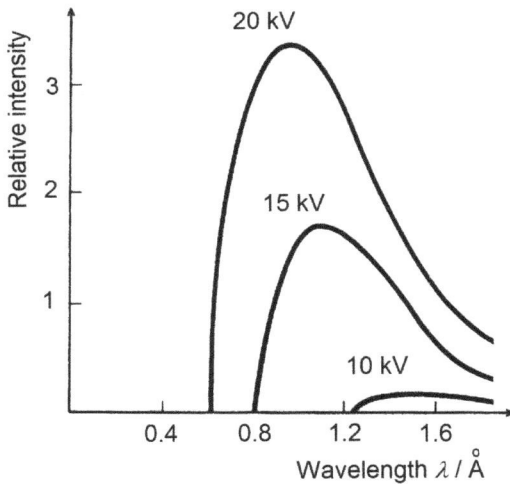

Fig. 5.2. Variation in intensity of X-radiation with wavelength for three applied voltages. As the voltage V increases, the wavelength corresponding to the maximum intensity moves to progressively shorter wavelengths, as expected from Eq. (5.2). The continuous spectrum is known as 'white' radiation, or Bremsstrahlung (Ger. 'braking radiation'). The sharp wavelength cut-off indicates insufficient excitation energy at that value of λ; the quantum condition is not met.

potential difference between the metal anode and filament cathode. The efficiency of X-ray generation is approximately 10% in terms of energy, the remainder being lost as heat to the water-cooled target.

The output from the X-ray tube is a function of the applied voltage, as shown in Fig. 5.2. An X-ray wavelength depends on two energy levels:

$$\lambda = hc/(E_2 - E_1) \qquad (5.3)$$

where E_1 and E_2 are energy levels of the atoms of the target material. The continuous 'white' radiation, and the sharp cut-off at the low wavelength (high energy) end of the spectrum correspond to the precise minimum value of λ given by Eq. (5.1).

At a sufficiently high value of the applied voltage, electrons incident upon the metal target excite electrons from the inner K shell of the metal target. As electrons fall back to those levels, the *characteristic* K spectrum is produced; the K_α and K_β spectra are always excited together:

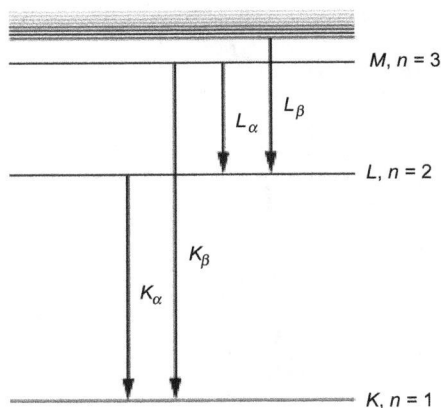

The K, L and M energy levels of an atom

The K_α spectrum comprises two components corresponding to two slightly different L level energies for $L \rightarrow K$ transitions, and the $M \rightarrow L$ transition gives rise to the K_β doublet spectrum, of mean wavelength of *ca.* 1.387 Å, as shown in Fig. 5.3. With a copper target, the K_{α_1} wavelength is 1.54056 Å and the K_{α_2} is 1.54439 Å in the ratio of 2:1, so that the mean value for K_α is 1.54184 Å.

5.3.1.   Monochromatic radiation

Most of the unwanted K_β component can be removed by passing the X-ray beam through a nickel *filter*. The *resonance level,* or *absorption*

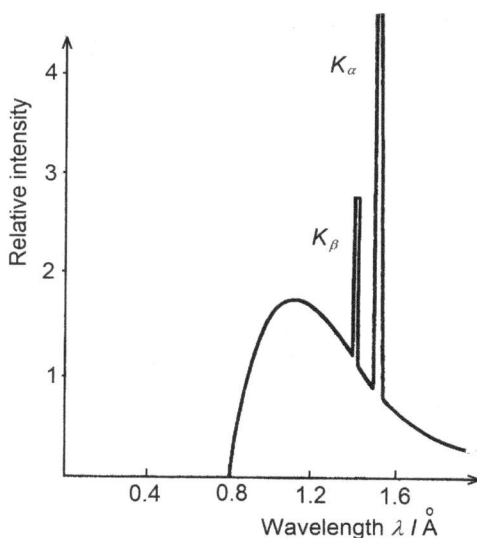

Fig. 5.3.   At a sufficiently high value of applied voltage, the high intensity *characteristic* K spectrum is emitted, superimposed on to the continuous radiation.

edge, for nickel occurs at 1.4886 Å, a value which lies between those of the K_α and K_β spectral wavelengths. Elemental nickel is effective in absorbing the higher energy K_β radiation so as to give a very nearly *monochromatic source* of X-rays.

The absorption of X-rays follows the exponential law

$$I = I_0 \exp(-\mu t) \tag{5.4}$$

where I is the transmitted intensity of X-rays and I_0 the corresponding incident intensity; μ is the linear absorption coefficient of the material and t its thickness in the X-ray path.

Self-assessment 5.1.   (a) What is the minimum wavelength and corresponding energy associated with X-rays generated in an X-ray tube by a potential difference of 20 kV? (b) What is the ratio of I/I_0 for a nickel filter of 0.23 mm thickness if $\mu_{Ni} = 0.76$ cm^{-1}?[a]

---

[a] Answers to all self-assessments are given at the end of the chapter.

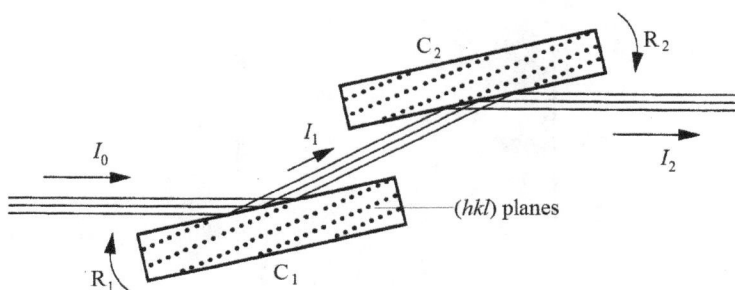

Fig. 5.4. Schematic diagram of a graphite double-crystal monochromator. The incident beam has intensity I_0 and the crystals C_1 and C_2 are set in positions to produce the strong (0002) reflection; the coupled rotors R_1 and R_2 allow different wavelengths to be selected. The twice-reflected beam of intensity I_2 is monochromatic for all practical purposes.

Although filtered radiation is a satisfactory source of X-rays for diffraction studies, a preferable method makes use of a crystal as a monochromator. Figure 5.4 illustrates the principle of a double-crystal monochromator. Incident X-rays of intensity I_0 impinge on the (0002) planes of a graphite crystal C_1 set at the correct angle for reflection, and pass as an essentially monochromatic beam of intensity I_1 on to a second crystal C_2 set at the same angle to improve further the monochromatization. The coupled rotators R_1 and R_2 allow different wavelengths to be selected. The ratio of peak to background radiation is high so that strongly monochromatic radiation I_2 is produced. Harmonics $\lambda/2, \lambda/3, \ldots$ are reflected at the same angles but are of very weak intensity from a graphite crystal.

Whatever the method of production of a monochromatic X-ray source, the angular limit of the beam is defined by collimation, which may be of pinhole or multi-fibre capillary type; a typical angle of beam divergence is approximately 0.02 rad.

5.3.2. Synchrotron radiation source

A synchrotron is a cyclic particle accelerator in which an accelerating electron beam travels around a fixed closed-loop circuit. A magnetic field bends the particle beam into its closed path; it increases with time during the accelerating process, and is synchronized to the increasing kinetic energy of the accelerated particles. The maximum energy of the electrons is *ca.* 10^9 eV, which implies a speed approaching that of light. The movement of the electrons is sufficiently fast for the emitted energy to be of X-ray wavelengths. The spectrum is smooth and continuous, and offers a choice between white radiation and a strictly monochromatic beam. The beam is horizontally polarized and its very small divergence means that it is well collimated.

Thus, the synchrotron provides an extremely powerful X-ray source, several orders of magnitude greater than that from other traditional sources. The importance of this feature is the ability to collect data from very small crystals in very short times, which is important in the examination of crystals that are readily attacked by the radiation. Whatever be the method used to obtain X-rays, the analysis of a crystal structure by X-ray diffraction requires, except for the Laue method, a strong, collimated, monochromatic X-ray beam. Many reports on the use of synchrotron radiation in structure analysis are available online.

5.4. X-ray Scattering

Electromagnetic radiation comprises plane self-propagating, transverse, oscillating waves of electric and magnetic fields. With respect to the direction of propagation of a wave, if the electric field is defined in a vertical plane, then the magnetic field is in a plane that is at right angles to that direction; these two fields are always in phase with each other. When X-rays interact with electrons, the alternating vector of the electric field imparts acceleration to the electrons. Electromagnetic theory shows that an accelerated particle emits radiation by an absorption and re-emission processes, with the emitted radiation travelling in all directions. Two forms of X-rays that are scattered by the electrons of atoms are important in X-ray diffraction.

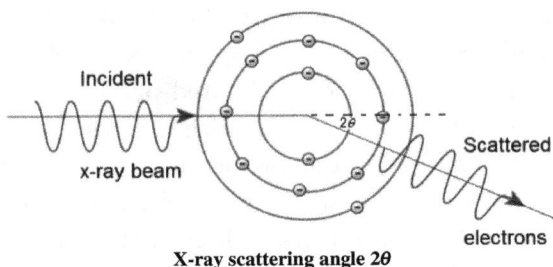

X-ray scattering angle 2θ

In *Thomson scattering*, which is elastic and normally coherent, the incident and scattered radiant waves have the same wavelength, and a definite phase relationship between them. A theoretical treatment of Thomson scattering shows that the intensity of radiation scattered at an angle 2θ is proportional to $(1 + \cos^2 2\theta)/2$ and inversely proportional to the mass m_e of the electron. The $1/m_e$ factor explains how electrons are the only effective scattering species for X-rays: even hydrogen, although it has the same magnitude of charge, is *ca.* 1840 times heavier than the electron. The term $(1 + \cos^2 2\theta)/2$ is a geometrical factor known as the *polarization factor p*.

Compton scattering is an incoherent process: the scattered radiation has a wavelength longer than that of the incident radiation, implying a transfer of energy in elastic collisions with the atomic nuclei (inelastic scattering). The wavelength change $\delta\lambda$ on scattering is given by

$$\delta\lambda = \frac{h}{m_e c}(1 - \cos 2\theta) \tag{5.5}$$

Self-assessment 5.2. An X-ray photon is scattered incoherently at a scattering angle of 40° to the forward direction of the incident beam. What is the change in wavelength?

5.4.1. Scattering by electrons in an atom

Let a plane monochromatic X-ray beam of amplitude Ψ_0 be incident upon an atom, assumed to be spherical, containing a single electron.

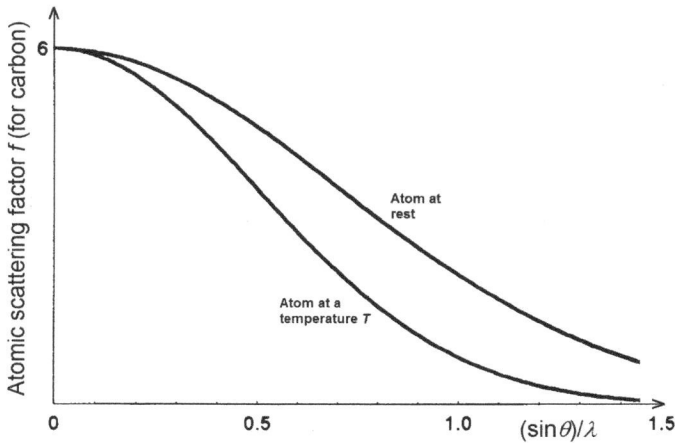

Fig. 5.5. Atomic scattering factor f for carbon as a function of $(\sin\theta)/\lambda$. At $\sin\theta = 0$, $f = Z$, the atomic number for the species. The upper curve refers to the atom at rest, while the lower curve includes a correction for thermal vibrations in the atom at ambient temperature.

The X-rays are scattered at an angle 2θ with respect to the direction of the incident beam, and the intensity of the scattered beam is then a function of θ given by

$$\Psi_\theta = f_\theta \Psi_0 \tag{5.6}$$

where f_θ is an *atomic scattering factor*.

In an atom of more than one electron, interference takes place in the scattering process within the atom. Theoretical calculation shows that the atomic scattering factor (aka atomic form factor) follows a curve of the type shown in Fig. 5.5. At $\sin\theta = 0$, there is no interference, so that f is then is equal to Z, the atomic number of the species. The atomic scattering factor may be defined as the ratio of the amplitude of coherent scattering from an atom to that of scattering by a single electron at the centre of the atom.

Consider an atom containing two electrons at points within it, such as O and A, as shown in Fig. 5.6. X-rays of wavelength λ are in phase along the incident wavefront OX and again at the diffracted wavefront AY. The path difference δ between the two wavefronts after scattering is represented by the distance $OY - AX$. Then the phase

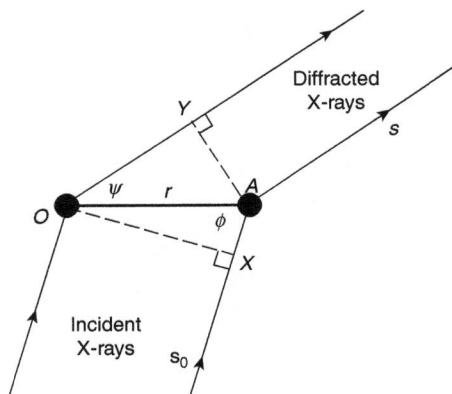

Fig. 5.6.   Scattering of X-rays from two electrons O and A in an atom; s_0 and s are unit vectors in the incident and diffracted X-ray beams, respectively. The angle of incidence is ϕ and the angle of scatter is ψ.

difference ϕ, or $(2\pi/\lambda)\delta$, at any point P in the scattered direction such that $OP \gg OA$ is given by

$$\phi = (2\pi/\lambda)(OY - AX) \qquad (5.7)$$

The distance OA is designated $r(\equiv |\mathbf{r}|)$ and the incident and diffracted waves may be defined by the vectors $\mathbf{s}_0$ and $\mathbf{s}$, respectively. Then, $AX = \mathbf{r} \cdot \mathbf{s}_0$ and $OY = \mathbf{r} \cdot \mathbf{s}$, so that

$$\phi = (2\pi/\lambda)(\mathbf{r} \cdot \mathbf{s} - \mathbf{r} \cdot \mathbf{s}_0) = 2\pi\mathbf{r} \cdot (\mathbf{s}/\lambda - \mathbf{s}_0/\lambda) = 2\pi\mathbf{r} \cdot \mathbf{S} \qquad (5.8)$$

The meaning of $\mathbf{S}$ is shown in Fig. 5.7; it is the normal to a plane through O which may be termed a *reflecting plane*. Hence,

$$|\mathbf{S}| = S = |\mathbf{s} - \mathbf{s}_0|/\lambda \qquad (5.9)$$

The amplitude of the wave scattered at the point P by two electrons, relative to the incident amplitude, is then the contribution given by Eq. (5.6) plus another such amount modified by a function of the phase factor in Eq. (5.8), which leads to the equation.

$$\Psi_\theta = f_\theta + f_\theta \exp(i2\pi\mathbf{r} \cdot \mathbf{S}) = f_\theta[1 + \exp(i2\pi\mathbf{r} \cdot \mathbf{S})] \qquad (5.10)$$

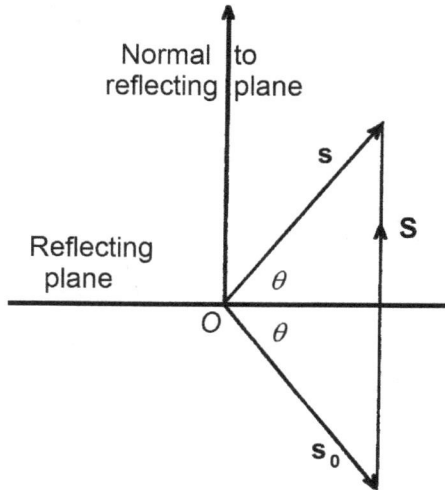

Fig. 5.7. Relationship of the scattering vector **S**, or $(\mathbf{s} - \mathbf{s}_0)/\lambda$, to its reflecting plane (hkl); the magnitude $|\mathbf{S}|$, or S, is $(2\sin\theta)/\lambda$.

If neither electron lies at the origin, then the unity term on the right-hand side of Eq. (5.10) would be replaced by another exponential function. In general, for n electrons, the amplitude expression becomes

$$\Psi_{\theta,n} = \sum_{j=1}^{n} f_{\theta,j} \exp(i2\pi \mathbf{r}_j \cdot \mathbf{S}) \qquad (5.11)$$

The use of the exponential term for the phase difference expression is discussed shortly.

5.4.2. Scattering by a crystal: Laue and Bragg equations

A crystal is a regular, three-dimensional array of atoms, and X-ray diffraction from a crystal is a combined scattering and interference process. It has been described by Laue and Bragg with different but equivalent treatments. In terms of the Laue treatment, consider a regular row of scattering centres of spacing **a** in the path of a parallel

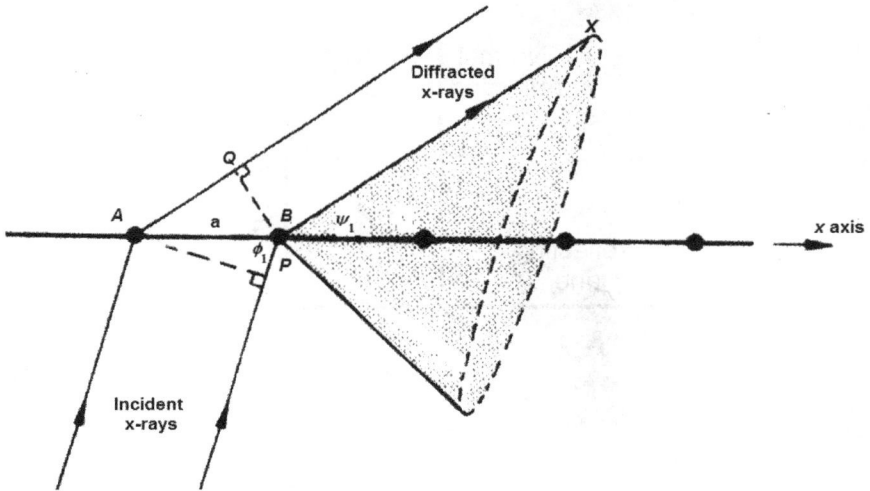

Fig. 5.8.   Diffraction from a one-dimensional (row) array of scattering centres of spacing a. The Laue equation (5.12) is satisfied by any generator of the cone of semi-angle ψ_1, but constructive interference takes place only for integral values of h.

X-ray beam. The X-rays are incident on the row at the angle ϕ_1 and are scattered at the angle ψ_1, as shown in Fig. 5.8.

The path difference for X-rays scattered by neighbouring scattering centres in the row is $AQ - BP$, or $a\cos\psi_1 - a\cos\varphi_1$. When this difference is equal to an integral number of wavelengths of the X-rays, constructive interference occurs, then

$$a(\cos\psi_1 - \cos\phi_1) = h\lambda = \mathbf{a}\cdot\mathbf{s} - \mathbf{a}\cdot\mathbf{s}_0 \qquad (5.12)$$

This equation is satisfied by any generator, such as BX of a cone, which is coaxial with the row of centres and has a semivertical angle ψ_1, which is the angle between the row and a diffracted beam.

Equation (5.12) may also be written, using Eq. (5.9), as

$$\mathbf{a}\cdot(\mathbf{s}-\mathbf{s}_0)/\lambda = \mathbf{a}\cdot\mathbf{S} = h \qquad (5.13)$$

In two dimensions, in-phase scattering occurs when two such cones as that shown in Fig. 5.8 intersect, which they do in two lines, say, OX and OY, at angles ψ_1 and ψ_2 to the rows of spacings a and b, respectively; a second Laue equation then arises, $b(\cos\psi_2 - \cos\phi_2) = k\lambda$,

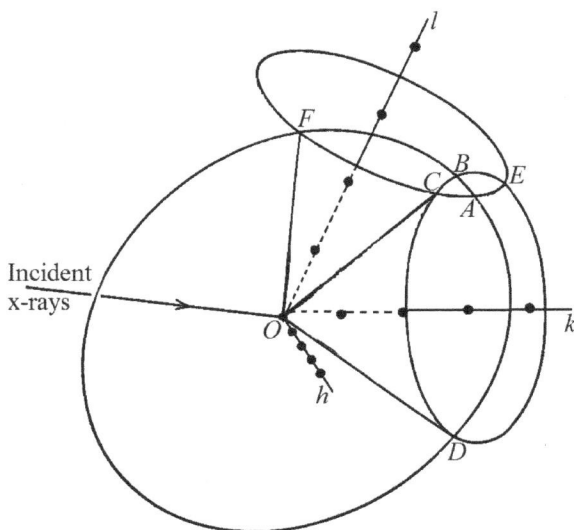

Fig. 5.9. Three cones representing diffraction from the rows of spacings a, b and c. Diffraction from the complete array of atoms occurs when the intersections OA, OB and OC coincide. [Reproduced with permission from Buerger MJ, *X-ray Crystallography*, John Wiley & Sons (1942).]

to be satisfied simultaneously with Eq. (5.12) for diffraction from the net of scattering points. Proceeding finally to the three-dimensional case, there are now three cones, as shown in Fig. 5.9, involving now a third equation, $c(\cos \psi_3 - \cos \phi_3) = l\lambda$.

When the three cones intersect mutually, six lines of intersection, OA, OB, OC, OD, OE and OF, are developed. Along the directions of pairs of these lines, cooperative scattering takes place from the corresponding nets of points; two Laue equations are satisfied in each case. When the lines OA, OB and OC coincide, the whole array, thus implying every point, scatters in phase according to the *three Laue equations*, which may be written concisely as

$$\mathbf{a} \cdot \mathbf{S} = h$$
$$\mathbf{b} \cdot \mathbf{S} = k \qquad\qquad (5.14)$$
$$\mathbf{c} \cdot \mathbf{S} = l$$

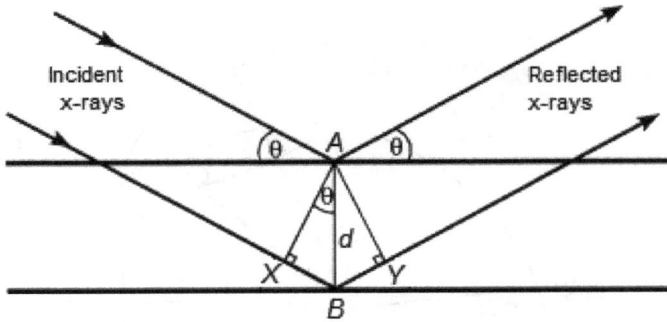

Fig. 5.10.   Bragg reflection from a family of crystal planes (hkl) of interplanar spacing $d(hkl)$. The path difference for two typical rays incident at a glancing angle θ is $XB + BY$. If the Bragg equation is fulfilled for these two planes, it will be fulfilled also for the remainder of planes in the family.

An alternative approach by Bragg was deduced from a comparison with reflection of light from a plane mirror, in which process a $\theta/2\theta$ relationship exists for reflection. The Bragg equation can be derived through Fig. 5.10 in which X-rays are incident on atomic planes of spacing d at the *glancing angle* θ; two planes of the family (hkl) are shown. The incident and *reflected* (diffracted) X-rays make the same angle θ with the planes; the scattering angle remains as 2θ.

The path difference δ for the two typical rays shown is $XB + BY$, so that

$$\delta = 2d \sin \theta \qquad (5.15)$$

For constructive interference, δ must be an integral number n of X-ray wavelengths λ. Hence,

$$2d \sin \theta = n\lambda \qquad (5.16)$$

which is the *Bragg equation* for reflection of X-rays from crystal planes. The exact mirror analogy breaks down since Eq. (5.16) must be satisfied for reflection to occur, but it led to the deduction of this important equation. The equation is written now as

$$2d_{hkl} \sin \theta_{hkl} = \lambda \qquad (5.17)$$

Table 5.1. Interplanar spacings and the Bragg equation: $a = 5$ Å, $hkl = (120)$.

Original notation			Current usage	
hkl	Order n	$d_{\text{old}}(hkl)/\text{Å}$	hkl	$d_{\text{new}}(hkl)/\text{Å}$
(120)	1	2.236	120	2.236
	2	2.236	240	1.118
	3	2.236	360	0.745
	4	2.236	480	0.559
	5	2.236	510,0	0.447

The relationship is $d(hkl)/n = d(nh, nk, nl)$, with h, k and l taking factors as necessary. Table 5.1 illustrates this relationship.

It is a common practice to refer to X-ray *reflection* from a crystal, in the context of the Bragg equation, rather than to X-ray diffraction. This treatment shows that X-ray reflections from a single crystal occur in specific directions governed by the Bragg equation. This inspirational formulation came to Bragg after walking past a plantation of trees and noting that every so often the trees seemed to line up exactly like the lattice points in a crystal. For further details on the scattering process in atoms and crystals, the reader is referred to the literature [2, 3].

5.4.2.1. *Argand diagram*

An Argand diagram provides a convenient illustration of the procedure for summing waves in terms of both amplitude and phase, and Eq. (5.11) is an expression of this type of summation. In Fig. 5.11, each of the two waves in the summation is of the form $f_j \exp(i\phi_j)$, where ϕ_j is given by Eq. (5.8). The subscript θ to f_j is dropped for convenience; f is always a function of θ. The exponential term can be expanded by Euler's equation, $\exp(\pm i\phi) = \cos\phi \pm i\sin\phi$, and following Eq. (5.11) in order to handle the real and imaginary components of the complex number $\mathbf{F}$ conveniently. For the two waves shown on their Argand diagram, with amplitudes f_1 and f_2 and phases ϕ_1 and ϕ_2, the addition is given by

$$\mathbf{F} = \mathbf{f}_1 + \mathbf{f}_2 = f_1 \exp(i\phi_1) + f_2 \exp(i\phi_2) = |\mathbf{F}| \exp(i\phi) \qquad (5.18)$$

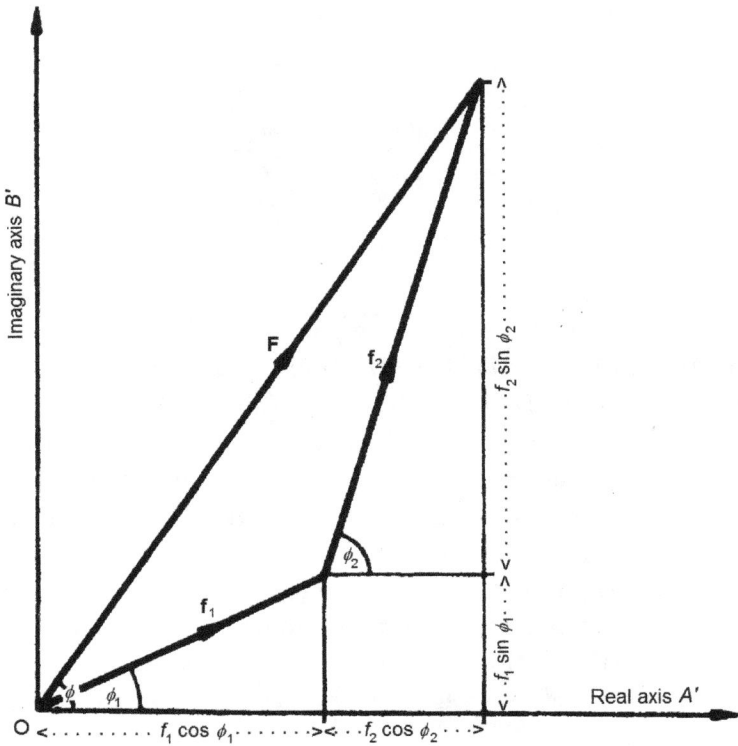

Fig. 5.11. Combination of two waves $\mathbf{f}_1 = f_1 \exp(\mathrm{i}\phi_1)$ and $\mathbf{f}_2 = f_2 \exp(\mathrm{i}\phi_2)$, as shown on an Argand diagram. The resultant wave is $\mathbf{F}$, with an amplitude $|\mathbf{F}|$, or F, and a phase angle ϕ.

where ϕ is the phase angle associated with the resultant amplitude $|\mathbf{F}|$. In general, for n waves (from n atoms), the expression is

$$\mathbf{F} = \sum_{j=1}^{n} f_j \exp(\mathrm{i}\phi_j) \tag{5.19}$$

and

$$F = |\mathbf{F}| = \sqrt{\mathbf{F}\mathbf{F}^*} \tag{5.20}$$

where $\mathbf{F}^*$ is the complex conjugate of $\mathbf{F}$, that is, $|\mathbf{F}| \exp(-\mathrm{i}\phi)$; thus, the complex conjugate is the reflection of $\mathbf{F}$ across the real axis.

Resolving $\mathbf{F}$ into its real A' and imaginary B' components leads to

$$|\mathbf{F}| = F = \sqrt{A'^2 + B'^2} \tag{5.21}$$

where

$$A' = \sum_{j=1}^{n} f_j \cos \phi_j \tag{5.22}$$

and

$$B' = \sum_{j=1}^{n} f_j \sin \phi_j \tag{5.23}$$

The phase angle ϕ of the resultant $\mathbf{F}$ is given by

$$\phi = \tan^{-1} \frac{B'}{A'} \tag{5.24}$$

Self-assessment 5.3. Two waves of amplitudes 100 and 50 have phases of $0°$ and $240°$, respectively. Calculate the resultant amplitude and phase angle after the waves have been combined.

5.4.2.2. *Linking $\mathbf{F}$ with the structure*

So far, the amplitudes have been linked with the structure because they are the properties of its component atoms. Now, it is necessary to link in the phases of each scattering component.

The scattering of the jth atom is, from Eq. (5.8), given by $f_{j,\mathbf{s}} \exp(\mathrm{i}2\pi\mathbf{r}_j \cdot \mathbf{S})$. The vector $\mathbf{r}_j$ from the origin of a unit cell governed by the basic translations $\mathbf{a}$, $\mathbf{b}$ and $\mathbf{c}$ the jth atom at a position with coordinates x_j, y_j, z_j is given, following Eq. (3.2), as

$$\mathbf{r}_j = x_j\mathbf{a} + y_j\mathbf{b} + z_j\mathbf{c} \tag{5.25}$$

and from Eq. (3.18),

$$\mathbf{d}^*_{hkl} = h\mathbf{a}^* + k\mathbf{b}^* + l\mathbf{c}^*$$

Since $2\sin\theta_{hkl}/\lambda$ is equal to d^*_{hkl}, then from Fig. 5.7, and using Eqs. (3.18) and (5.8),

$$\mathbf{r}_j \cdot \mathbf{S} = (x_j\mathbf{a} + y_j\mathbf{b} + z_j\mathbf{c}) \cdot (h\mathbf{a}^* + k\mathbf{b}^* + l\mathbf{c}^*) = hx_j + ky_j + lz_j \tag{5.26}$$

since $\mathbf{a} \cdot \mathbf{a}^* = 1$ and $\mathbf{a} \cdot \mathbf{b}^* = 0$, and similarly for other such pairs. Thus, from Eq. (5.8), it follows that the phase angle in terms of the coordinates is given by

$$\phi_j = 2\pi(hx_j + ky_j + lz_j) \tag{5.27}$$

and the phase contribution of the jth atom is, therefore,

$$\exp[i2\pi(hx_j + ky_j + lz_j)] \tag{5.28}$$

Then, following Eq. (5.18), the *structure factor equation* is

$$\mathbf{F}(hkl) = \sum_{j=1}^{N} f_j \exp[i2\pi(hx_j + ky_j + lz_j)] \tag{5.29}$$

and applies to a crystal of N atoms at rest. At ambient temperature, a correction for the thermal vibration of the atoms must be applied. It may be isotropic or anisotropic and the order of efficacy is individual anisotropic > individual isotropic > overall isotropic, more of which will be discussed shortly.

Self-assessment 5.4. Show that the Bragg and Laue equations are equivalent ways of representing X-ray diffraction in terms of direction.

5.4.3. The structure factor F in practice

This section will consider some typical practical applications of the structure factor equation, Eq. (5.29). The trigonometrical identities in Appendix A7 may be helpful in this discussion.

5.4.3.1. *Friedel's law*

X-ray diffraction spectra may be treated as a representation of the reciprocal lattice of a crystal (Section 3.7) in which each hkl spectrum is weighted in proportion to its structure amplitude $|\mathbf{F}(hkl)|$. In applications of the structure factor equation in X-ray crystallography, wherein the effects of anomalous scattering (Section 6.16) are not significant, the relationship

$$I(hkl) = I(\bar{h}\bar{k}\bar{l}) \tag{5.30}$$

holds, within the limits of experimental error, and expresses the centrosymmetric nature of an X-ray diffraction record; Eq. (5.30) is an expression of *Friedel's law*. Alternatively, it may be written as

$$|\mathbf{F}|(hkl) = |\mathbf{F}|(\bar{h}\bar{k}\bar{l}) \tag{5.31}$$

with the same qualifications.

The atomic scattering factor, being a function of $\sin\theta$, has the same value for the hkl and the $\bar{h}\bar{k}\bar{l}$ reflections. Thus, $f_\theta = f_{-\theta}$ because the hkl and $\bar{h}\bar{k}\bar{l}$ reflections are opposite sides of the plane and so reflect at the same Bragg angle.

From Eq. (5.29),

$$\mathbf{F}(\bar{h}\bar{k}\bar{l}) = \sum_{j=1}^{n} f_j \exp[-i2\pi(hx_j + ky_j + lz_j)] \tag{5.32}$$

From Fig. 5.12, the following relations may be seen:

$$\begin{aligned}
\mathbf{F}(hkl) &= A'(hkl) + iB'(hkl) \\
\mathbf{F}(\bar{h}\bar{k}\bar{l}) &= A'(\bar{h}\bar{k}\bar{l}) + iB'(\bar{h}\bar{k}\bar{l}) = A'(hkl) - iB'(hkl)
\end{aligned} \tag{5.33}$$

The same result can be reached working from the exponential form in Eq. (5.32):

$$|\mathbf{F}|(hkl) = |\mathbf{F}|(\bar{h}\bar{k}\bar{l}) = \sqrt{A'^2(hkl) + B'^2(hkl)} \tag{5.34}$$

which is similar to Eq. (5.21); also, since $\phi = \tan^{-1}\frac{B'}{A'}$ it, follows that

$$\phi(hkl) = -\phi(\bar{h}\bar{k}\bar{l}) \tag{5.35}$$

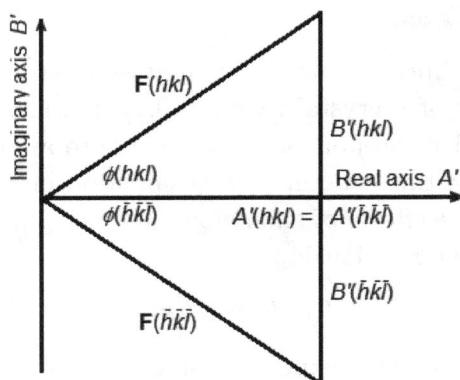

Fig. 5.12.   The structure factor $\mathbf{F}(hkl)$ and its components: $A'(hkl)$, $B'(hkl)$ and $\phi(hkl)$.

5.4.4.   Typical applications of the structure factor equation

This section will examine the structure factor in three situations: a centrosymmetric unit cell, a centred unit cell and a unit cell with glide planes and screw axes.

5.4.4.1.   *Structure factor for a centrosymmetric crystal*

In a centrosymmetric structure, it is almost always an advantage to have the centre of symmetry at the origin, the point 0, 0, 0. In these discussions, the interest centres upon the *geometrical structure factor*, the components of which are Eqs. (5.22) and (5.23) without the inclusion of the atomic scattering factors. Thus, the geometrical structure factor depends only upon the symmetry of the structure. The two functions are

$$A(hkl) = \sum_{j=1}^{N} \cos[2\pi(hx_j + ky_j + lz_j)] \tag{5.36}$$

and

$$B(hkl) = \sum_{j=1}^{N} \sin[2\pi(hx_j + ky_j + lz_j)] \tag{5.37}$$

where N is the total number of atoms in the unit cell.

In centrosymmetric crystals, the atoms occur in the pairs $\pm(x, y, z)$ in the structure. Then,

$$A(hkl) = \sum_{j=1}^{N/2} \cos[2\pi(hx_j + ky_j + lz_j)] + \cos[2\pi(-hx_j - ky_j - lz_j)]$$

$$(5.38)$$

$$= 2\sum_{j=1}^{N/2} \cos[2\pi(hx_j + ky_j + lz_j)] \qquad (5.39)$$

since $\cos(-\theta) = \cos\theta$. For the B component, since $\sin(-\theta) = -\sin\theta$,

$$B(hkl) = \sum_{j=1}^{N/2} \sin[2\pi(hx_j + ky_j + lz_j)] + \sin[2\pi(-hx_j - ky_j - lz_j)]$$

$$= 0 \qquad (5.40)$$

The advantage in placing the centre of symmetry at the origin is now evident. In any other setting of the origin, there will be a non-zero value of the B component. Another important feature is that, from Eq. (5.24), the phase angle ϕ is either 0 or π, which is of considerable help in a crystal structure analysis.

5.4.4.2. *Structure factor for a C-centred unit cell*

In a C-centred unit cell, the atoms are arranged in the pairs $(x, y, x; 1/2 + x, 1/2 + y, z)$. Then from Eqs. (5.36) and (5.37),

$$A(hkl) = \sum_{j=1}^{N/2} \cos[2\pi(hx_j + ky_j + lz_j)$$

$$+ \cos[2\pi(hx_j + ky_j + lz_j) + (h + k)/2)]$$

$$= 4\sum_{j=1}^{N/2} \cos[2\pi(hx_j + ky_j + lz_j) \cos\pi(h + k)/2]$$

$$(5.41)$$

From this equation, it is evident that the term $\cos \pi (h+k)/2$ is unity if $h + k = 2n$, that is, an even integer, and zero if $h + k$ is odd. For the B component,

$$
\begin{aligned}
B(hkl) &= \sum_{j=1}^{N/2} \sin[2\pi(hx_j + ky_j + lz_j)] \\
&\quad + \sin[2\pi(hx_j + ky_j + lz_j) + (h+k)/2)] \\
&= 4\sum_{j=1}^{N/2} \sin[2\pi(hx_j + ky_j + lz_j) \cos \pi (h+k)/2] \quad (5.42)
\end{aligned}
$$

The same condition applies with respect to $h + k$. Thus, the geometrical structure factor may be written for $(h + k)$ even $(2n)$ and odd $(2n+1)$ integral numbers as

$h + k = 2n$

$$
A(hkl) = 4\sum_{j=1}^{N/2} \cos 2\pi(hx_j + ky_j + lz_j) \quad (5.43)
$$

$$
B(hkl) = 4\sum_{j=1}^{N/2} \sin 2\pi(hx_j + ky_j + lz_j) \quad (5.44)
$$

$h + k = 2n + 1$

$$
A(hkl) = B(hkl) = 0 \quad (5.45)
$$

It may be seen that Eq. (5.43) has twice the magnitude of Eq. (5.39), on account of the C-centring. The multiplier, usually denoted by the G-factor, depends upon the form of centring:

Unit Cell	G
P	1
A, B, C, I	2
F	4

The conditions on $h + k$ derived above are known as *limiting conditions* (aka reflection conditions). Note that energy is not destroyed by absences: in the case of the C-centred unit cell, the energy apparently lost for reflections with $h + k = 2n + 1$ is redistributed to those reflections for which $h + k = 2n$; it is embodied in the G-factor.

5.4.4.3. *Reflection conditions*

Reflection conditions, or limiting conditions, are associated with space group symmetry; they will have been noted in the description of space groups in Chapter 4. In general, they apply irrespective of the chemical contents of the unit cell. They govern the possible types of *hkl* spectra that are observable on the X-ray diffraction pattern of a crystal of a given space group. Looked at in reverse, a knowledge of the limiting conditions derived by experiment is a guide to the determination of the space group of a crystal. For example, if a set of diffraction data when *indexed*, that is, when the h, k and l indices have been assigned to the reflection data, shows throughout that $h + k$ is an even integer, it implies that the structure is based on a C-centred unit cell. Similar conditions apply to all forms of centring and to glide planes and screw axes; these conditions are derivable from the structure factor equation.

5.4.4.4. *Screw axes and glide planes*

In order to highlight the convenient separation of the geometry and contents of the unit cell, as in the examples above, let N be the total number of atoms in the unit cell, n be the number of atoms in an asymmetric unit and m be the number of asymmetric units. Then, symbolically,

$$\sum_{j=1}^{N} = \sum_{r=1}^{n} \sum_{s=1}^{m} \tag{5.46}$$

where r counts the symmetry-independent atoms and s counts to the asymmetric units. Thus, the structure factor equation may be considered in two parts. For the symmetry-related atoms,

$$A_r(hkl) = \sum_{s=1}^{m} \cos 2\pi(hx_s + ky_s + lz_s)$$

$$B_r(hkl) = \sum_{s=1}^{m} \sin 2\pi(hx_s + ky_s + lz_s)$$

(5.47)

Then, extending to the n atoms of the asymmetric unit,

$$A'(hkl) \sum_{r=1}^{n} f_r A_r(hkl)$$

$$B'(hkl) = \sum_{r=1}^{n} f_r B_r(hkl)$$

(5.48)

where the terms $A_r(hkl)$ and $B_r(hkl)$ are the components of the geometrical structure factor as discussed above; they are governed by the space group symmetry and are independent of the nature of the atoms in the crystal structure; they will often be written as just A and B in this context.

In order to demonstrate the effect of screw axes and glide planes, consider the space group $P2_1/c$. The general equivalent positions are $x, y, z; \ \bar{x}, \bar{y}, \bar{z}; \ x, 1/2 - y, 1/2 + z; \ \bar{x}, 1/2 + y, 1/2 - z$.

The structure is centrosymmetric with $\bar{1}$ at the point 0, 0, 0. From Eq. (5.47), $B' = 0$. Then, from Eq. (A7.1) and dropping the subscript s,

$$A(hkl)/2 = \cos 2\pi(hx + ky + lz)$$

$$+ \cos 2\pi \left(hx - ky + lz + \frac{k+l}{2} \right)$$

(5.49)

$$A(hkl)/4 = \cos 2\pi \left(hx + lz + \frac{k+l}{4} \right) \cos 2\pi \left(ky - \frac{k+l}{4} \right)$$

Next, applying Eq. (A7.5), noting that $\cos 2\pi(\frac{k+l}{4}) \sin 2\pi(\frac{k+l}{4}) = 1/2 \sin \pi(k+l) = 0$, and separating for $k+l$ even or odd:

$$k + l = 2n : A(hkl) = 4\cos 2\pi(hx + lz)\cos 2\pi ky; \ B(hkl) = 0$$

(5.50)

$$k + l = 2n + 1 : A(hkl) = -4 \sin 2\pi(hx + lz) \sin 2\pi ky; \quad B(hkl) = 0$$
$$(5.51)$$

Self-assessment 5.5. Carry out the manipulation on Eqs. (5.50) and (5.51) that leads to Eq. (5.49).

These two equations apply generally for this space group. However, it may be noted that $A_{hkl} = B_{hkl} = 0$ if $h = l = 0$ or if $k = 0$. Then, for the special reflections, $h0l$, other conditions exist. In total, then:

hkl reflections
  No systematic conditions

h0l reflections

$$k + l = 2n : \quad A(h0l) = 4 \cos 2\pi(hx + lz) \cos 2\pi(0)y; \quad B(h0l) = 0$$
$$(5.52a)$$

$$k + l = 2n + 1 : A(h0l) = -4 \sin 2\pi(hx + lz) \sin 2\pi(0)y;$$
$$B(h0l) = 0 \qquad (5.53a)$$

Thus, the $h0l$ reflections are absent for l odd, which indicates a *c-glide plane* normal to y.

0k0 reflections

$$k + l = 2n : \quad A(0k0) = 4 \cos 2\pi(ky); \quad B(0k0) = 0 \qquad (5.52b)$$
$$k + l = 2n + 1 : A(0k0) = -4 \sin 2\pi(ky); \quad B(0k0) = 0 \qquad (5.53b)$$

Thus, the $0k0$ reflections are absent for k odd, which indicates a 2_1 *screw axis* parallel to y.

These results are specific for space group $P2_1/c$; however, not all space groups are determined through the limiting conditions. The important reflections for monoclinic crystals are the hkl, $h0l$ and $0k0$. For example, if the limiting conditions of a monoclinic crystal are found to be

$$hkl : \text{None}, \quad 0k0 : k = 2n$$

then the space group could be either $P2_1$ or $P2_1/m$, since no limiting conditions arise for m symmetry.

Certain reflections can be absent because they are too weak in intensity to be observed; they are structure-dependent and are termed *accidental absences*. The limiting conditions must be systematic to be meaningful; limiting conditions are often referred to also as *systematic absences*. Note, however, that $h = 2n$ for a limiting condition implies $h = 2n + 1$ for a systematic absence. Relationships such as Eqs. (5.50) and (5.51) for the 230 space groups may be found in the literature [4, 5].

Self-assessment 5.6. What information about space groups can be obtained from each of the following sets of crystal reflection data on limiting conditions?

(a) Monoclinic. $hkl : h + k = 2n$; $h0l : l = 2n, h = 2n$; $0k0 : k = 2n$.

(b) Orthorhombic. hkl: None; $0kl : k = 2n$; $h0l : h = 2n$; $hk0$: None; $h00 : h = 2n$; $0k0 : k = 2n$; $00l$: None.

(c) In the space group for (b), two sets of special equivalent positions lie at the coordinates $0, 1/2, z$; $1/2, 0, z$; and $0, 0, z$; $1/2, 1/2, z$. What extra conditions arise for atoms on these sites in a crystal?

5.4.5.   Diffraction symbols

Self-assessment 5.6 indicates the application of a hierarchy in the interpretation of the information on a space group from the limiting conditions. In the monoclinic system, with the y-axis unique, the relevant reflections are, as has been shown:

hkl for unit cell type
h0l for the planes normal to y
0k0 for the axes parallel to y

Any observed condition that is dependent upon one higher in the list is redundant from the point of view of space group determination. As an example, consider space group $C2$. The limiting conditions are *hkl*: $h + k = 2n$; *h0l*: $(h = 2n)$; *0k0*; $(k = 2n)$, as shown in Fig. 4.12. The parenthetical conditions relating to the *h0l* and *0k0* reflections are a feature of space group $C2$; here, they do not *determine* an *a*-glide plane or a 2_1 screw axis even though the latter symmetry is present. The screw axis arises through the C centring, as the space group diagram makes clear.

In the orthorhombic system, the hierarchy is more extensive:

hkl for unit cell type
0kl for the planes normal to x
h0l for the planes normal to y
hk0 for the planes normal to z
h00 for the axes parallel to x
0k0 for the axes parallel to y
00l for the axes parallel to z

This hierarchy must be observed when deducing space group information from limiting conditions.

The symmetry information for a crystal comprises the system, the point group and the limiting conditions. It is not always possible to determine the point group at an early stage in the examination of a crystal, but the Laue class (Section 2.7.1) is usually known. The collated information on Laue class, unit-cell type and extinction conditions comprises a *diffraction symbol*. Compilations of International Tables [4, 5] include tables of diffraction symbols for all space groups. A condensed form of one such table for the orthorhombic system is listed in Table 5.2.

Table 5.2.   Diffraction symbols for the orthorhombic system: Laue class *mmm*.

Diffraction symbol	Point group and settings				
	222	2mm or	m2m or	mm2	mmm
mmm P - - -	$P222$	$P2mm =$ $Pmm2$	$Pm2m =$ $Pmm2$	$Pmm2$	$Pmmm$
mmm P - -2_1	$P222_1$				
mmm $P\,2_1\,2_1-$	$P2_12_12$				
mmm $P\,2_1\,2_1\,2_1$	$P2_12_12_1$				
mmm $P\ c$ - -			$Pc2m =$ $Pma2$	$Pcm2_1 =$ $Pmc2_1$	$Pcmm =$ $Pmma$
mmm $P\ n$ - -			$Pn2_1m =$ $Pmn2_1$	$Pnm2_1 =$ $Pmn2_1$	$Pnma =$ $Pmmn$
mmm $P\ c\ c$ -				$Pcc2$	$Pccm$
mmm $P\ c\ a$ -				$Pca2_1$	$Pcam =$ $Pbcm$
mmm $P\ b\ a$ -				$Pba2$	$Pbam$
mmm $P\ n\ c$ -				$Pnc2$	$Pncm =$ $Pnma$
mmm $P\ n\ a$ -				$Pna2_1$	$Pnam =$ $Pnma$
mmm $P\ n\ n$ -				$Pnn2$	$Pnnm$
mmm $P\ c\ c\ a$					$Pcca$
mmm $P\ b\ c\ a$					$Pbca$
mmm $P\ c\ c\ n$					$Pccn$
mmm $P\ b\ a\ n$					$Pban$
mmm $P\ b\ c\ n$					$Pncn$
mmm $P\ n\ n\ a$					$Pnna$
mmm $P\ n\ n\ n$					$Pnnn$
mmm C - --	$C222$	$C2mm =$ $Amm2$	$Cm2m =$ $Amm2$	$Cmm2$	$Cmmm$
mmm C - -2_1	$C222_1$				

(Continued)

Table 5.2.   (*Continued*)

Diffraction symbol	Point group and settings						
	222	2mm	or	m2m	or	mm2	mmm
mmm C - c -		C2cm = Ama2				Cmc2₁	Cmcm
mmm C - - a		C2ma = Abm2		Cm2a =Abm2			Cmma
mmm C - c a		C2ca = Aba2					Cmca
mmm C c c -						Ccc2	Cccm
mmm C c c a							**Ccca**
mmm I - - -	$\begin{bmatrix} I222 \\ I2_12_12_1 \end{bmatrix}$	I2mm = Imm2		Im2m = Imm2		Imm2	Immm
mmm I - a -		I2am = Ima2		Im2m =Imm2		Ima2	Imam = Imma
mmm I b a -						Iba2	Ibam
mmm I b c a							**Ibca**
mmm F - - -	F222	F2mm = Fmm2		Fm2m =Fmm2		Fmm2	Fmmm
mmm F d d -						**Fdd2**	
mmm F d d d							**Fddd**

Notes:

Owing to space limitations, a set of lines beginning with a point-group symbol, such as:

mmm P - - -	P222	P2mm = Pmm2	Pm2m = Pmm2	Pmm2	Pmmm

should be read as

mmmP - - -	P222	P2mm = Pmm2	Pm2m = Pmm2	Pmm2	Pmmm

1. Entries in **bold** type indicate space groups that are uniquely determined by the diffraction symbol.

2. Possible space groups are listed for different settings of the mm2 class, the mm2 entries being the standard symbols.

3. Two equivalent symbols reflect the way in which the diffraction symbol is first interpreted. For example, if the diffraction symbol is mmmPca - and the point group is known to be mmm, then the obvious (and not incorrect) choice would be Pcam. However, this symbol corresponds to the ba$\bar{\text{c}}$ setting for this group. The standard (**abc**) setting is Pbcm, as indicated in the table by the entry Pcam = Pbcm. Alternative settings have been discussed in Section 4.9.

4. The two chiral space groups I222 and I2₁2₁2₁ listed in parentheses cannot be distinguished by Laue symmetry; each space group contains three two-fold axes and three two-fold screw axes. The systematic absences of the I-centring cause the absences from the 2₁ axes to be non-independent. A similar situation occurs for the cubic space groups I23 and I2₁3. [see also Henry NFM and Lonsdale K. *International Tables for X-ray Crystallography*, Vol 1, Section 4.4, Kynoch Press, 1969.]

Self-assessment 5.7. What standard space group symbols are indicated by the following diffraction symbols?
(a) $\frac{2}{m}C$ - (b) $mmmPn$ -- (c) $mmmI$ - a -

5.5. Collecting X-ray Intensity Data

The early collection of X-ray diffraction data from crystals was carried out by methods involving cameras and photographic film, using monochromatic (filtered) X-rays. It follows from Section 5.4.3.1 that an X-ray diffraction photograph of a crystal is a picture of its reciprocal lattice (Section 3.7) with each *hkl* lattice point weighted by a function of its intensity.

The first X-ray photographs of crystals were obtained in 1912 by von Laue and his associates, and the type of photograph obtained then now carries his name. Figure 5.13 is a *Laue photograph* of a crystal taken with the incident X-ray beam passing through the crystal and normal to a flat film.

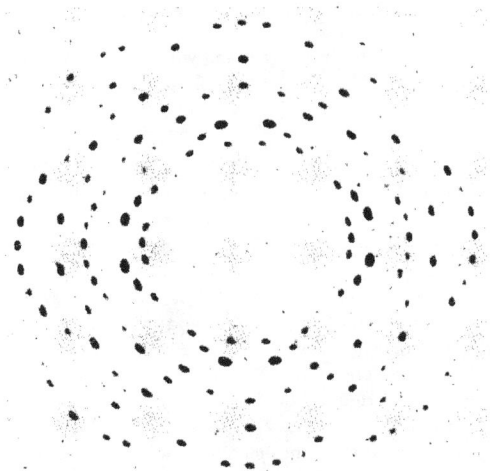

Fig. 5.13. Laue X-ray photograph, showing four-fold symmetry in the crystal along the direction of the X-ray beam (normal to the film) and reflection (mirror) symmetry in horizontal and vertical planes in the crystal.

It shows clear evidence of four-fold symmetry in the direction of the X-ray beam. The detection of symmetry elements is one of the important features of the Laue method. Unfiltered X-rays and a stationary crystal are used in order to obtain this type of photograph; the crystal effectively selects the correct wavelength from the 'white' radiation for reflection in accordance with the Bragg law, Eq. (5.17). It was not used in early structure analysis because of the difficulty then in assigning the *hkl* indices to the reflections.

A technique more suited to structure analysis was developed that involved a *rotating crystal* within a cylindrical film. The method uses effectively monochromatic radiation and data are collected in layers, usually with respect to a principal axis. A photograph with a crystal oscillating about the *a*-axis direction in the crystal is shown in Fig. 5.14; each layer line has a constant value of *h*. Each reflection must be indexed and its intensity measured in order to proceed with a structure analysis, and there is always the possibility of different reflections overlapping at a given point on the film. The intensities were measured either by a photometric device that measures the extent of the blackening of a spot on the film, or visually by comparison with a calibrated spot scale. As a picture of the reciprocal lattice, however, it suffers from both collapse and distortion.

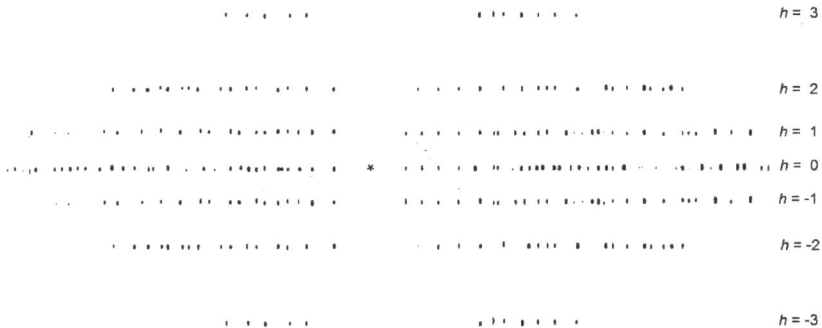

Fig. 5.14. X-ray photograph of a crystal oscillating about its *a*-axis; each layer line has a constant value for the index *h*. The photograph can be used *inter alia* to calculate the *a* spacing in the crystal.

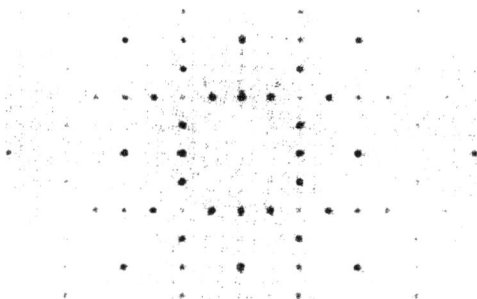

Fig. 5.15. X-ray precession photograph of a cubic crystal; the two central rows correspond to a^* (vertical) and b^* (horizontal) in the crystal, and systematic absences can be detected.

An improvement was made with the introduction of moving-film methods, in which the film holder traversed linearly and synchronously with the rotation of the crystal on its chosen axis. One of these methods, the *Weissenberg* technique, gave each reflection its own place on the film record. The reciprocal lattice was no longer collapsed by this method but remained distorted. Nevertheless, it was used for many years to collect single-crystal diffraction data.

An ingenious device was developed by Buerger which collected intensity data without distortion of the geometry of the reciprocal lattice. In his *precession method*, a flat film is linked to the crystal rotation, or oscillation, axis. With the instrument set at its zero, the X-ray beam strikes the crystal parallel to a real axis that is normal to the film. The crystal (with the film) is tilted by an angle of *ca.* 30° and allowed to precess in such a manner that a real crystal axis traces a cone about the direction of the X-ray beam. The mechanism and motion of the assembly is complex, but the result is an undistorted picture of the weighted reciprocal lattice of the crystal, as shown in Fig. 5.15. The data can be indexed by inspection and the intensities measured photometrically.

5.5.1. Geometry of X-ray data collection

The Bragg construction for X-ray diffraction is based on the idea of reflections from planes, but the geometry of a distribution of planes

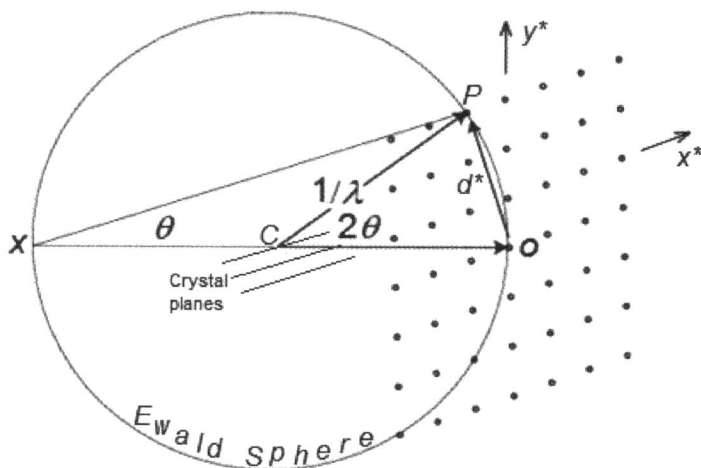

Fig. 5.16.   The Ewald sphere, or sphere of reflection, of centre C and radius $1/\lambda$. The incident X-ray beam is along the diameter XO, where O is the origin of the reciprocal lattice through which the rotation axis normal to the x^*–y^* plane passes; the reflected ray is along CP. A reflection, $\bar{1}30$ in the example here, occurs as the reciprocal lattice point P intersects the Ewald sphere.

in a crystal is not always an easy concept to study. The reciprocal lattice concerns the distribution of points with weights proportional to diffracted intensity, and this picture makes for an easier appreciation of the geometry of X-ray diffraction. This use of the reciprocal lattice was introduced by Bernal [6] and developed fully by Ewald [7].

In Fig. 5.16, the concept of the *Ewald sphere* is illustrated. This sphere (aka *sphere of reflection*) has a radius $1/\lambda$ and lies on the X-ray beam direction XO as a diameter, where O is the origin of the reciprocal lattice; the crystal lies at the centre C of the sphere. For a reflection hkl, the construction shows that $XO = 2/\lambda$ and $\widehat{XPO} = 90°$. Using the Bragg equation and the definition $d^* = 1/d$,

$$OP = XO \sin \theta_{hkl} = (2/\lambda) \sin \theta_{hkl} = 1/d_{hkl} = d^*_{hkl} \qquad (5.54)$$

The size of the sphere and, hence, the size of the reciprocal lattice is determined by the wavelength of the X-radiation. In the diagram, P is the reciprocal lattice point $\bar{1}30$ that represents the family of parallel, equidistant ($\bar{1}30$) planes in the crystal; CP is the direction

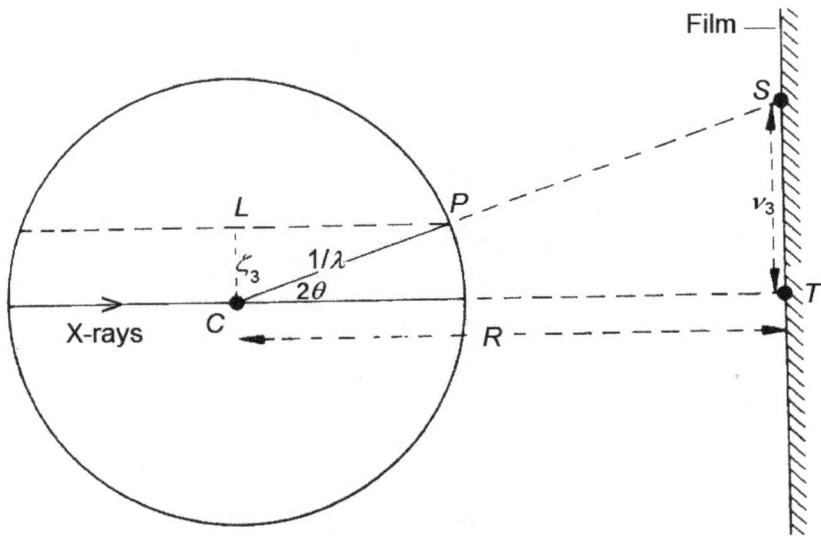

Fig. 5.17.   Determining a unit-cell dimension from an oscillation-type photograph with the crystal oscillating about the $a(x)$-axis. A reciprocal lattice layer spacing ζ_n is given by $\zeta_n^2 = (\nu_n/R)^2/[1+(\nu_n/R)^2]$, and $a = n\lambda/\zeta_n$, where n is the number of the layer from which ν_n is measured — layer number 3 in this example.

of the diffracted beam from that family. Thus, the Ewald construction shows that an X-ray reflection from the (hkl) planes occurs when the hkl reciprocal lattice point intersects the surface of the Ewald sphere, and the path CP of the diffracted beam is from the crystal through the point hkl. The possible reflections in any experiment are those lying within a sphere of radius $2/\lambda$, which is known as the *limiting sphere*.

The repeat distance along a rotation or oscillation axis is readily determinable from a photograph such as that in Fig. 5.14. The relevant geometry is shown in Fig. 5.17. The spot S on layer 3 of the upper half of the film is at a height ν_3 from the zero layer (half the distance ν_3 to $\nu_{\bar{3}}$ for better precision). By similar triangles, ν_3 corresponds to the reciprocal lattice spacing ζ_3. Then, $\nu_3/R = \zeta_3/\sqrt{1-\zeta_3^2}$ and $a = n\lambda/\zeta_n$, where $n = 3$ in this example.

The determination of unit-cell dimensions is even more straightforward with a precession photograph. In order to illustrate it clearly

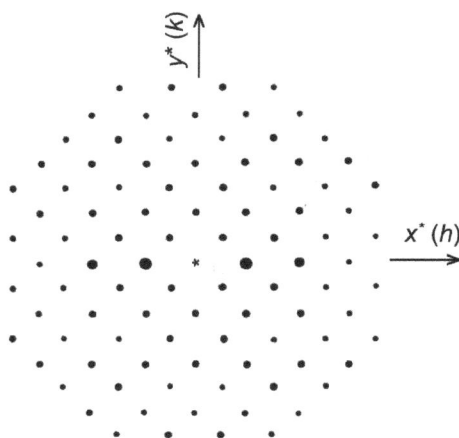

Fig. 5.18. Idealized precession photograph of the $hk0$ layer of an orthorhombic crystal; Cu $K\alpha$ X-radiation, $\lambda = 1.5418$ Å, was assumed for its construction.

and to highlight another feature of the precession method, an idealized photograph of the $hk0$ layer of the reciprocal lattice of an orthorhombic crystal has been constructed, as shown in Fig. 5.18. The crystal-to-film distance is 60.0 mm and the crystal is mounted on the camera so as to oscillate about the c-axis; Cu $K\alpha$ X-radiation ($\lambda = 1.5418$ Å) was assumed.

On the $hk0$ photograph, measurements are made of the lengths of the rows of fourteen spots in both the x^* and y^* directions. The measurements on the actual photograph are 43.5 mm along x^* and 42.5 mm along y^*; the photograph has a multiplying scale factor of 4.00. Then, by the same geometry as above, except that now the reciprocal lattice spacings are measurable directly:

$$a* = \frac{43.5 \text{ mm}/14}{60.0 \text{ mm}} \times 4.00 = 0.2071, \text{ so that}$$
$$a = 1.5418 \text{ Å}/0.2071 = 7.44 \text{ Å}$$

The precession photograph is useful also in revealing information on the space group of a crystal.

Self-assessment 5.8. (a) Use the data on the precession photograph given above to determine the value of the unit-cell dimension b for this crystal. (b) Index the precession photograph of
Fig. 5.18 and note the systematic absences. From other investigations, it was determined that no systematic absences occur for
the general hkl reflections. Assuming that the systematic absences
on the photograph are representative of the entire given layer of
the reciprocal lattice, what is indicated for the probable space
group?

5.5.1.1. *Possible number of reflections in a data set*

The volume V_s of the limiting sphere is $\frac{4}{3}\pi(2/\lambda)^3$ and the number
N of possible reflections in a data set is, in the most general case,
V_s/V^*, where V^* is the volume of the reciprocal unit cell ($= 1/V$).
The practical value is given by

$$N = \frac{33.510}{\lambda^3} \times V \times \frac{\sin^3 \theta_{max}}{mG} \tag{5.55}$$

where θ_{max} is the upper experimental limit of θ, m is the order of
the crystal point group and G is the unit-cell translation factor (Section 5.4.4.2).

Self-assessment 5.9. An orthorhombic crystal of space group
$Abm2$ has the unit-cell dimensions $a = 5.000$ Å, $b = 9.000$ Å,
$c = 22.00$ Å. How many unique reflections could be measured
with Cu $K\alpha$ radiation ($\lambda = 1.5418$ Å) if θ max $= 85°$?

5.5.2. X-ray diffractometry

While single-crystal photographic methods still have an application
in some X-ray structure determinations, the majority of crystal structures are solved today with data collected by X-ray diffractometry.
Figure 5.19 shows schematically basic features of a four-circle single-
crystal X-ray diffractometer.

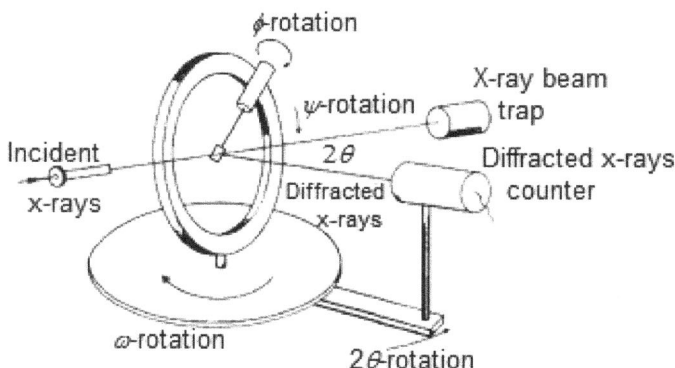

Fig. 5.19. Basic components of an X-ray single-crystal goniometer, showing the four circles, ϕ, ψ, ω and 2θ, that are used to bring the crystal into a reflecting position; also shown are the collimated incident X-ray beam direction, and the collecting and measuring device for the diffracted spectra.

A goniometer is an instrument that allows an object to be rotated to a precise angular position; it is the integral part of any X-ray diffractometer. The X-ray diffractometer is the basic goniometer assembled together with an X-ray source and a detector for collecting and measuring the intensity of the diffracted X-ray beams. The Enraf-Nonius CAD4 kappa X-ray goniometer is shown schematically in Fig. 5.20.

The goniometer itself carries the goniometer head on which the crystal is mounted. It has angular and linear adjustments; a typical goniometer head is shown in Fig. 5.21. A small single crystal on a glass fibre is mounted on to the goniometer head and exposed to a beam of monochromatic X-radiation, obtained normally by using a crystal monochromator. Molybdenum $K\alpha$ X-radiation ($\lambda = 0.71069$ Å) or copper $K\alpha$ X-radiation ($\lambda = 1.5418$Å) is used generally for single-crystal diffraction. The X-ray beam is collimated and directed on to the sample, while the crystal is retained at the centre of the goniometer during the data-collection process. The X-ray diffractometer software allows the unit-cell dimensions to be determined. High-angle data are used for the refinement, since the

The Essence of Crystallography

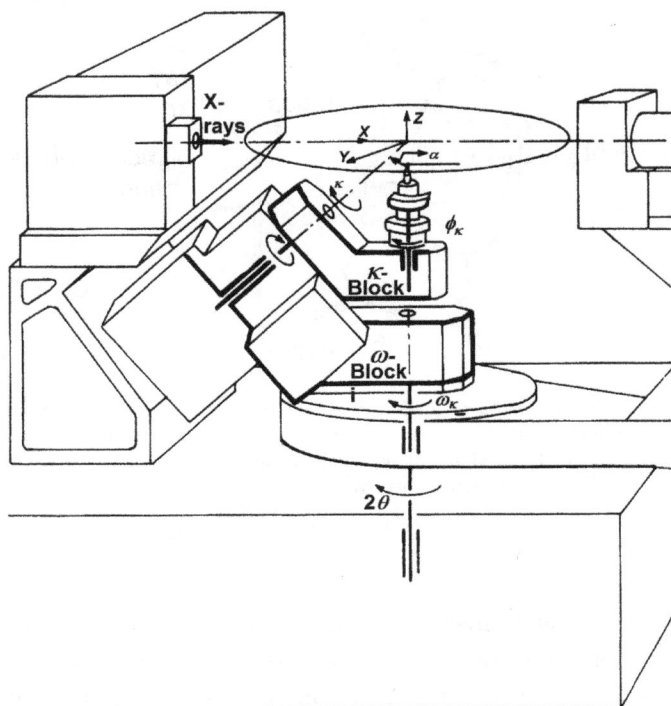

Fig. 5.20. Schematic diagram of the Enraf-Nonius kappa X-ray diffractometer that formed the basis of the X-ray diffractometer in Fig. 5.22; the X, Y and Z instrumental coordinate system and the rotation directions are indicated. [*Structure Determination by X-ray Crystallography*, Mark Ladd and Rex Palmer, 5th edn. (2013). Reproduced by permission of Springer Science+Business Media, NY.]

variation in $\sin\theta$ with θ is less at high angles. Refinement of the unit-cell parameters is carried out by the software, using the method of least squares.

A suitable axis is chosen for the rotation of the crystal during the period of data collection, and the collimated, monochromatic X-ray beam is directed on to it. The X-ray goniometer software orients the crystal and moves it to successive reflection positions within the experimental recording range. A detector records and processes the diffraction spectra, converting the signal to an appropriate output format.

Fig. 5.21.   Bruker goniometer head mount for single-crystal X-ray diffraction. Two angular adjustments and two linear adjustments are provided, and the goniometer head retains the crystal automatically in the centred position of the diffractometer circles. See also ⟨https://my.bruker.com/acton/attachment/2655/f-0ade/1/-/-/-/-/42-08%20-%20PS%20SC-XRD%2057%20Automated%20Goniometer%20Head%20-%20Optimized%20Sample%20Centering.pdf?utm_term=PDF%20SCD%20AGH&&utm_content=landing+page&utm_source=Act-On+Software&utm_medium=landing+page&cm_mmc=Act-On%20Software-_-Landing%20Page-_-_-PDF%20SCD%20AGH&sid=TV2:7IG7RS1Z6⟩ [Reproduced with permission from Bruker AXS GmbH, Karlsruhe.]

A diffractometer can be used in conjunction with synchrotron radiation, such as the Diamond light source at Harwell UK, which provides a very intense and tuneable beam. This system enables very rapid data collections with small crystals, of size even less than 0.1 mm, and with a choice of radiation wavelength. These features are particularly valuable for macromolecular crystals, which are prone to damage by irradiation. It is also quite usual for these crystals to be flash frozen in a continuous stream of vapour from a liquid nitrogen source. This stabilizes the sample and improves the quality of the diffracted X-ray data.

A renewed version of the very successful Enraf-Nonius CAD4 kappa goniometer shown in Fig. 5.20 can be seen today in the Bruker D8 VENTURE single-crystal kappa diffractometer system which is

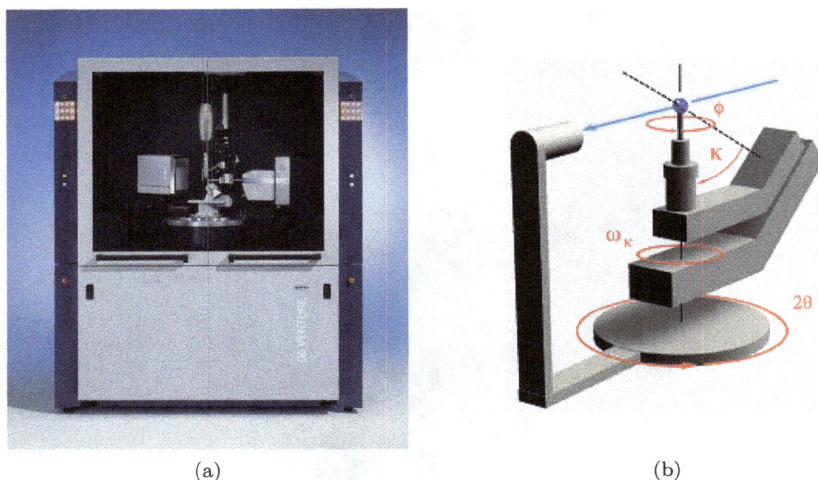

(a) (b)

Fig. 5.22. X-ray diffractometry: (a) The Bruker D8 VENTURE X-ray kappa diffrac-
tometer enclosure, equipped with the ImuS DIAMOND Cu ImuS X-ray source, PHO-
TON III detector and cryogenic cooling system. (b) The kappa goniometer of the D8
VENTURE, showing the four circles ϕ, κ, ω_κ and 2θ. [Reproduced with permission from
Bruker AXS GmbH, Karlsruhe.]

illustrated in Fig. 5.22. Diffracted X-rays are detected and measured
by the 'Photon III' CPAD detector (charge-integrating pixel array
detector), which is a highly efficient quantum counter with single
pixel sensitivity.

As well as the measurement of the intensity of the diffracted
spectra, the background scattered radiation is also monitored so as
to obtain a true intensity of each *hkl* spectrum. The accompanying
diffractometer software package carries out all operations concerned
with collecting and measuring the X-ray diffraction data set, and
applying the necessary corrections (Section 5.5.3); it also handles
the problems associated with twinned crystals. The diffractometer
is provided with a facility for collecting data at low temperatures,
which is useful in reducing the effect of thermal vibrations of the
atoms, thereby sharpening the diffraction data. The Bruker APEX3
GUI software package for the Bruker D8 system is based on the well-
known and highly successful SHELX program systems for solving

crystal structures, and processes all stages of the crystal structure analysis.

Another important X-ray diffraction system for both small molecule and protein crystallography is the Rigaku *SuperNova*. It uses a hi-flux X-ray microfocus source and has a fast, high performance charge-coupled device (CCD) as detector. This combination permits a rapid collection of intensity data. The radiation is easily switchable between copper and molybdenum target materials [9].

X-ray powder diffraction is also employed to determine the structures of crystalline materials, especially those substances that are difficult to prepare in a sufficiently large single-crystal form. This method has been made feasible mainly by the advent of the X-ray powder diffractometer. A problem with data from powder cameras arises from the overlapping of diffraction lines, which renders indexing of the reflections uncertain. The powder diffractometer is capable of a very high resolution, and many structures have been solved from powder data.

An example of the increased sharpness of the resolution of diffractometer data over that of the powder data may be seen by comparing the diffractometer trace for powdered sodium chloride in Fig. 5.23 with the powder pattern, albeit an early example, of this substance that appears with Problem 5.7. Another intensity pattern showing a very high degree of resolution is that for sodium thiosulphate pentahydrate, which is monoclinic with space group $P\frac{2_1}{c}$; the diffractometer trace illustrated in Fig. 5.24 was obtained with monochromatic $CuK\alpha_1$ radiation.

The basic geometry of a powder diffractometer working in the Debye–Scherrer mode is shown in Fig. 5.25, and an example of an instrument that operates in this mode is the Bruker D8 Advance, shown in Fig. 5.26. The accompanying computer software includes the Rietveld profile analysis and least-squares algorithms that fit a predicted pattern to the observed pattern. Crystal structures of proteins have been determined and refined from both X-ray and neutron powder data [10].

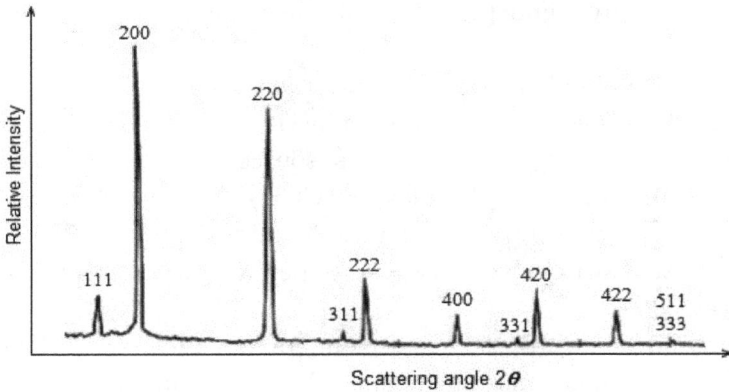

Fig. 5.23. X-ray goniometer trace of intensity *vs.* scattering angle from powdered sodium chloride. The weakness in intensity of reflections with $(h + k + l)$ odd as compared to those with the sum $(h + k + l)$ even is evident. The superposition of reflections 333 and 511 is unavoidable in this example since the sum $(h^2 + k^2 + l^2)$ is 27 for both reflections: note that from Eqs. (5.25) and (5.26) and for $a = b = c$ (cubic), $\sin\theta = (\lambda/2a)\sqrt{h^2 + k^2 + l^2}$.

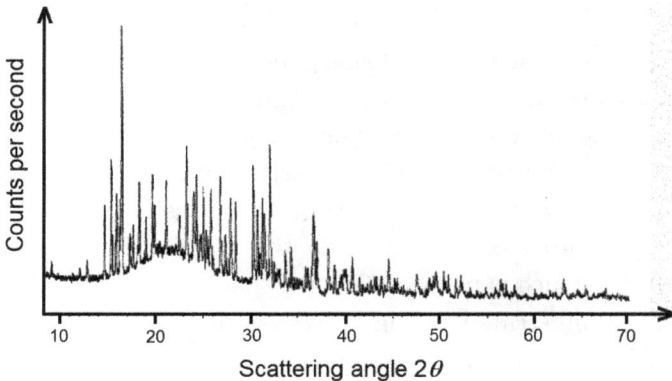

Fig. 5.24. Highly resolved X-ray diffractometer trace from powdered sodium thiosulphate $Na_2S_2O_3$, taken with monochromatized $Cu\,K\alpha_1$ X-radiation.

5.5.3. Corrections to measured intensities

The intensity of an *hkl* reflection at an ideal Bragg angle θ_0 is, in practice, an integrated value over a small range $\pm\delta\theta_0$, as indicated in Fig. 5.27. The measured quantities are diffracted X-ray intensities

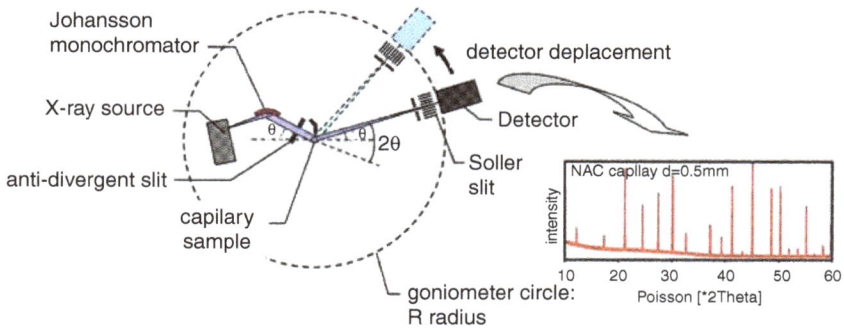

Fig. 5.25. The geometry of an X-ray powder goniometer (XRD) operating in the Debye–Scherrer mode. [Reproduced with permission from Bruker AXS GmbH, Karlsruhe.]

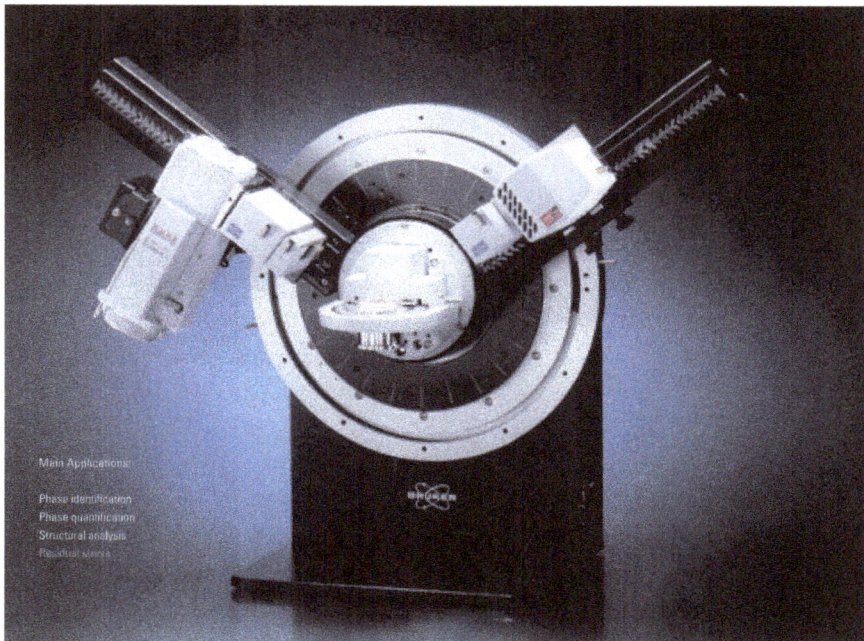

Fig. 5.26. Bruker D8 Advance X-ray powder goniometer (XRPD). The incident X-rays are monochromatized to $K\alpha_1$ radiation from cobalt, copper or molybdenum by a Johansson monochromator. [Reproduced with permission from Bruker AXS GmbH, Karlsruhe.]

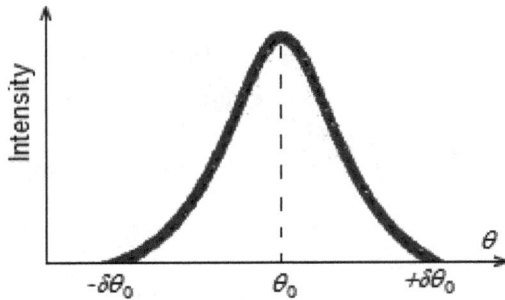

Fig. 5.27. Variation in diffracted X-ray intensity with Bragg angle θ. The measured intensity is the reflecting power at the value of θ_0 integrated over the small range $\pm\delta\theta_0$.

$I(hkl)$, and diffractometer software applies a number of corrections in order to obtain the amplitudes of structure factors on a common, relative scale. The required corrections are discussed briefly in the following sections.

5.5.3.1. *Multiplicity of planes*

Two factors determine the multiplicity of a given plane in a crystal, namely, the Laue group and whether or not the reflection lies on a symmetry element of the group. Consider a monoclinic crystal of Laue symmetry $\frac{2}{m}$. The symmetry operators for this point group applied to a general reflection (hkl) generate the set (hkl), $(\bar{h}k\bar{l})$, $(h\bar{k}l)$ and $(\bar{h}k\bar{l})$, so that the multiplicity of (hkl) is 4. If the plane is a special form, such as $h0l$ in this group, then the multiplicity is 2. In the orthorhombic mmm class, the multiplicities are 8 for hkl, 4 for $0kl$, $h0l$ and $hk0$, and 2 for $h00$, $0k0$ and $00l$. All other cases can be deduced in the same manner.

Consider an orthorhombic crystal of class mmm. The reciprocal spacing $d^*(hkl)$ is given by the expression $d^*(hkl) = \left(\frac{h^2}{a^2} + \frac{k^2}{b^2} + \frac{l^2}{c^2}\right)^{1/2}$ and $d^*(hkl)$ is proportional to $\sin\theta(hkl)$. For each of the eight planes $\{123\}$, $h^2 = 1$, $k^2 = 4$ and $l^2 = 9$, so that $\theta(123)$ has the same value for each plane in the family, and the multiplicity $m(hkl)$ is 8 in this example, the order of point group mmm.

5.5.3.2. *Lorentz and polarization correction factors*

The Lorentz factor L is a function of the geometry of the experiment, and represents a time-of-reflection opportunity for a reflection. With a rotating crystal and the X-ray beam normal to a reflecting plane, $L = 1/\sin 2\theta$.

The polarization factor p takes into account the fact that the incident X-ray beam is unpolarized radiation, whereas after reflection from crystal planes, it is polarized and of decreased intensity, the decrease being a function of the scattering angle 2θ, namely, $1/2(1 + \cos^2 2\theta)$. For a beam after reflection from a crystal monochromator, the polarization factor is $(1 + \cos^2 2\theta \cos^2 2\theta_m)/(1 + \cos^2 2\theta_m)$, where $2\theta_m$ is the angle between the reflecting plane and scattered beams at the monochromator. These two corrections are taken together normally, and known as the Lp factor.

5.5.3.3. *Extinction*

A crystal immersed in an X-ray beam at the Bragg angle for a given family of planes suffers Bragg reflection from a second plane in the family since it is incident at the same angle θ, as shown in Fig. 5.28. For example, a first-reflected ray QT is reflected within the crystal along the path TS, and so on. As there is always an inherent phase change of $\pi/2$ on reflection, a double-reflected ray has a phase change of π with respect to the incident ray PQ. The phase change of π is not considered in crystal-structure analysis because it is common to all reflections. There are two extinction processes to examine.

5.5.3.3.1. Primary extinction

X-rays that are reflected n and $(n + 2)$ times differ in phase by π, which leads to a reduction in the intensity of diffraction. The beam energy is conserved: each beam that is depleted in energy by scattering into another beam is enhanced by a similar process of scattering from other beams [11]. This effect is known as *primary extinction*, and it is much reduced if the crystal is not geometrically perfect.

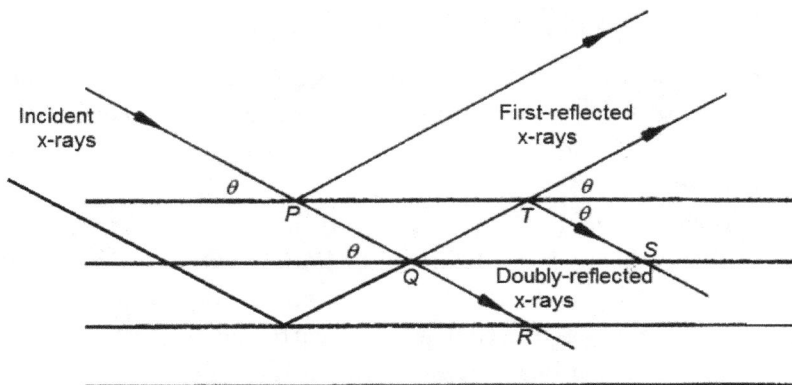

Fig. 5.28.   Primary extinction: the phase changes by reflection at Q and T are $\pi/2$ each, so that between the directions QR and TS, the total phase change is π. Thus, the X-ray beam is attenuated as it penetrates further into the crystal.

Few crystals are perfect: they are composed of very slightly mis-aligned blocks of structure, forming the so-called *mosaic structure* shown below.

Mosaicity in a crystal

The ranges of perfection are 50–100 μm. For an ideally perfect crystal, $I \propto |\mathbf{F}|$, whereas for the ideally imperfect crystal, $I \propto |\mathbf{F}|^2$. Imperfection is the state of most crystals and its degree of imperfection may be increased by thermal shock, such as immersion in liquid nitrogen, which leads also to a more uniform state.

5.5.3.3.2.   Secondary extinction

The first few planes of a crystal that encounter incident X-rays reflect a high proportion of the X-ray beam. Planes deeper in the crystal receive less incident intensity and, therefore, reflect less than would be expected. The effect is particularly noticeable with intense, low-order reflections. If the degree of imperfection is large, secondary extinction is small and may be even negligible. Nevertheless, it can be brought into the refinement stage of a crystal structure determination in terms of an *extinction parameter* ζ that modifies the scale factor for the structure factors.

5.5.3.4. *Absorption*

Matter absorbs X-rays and crystals are no exception. Absorption follows the exponential law in Section 5.3.1, recalled here for convenience:

$$I = I_0 \exp(-\mu t) \qquad (5.56)$$

If the material is a single substance, the *atomic absorption coefficient* μ_a is given by $\mu_a = M_r \mu / \rho L$, where M_r is the relative atomic mass of the material and ρ is its density. More general is the *mass absorption coefficient* μ_m given through the equation

$$\mu = \sum_j \mu_{m,j} \rho_{m,j} \qquad (5.57)$$

Example 5.1. The density of sodium chloride is 2.165 g cm^{-3} and the relative atomic masses are Na 22.990, and Cl 35.453; the mass absorption coefficients for Cu$K\alpha$ X-radiation are 30.1 cm^2 g^{-1} and 106 cm^2 g^{-1} for sodium and chlorine, respectively. Then, the linear absorption coefficient is given as

$$\mu = 2.165 \text{ g cm}^{-3} \times \left[\frac{\begin{array}{c}(30.1 \text{ cm}^2 \text{ g}^{-1} \times 22.990) \\ +(106 \text{ cm}^2 \text{ g}^{-1} \times 35.453)\end{array}}{58.443} \right] = 164.8 \text{ cm}^{-1}$$

If the X-ray beam passes through a sodium chloride crystal of thickness 0.2 mm, then the *attenuation factor* I/I_0 is $\exp(-164.8 \text{ cm}^{-1} \times 0.02 \text{ cm}) = 0.037$.

In correcting for absorption, it is necessary to take into account the path t_i in the crystal taken by the incident beam before reflection and the path t_r after reflection. These two parameters define a *transmission factor* T, given as

$$T = \exp[-\mu(t_i + t_r)] \qquad (5.58)$$

If the shape of the crystal is known exactly, then

$$T = \frac{1}{V} \int_V \exp[-\mu(t_i + t_r)]dV \qquad (5.59)$$

A satisfactory empirical correction is possible with diffractometer data collection, and the diffractometer software can be used to apply this correction. In some cases, a crystal can be ground to a spherical shape, in which case the total path length is a constant for all reflections.

5.5.3.5.   *Scale and temperature factors*

The corrected intensities present a set of relative values with a precision that depends upon the statistics of the detection and counting processes. In addition, the atomic scattering factors (Section 5.4.1) listed in the literature sources apply to atoms at rest, that is, at 0 K. A first approximation to scale and temperature factors may be obtained by Wilson's method [12]. The *ideal intensity* $|\mathbf{F}|^2$ may be defined as

$$|\mathbf{F(h)}|^2 = \sum_i (g_{i,\theta})^2 + \sum_i \sum_j (g_{i,\theta})(g_{j,\theta}) \exp(i2\pi \mathbf{h} \cdot \mathbf{r}_{i,j}) \qquad (5.60)$$

where g is the atomic scattering factor at an ambient temperature, $\mathbf{h}$ stands for *hkl*, and $\mathbf{r}_{i,j}$ is the interatomic distance $\mathbf{r}_j - \mathbf{r}_i$. Wilson showed that, over a sufficient number of terms, the double summation term tends to a very small value, so that the average intensity is given as

$$\langle |\mathbf{F}(\mathbf{h})|^2 \rangle = \sum_j (g_{j,\theta})^2 \tag{5.61}$$

provided that the averaging is carried out in small, local ranges of reciprocal space. Then, using the Debye–Waller temperature correction [13, 14]

$$f_T = f_0 \exp[-B(\sin^2 \theta)/\lambda^2] \tag{5.62}$$

where the *overall isotropic* temperature factor B is equal to $8\pi^2$ $\langle u^2 \rangle$, $\langle u^2 \rangle$ being the mean square displacement of an atom from its average position. Then, applying a scale factor K to $|\mathbf{F}|$ and assuming n ranges of data,

$$K^2 \langle |\mathbf{F}|^2 \rangle = \exp[-B(\sin^2 \theta_n)/\lambda^2]^2 \sum_j f_{j,\theta_n}^2$$

where j includes all atoms in the unit cell. Setting $q_n = \sum_j f_{j,\theta_n}^2 / \langle |\mathbf{F}|^2 \rangle$ and taking natural logarithms, it follows that

$$\ln q_n = 2 \ln K + 2B(\sin^2 \theta_n)\lambda^2 \tag{5.63}$$

Thus, a graph of $\ln q_n$ vs. $(\sin^2 \theta_n)\lambda^2$ should be linear, with a slope of $2B$ and an intercept of $2 \ln K$. For a variety of reasons [15, 16], a Wilson plot may deviate from linearity, but it can provide a useful starting value of the scale and temperature factor for a set of diffraction data. The important effect of a temperature correction is illustrated by the data in Table 5.3, which compares the rest of values of the form factor f for oxygen with those to which B-values of 2 Å and 4 Å have been applied.

The success of the Wilson plot depends largely on a distribution of structure factor amplitudes (ideal intensities) that changes smoothly with $\sin \theta$. Also, the space group may not yet be determined unambiguously. For both of these situations, a discussion of the statistics of intensity distributions may prove revealing.

At this stage, a set of *hkl* intensity data has been collected and the necessary corrections applied so as to obtain a set of observed

Table 5.3.   Debye–Waller correction for oxygen: $\lambda = 1.5418\,\text{Å}$.

$(\sin\theta)/\lambda$	$(\sin^2\theta)/\lambda^2$	f_0	$f_T(B = 2\,\text{Å}^2)$	$f_T(B = 4\,\text{Å}^2)$
0.00	0.00	8	8	8
0.10	0.01	6.83	6.69	6.56
0.20	0.04	4.77	4.40	4.06
0.30	0.09	3.34	2.79	2.33
0.40	0.16	2.60	1.89	1.37
0.50	0.25	2.25	1.36	0.83
0.60	0.36	2.05	1.00	0.71
0.70	0.49	1.90	0.71	0.27

structure amplitudes that will be denoted as $|\mathbf{F}_{\text{obs}}|$ and used in a crystal structure analysis.

Self-assessment 5.10. A tetragonal crystal of point group $mm2$ has the dimensions $a = b = 5.000$ Å, $c = 10.00$ Å. (a) What is the multiplicity factor m for the 110 reflection? (b) Calculate the Lorentz and polarization corrections for $I(110)$; the X-radiation wavelength is 1.5418 Å.

5.6.   Statistics of a Diffraction Pattern

At this stage, the space group of the crystal under examination may be uncertain on account of the Friedel ambiguity. Several features are yet to be considered in this context.

5.6.1.   Accidental absences

A relatively small number of data may be absent because they are too weak to be recorded; the sum $\sum_j f_j \exp[(\mathrm{i}2\pi(hx_j + ky_j + lz_j)]$ may tend to a vanishingly small quantity for some hkl reflections. Such reflections are known as *accidental absences*, but they are still a feature of the structure. It has been shown that such 'unobserved' but possible reflections should be allocated a value in the region of $0.6|\mathbf{F}_{\text{obs, min}}|$, where $|\mathbf{F}_{\text{obs, min}}|$ is the minimum value of $|\mathbf{F}_{\text{obs}}|$ in the locality of the unobserved reflection. Alternatively, they may be measured again over a longer time period. These procedures ensure that all possible reflections are brought into the data set for analysis.

There may be other reasons why a particular reflection has not been measured well. To facilitate recognition of such data, crystallographic software such as SHELXL produces a list of $|\mathbf{F}_{\text{obs}}|$ and the corresponding calculated values of $|\mathbf{F}_{\text{calc}}|$ that show exceptionally poor agreement. These data terms can, with discretion, be omitted from the refinement procedure.

5.6.2. Laue symmetry

On account of Friedel's law (Section 5.4.3.1), the diffraction pattern, in the absence of anomalous (resonance) scattering, conforms to one of the 11 Laue classes. Anomalous behaviour occurs most strongly with atoms that absorb at a wavelength close to that of the incident radiation. For example, copper radiation of wavelength 1.5418 Å is unsuitable for use with compounds containing iron which has an absorption edge wavelength of 1.7433 Å.

5.6.3. Systematic absences

In a structure with a body-centred unit cell, for example, the structure factor takes the form

$$\mathbf{F}_{hkl} = 2\cos^2 2\pi \left(\frac{h+k+l}{4}\right) \sum_{j=1}^{n/2} f_j \exp[\mathrm{i}2\pi(hx_j + ky_j + lz_j)]$$

(5.64)

which is twice the value that it would have had in a corresponding primitive unit cell. Possible reflections with $(h + k + l) = 2n + 1$, where n is an integer, are systematically absent. The associated energy is redistributed to those reflections with $(h + k + l) = 2n$, as shown by the factor 2 in Eq. (5.64).

5.6.4. Abnormal averages

The symmetry of a crystal links the vectors, $\mathbf{r}_j$, given by

$$\mathbf{r}_j = x_j\mathbf{a} + y_j\mathbf{b} + l_j\mathbf{c} \qquad (5.65)$$

in groups of two or more, and Eq. (5.39) is one such example. The term $\sum_j g_j^2$, written conventionally as $\sum$, is termed a *distribution parameter*. It may be enhanced by the value of the epsilon factor ε, which appears in the following discussion and is listed in Table 5.4 for the 32 crystal classes.

Consider space group Pm with the y-axis unique, for which the general equivalent positions are x, y, z, and x, $\bar{y}$, z. It is straightforward to obtain the equations

$$A'(hkl) = 2\sum_j^{n/2} g_j \cos 2\pi(hx_j + lz_j)\cos 2\pi ky_j \qquad (5.66)$$

and

$$B'(hkl) = 2\sum_j^{n/2} g_j \sin 2\pi(hx_j + lz_j)\cos 2\pi ky_j \qquad (5.67)$$

The *central limit theorem* states that, in a sequence of independent random variables $x_j, (j = 1, 2, \ldots, n)$ for which the mean values are m_j and variances σ_j^2, the sum $x = \sum_j x_j$ tends to a normal (Gaussian) distribution, with a mean $m = \sum_j m_j$ and a variance $\sigma^2 = \sum_j \sigma_j^2$, as the number n of terms tends to infinity. In the example of Pm, the mean values $\langle A'(hkl)\rangle$ and $\langle B'(hkl)\rangle$ tend to zero since positive and negative values of $A'(hkl)$ and $B'(hkl)$ are equally probable. The variance of a large sample is given by

$$\sigma^2 = \frac{1}{n}\sum_j(x_j - \langle x\rangle)^2 = \frac{1}{n}\sum_j x_j^2 - \langle x^2\rangle \qquad (5.68)$$

Table 5.4. Centric reflections and ideal-intensity multiples (ε-factors) in the 32 crystal classes.

Crystal class	New diffraction symbol	Centric sets	Multiples (ε)
1	$1P$	None	1/1
$\bar{1}$	$\underline{1P}$	All	1/1
m	$2/mP-/-$	$(0k0)$	1/2
2	$2/mP-/\underline{-}$	$(h0l)$	2/1
$2/m$	$2/m\underline{P}-/-$	All	2/2
$mm2$	$mmmP---$	[$(hk0)$ masks $(h00)$, $(0k0)$]	2/2; 2/2; 4/1
222	$mmmP\underline{-}\ \underline{-}\ \underline{-}$	3 principal zones only	2/1; 2/1; 2/1
mmm	$mmm\underline{P}-\ -\ -$	All	4/2; 4/2; 4/2
4	$4/mP-/\underline{-}$	$(hk0)$	4/1
$\bar{4}$	$4/mP-/\underline{-}$	$(hk0)$; $(00l)$	2/1
$4/m$	$4/m\underline{P}-/-$	All	4/2
$\bar{4}2m$	$4/mmmP-/\underline{-}\ -\ -\ -$	[$(hk0)$, $\{hh0\}$]; [$\{h0l\}$, $(00l)$]	4/1; 2/1; 2/2
$4mm$	$4/mmmP-/\underline{-}\ --$	[$(hk0)$, $\{h00\}$, $\{hh0\}$]	8/1; 2/2; 2/2
422	$4/mmmP-/\underline{-}\ -\ -$	$(hk0)$; $\{h0l\}$ $\{hh\}$	4/2; 2/1; 2/1
$4/m\ mm$	$4/mmm\underline{P}-/-\ -\ -$	All	8/2; 4/2; 4/2
3	$\bar{3}P\ -$	None	3/1
$\bar{3}$	$\bar{3}\underline{P}\ -$	All	3/1
$3m(1)$	$\bar{3}m1P\ -\ -\ -$	$\{h0\bar{h}0\}$	6/1; 1/2; 2/1
$32(1)$	$\bar{3}m1P-\ \underline{-}\ -$	$\{h0\bar{h}l\}$	3/1; 2/1; 1/1
$\bar{3}m(1)$	$\bar{3}m1\underline{P}\ -\ -\ -$	All	6/1; 2/2; 2/1
6	$6/mP-/\underline{-}$	$(hk0)$	6/1
$\bar{6}$	$6/m\underline{P}-/-$	$(00l)$	3/2
$6/m$	$6/m\underline{P}-/-$	All	6/2
$\bar{6}m2$	$6/mmmP-/---\underline{-}$	[$\{hhl\}$, $\{hh0\}$, (001)]	6/2; 2/2; 4/1
$6mm$	$6/mmmP-/---\underline{-}$	[$(hk0)$, $\{hh0\}$, $\{h00\}$]	12/1; 2/2; 2/2
622	$6/mmmP-/\underline{-}\ -\ -$	$(hk0)$; $(h0l)$; (hh)	6/1; 2/1; 2/1
$6/m\ mm$	$6/mmm\underline{P}-/-\ -\ -$	All	12/2; 4/2; 4/2
23	$m\bar{3}P\underline{-}\ -$	$\{hk0\}$	2/1; 3/1; 1/1
$m\bar{3}$	$m\bar{3}\underline{P}--$	All	4/2; 3/1; 2/1
$\bar{4}3m$	$m\bar{3}mP\underline{-}\ --$	[$\{hk0\}$, $\{hh0\}$]	4/1; 6/1; 2/2
432	$m\bar{3}mP\underline{---}$	$\{hk0\}$; $\{hhl\}$	4/1; 3/1; 2/1
m3m	$m\bar{3}m\underline{P}-\ -\ -$	All	8/2; 6/1; 4/2

Column 1: Crystal class (also, point-group symbol).

Column 2: Diffraction symbols: centric zones are bold-underlined; the unit-cell symbol is underlined where the point group is centrosymmetric.

Note that the centric distribution occurs (i) for all hkl if the lattice is centrosymmetric; (ii) for a zone if the corresponding two-dimensional projection is centrosymmetric; (iii) for a central lattice row if the corresponding one-dimensional projection is centrosymmetric.

Column 3: Centric reflections are listed explicitly.

Column 4: Average ideal-intensity, crystal-class-dependent multiples (ε-factors). Each p/q symbol gives the multiple p (ε-factor) for a reciprocal lattice row, and q (ε-factor) for the zone normal to that row. (It may be helpful to recall the full symbols $1m1$ and 121 for monoclinic m and 2.) [see also Rogers D. *Statistical Properties of Reciprocal Space* in Computing Methods in Crystallography, ed Rollett JS, Pergamon, 1965.]

Applying this result to Eq. (5.66), the variance of $A'(hkl)$ is given by

$$\langle A'(hkl)^2 \rangle = 4 \sum_j^{n/2} g_j^2 \langle \cos^2 2\pi(hx_j + lz_j) \rangle \langle \cos^2 2\pi k y_j \rangle \qquad (5.69)$$

It is straightforward to show that $\langle \cos^2 2\theta \rangle = (1/\pi) \int_0^\pi \cos^2 \theta d\theta = 1/2$, so that from Eq. (5.65)

$$\langle A'(hkl)^2 \rangle = \sum_{j=1}^{n/2} g_j^2 = 1/2 \sum \qquad (5.70)$$

By a similar argument, $\langle B'(hkl)^2 \rangle = \sum_j g_j^2 = 1/2 \sum$, so that

$$\langle |\mathbf{F}_{\text{obs}}(hkl)|^2 \rangle = \sum \qquad (5.71)$$

However, if the reciprocal lattice level for $k = 0$ is considered, then a similar analysis shows that

$$\langle |\mathbf{F}_{\text{obs}}(h0l)|^2 \rangle = 2 \sum$$

In this example, the ε-factor for the $h0l$ zone in Pm is 2. The ε-factor depends upon the *crystal class*, and the values for all classes are listed in Table 5.4 [17].

5.6.5. Centric and acentric distributions of intensities

The scaled ideal intensities $I(hkl) = |\mathbf{F}_{\text{obs}}(hkl)|^2$ or of certain nets or rows of amplitudes conform to an acentric, a centric or a hypercentric distribution. Certain mean values may help to decide if a crystal or a crystal zone is centrosymmetric. It can be shown (Appendix A8) that the ratio of the square of the average $|\mathbf{F}_{\text{obs}}|$ to the average of $|\mathbf{F}_{\text{obs}}|^2$ has characteristic values. Thus, for the acentric distribution

$\langle |\mathbf{F}_{\text{obs}}| \rangle = \frac{1}{2}\sqrt{\pi \sum}$ and $\langle |\mathbf{F}_{\text{obs}}|^2 \rangle = \frac{1}{4}\pi \sum$. Hence,

$$M_{\text{a}} = \langle |\mathbf{F}_{\text{obs}}| \rangle^2 / \langle |\mathbf{F}_{\text{obs}}|^2 \rangle = \frac{1}{4}\pi \sum \Big/ \sum = \pi/4 = 0.785 \qquad (5.72)$$

In a similar analysis for a centric distribution,

$$M_{\text{c}} = \langle |\mathbf{F}_{\text{obs}}| \rangle^2 / \langle |\mathbf{F}_{\text{obs}}|^2 \rangle = \left(2 \sum /\pi \right) \Big/ \sum = 2/\pi = 0.637$$
$$(5.73)$$

These results can be useful in identifying a centre of symmetry, and so complete the determination of the space group of a crystal.

5.6.6. Normalized structure factors

Statistics based on $|\mathbf{F}_{\text{obs}}|$ contain a dependence on the structure itself, through the atomic scattering factors implicit in the intensity values. Improved statistical results may be obtained by using unitary structure factors $|\mathbf{U}(hkl)|$ or normalized structure factors $|\mathbf{E}(hkl)|$. In this discussion, and in Appendix A8, normalized structure factors (E-values) will be used. The *normalized structure factor* is given by the equation

$$|\mathbf{E}|^2 = \frac{|\mathbf{F}_{\text{obs}}|^2}{\varepsilon \sum_j g_j^2} \qquad (5.74)$$

It is implicit that the $|\mathbf{F}_{\text{obs}}|$ data are on an approximately *absolute scale* (Section 5.5.3.5) and that g_j implies, as before, that a temperature correction has been applied to f_j.

A set of $|\mathbf{E}|$-values for a crystal exhibits important distinctions between centric and acentric distributions; some of the more useful relations are listed in Table 5.5.

Example 5.2. The following E-values statistics were determined for two crystals. The intensity statistics of Table 5.5 can be used to examine the centricity of the distributions. Each individual result is not necessarily clear cut.

	Crystal I	Crystal II
	Mean values	
$\langle |\mathbf{E}| \rangle$	0.85 A/C	0.84 A/C
$\langle |\mathbf{E}|^2 \rangle$	0.99	0.98
$\langle |\mathbf{E}|^2 - 1 \rangle$	0.91 C	0.82 A
	Distributions/ (%)	
$|\mathbf{E}| > 3.00$	0.20 C	0.05 A
$|\mathbf{E}| > 2.50$	0.90 C	0.98 C
$|\mathbf{E}| > 2.00$	2.70 A/C	2.84 A/C
$|\mathbf{E}| > 1.75$	7.14 C	6.21A/C
$|\mathbf{E}| > 1.50$	12.9 C	10.5 A
$|\mathbf{E}| > 1.00$	33.7 C	37.1

Although there are some uncertainties among the two sets of results, which is not an unusual situation with these types of tests, the evidence is that Crystal I is centric (C) and Crystal II is acentric (A). This assignment was confirmed by the complete structure analyses.

5.6.6.1. *Cumulative distributions of $|\mathbf{E}|$-values*

As well as the results in Table 5.5, a cumulative distribution of $|\mathbf{E}|$ is often significant. The function $N(\mathbf{E})$ is the fractional number of $|\mathbf{E}|$-values less than or equal to a given value of $|\mathbf{E}|$, and has significantly different characteristics for acentric and centric distributions, as shown in Fig. 5.29.

Table 5.5. $|E|$-value functions for acentric and centric distributions.

$|\mathbf{E}|$-value function	Acentric	Centric
Mean values		
$\langle|\mathbf{E}|\rangle$	0.886	0.798
$\langle|\mathbf{E}|^2\rangle$	1	1
$\langle|\mathbf{E}|^2 - 1\rangle$	0.736	0.968
$\langle(|\mathbf{E}|^2 - 1)^2\rangle$	1	2
Distributions	%	%
$|\mathbf{E}| > 3.00$	0.01	0.30
$|\mathbf{E}| > 2.50$	0.19	1.24
$|\mathbf{E}| > 2.00$	1.80	4.60
$|\mathbf{E}| > 1.75$	4.71	8.00
$|\mathbf{E}| > 1.50$	10.5	13.4
$|\mathbf{E}| > 1.00$	36.8	32.0

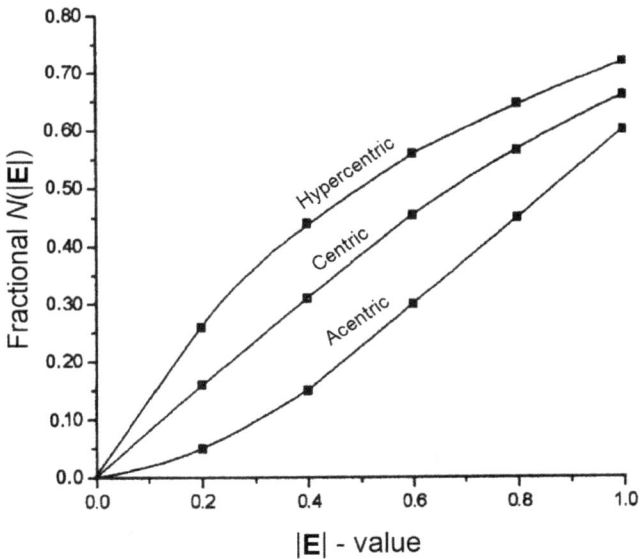

Fig. 5.29. Cumulative distributions of fractional $N(E)$ values for the acentric, centric and hypercentric intensity distributions.

5.6.6.2. *Hypersymmetry*

In the presence of non-crystallographic centres of symmetry in structural entities, enhanced values of the centric distribution may be observed. The crystal structures of pyrene [18] and benzo[a]pyrene [19] are examples of centrosymmetric molecules that crystallize in centrosymmetric space groups.

The degree of hypersymmetry depends upon the number of additional centres of symmetry, and the curve for a hypersymmetric distribution with one additional centre of symmetry is compared with that for a centric distribution in Fig. 5.29. Further discussions on intensity statistics can be found in the literature [17, 20].

Self-assessment 5.11. A monoclinic crystal shows systematic absences only for obs $|\mathbf{F}_{\text{obs}}(0k0)|$ with $k = 2n + 1$. In addition, the average value of $|\mathbf{F}_{\text{obs}}(hkl)|$ was 77 while the average value of $|\mathbf{F}_{\text{obs}}(hkl)|^2$ was 99. What is the probable space group?

Problems

5.1. Space group $P\frac{2_1}{c}$ was studied in Section 5.4.4.4. What equalities exist between the structure factors for the eight combinations of positive and negative signs for the hkl indices of a general reflection?

5.2. Determine the geometrical structure factors, A and B, for space group $Pma2$. From the results, list the conditions for systematic absences in this space group. The coordinates of the general equivalent positions are x, y, z; $\bar{x}, \bar{y}, z$; $1/2 - x, y, z$; $1/2 + x, \bar{y}, z$.

5.3. Three atoms have the amplitudes and phases 10, 15°; 20, 150°; 40, −110° on an Argand diagram. Calculate the amplitude and phase of the resultant.

5.4. (a) Evaluate the geometrical structure factor equation for an A-centred unit cell containing N atoms, using an exponential expression for $\mathbf{F}(hkl)$. (b) How does the result affect the possible reflections?

5.5. The following averaged data were obtained from a set of 150 $|\mathbf{F}_{obs}|$ reflections over six regions in reciprocal space. Determine a scale factor and a temperature factor by the Wilson method.

Range, n	1	2	3	4	5	6
$\langle(\sin^2\theta_n)/\lambda^2\rangle$	0.091	0.149	0.192	0.256	0.297	0.362
$\sum_j f_j^2\langle|\mathbf{F}_{obs,\,\theta_n}|^2\rangle$	1.215	1.592	1.934	2.703	3.236	4.016

5.6. α-Uranium crystallizes in the orthorhombic system in the orthorhombic system with $Z = 4$. The coordinates of the uranium atoms are $0, y, 1/4$; $1/2, 1/2 + y, 1/4$; $0, \bar{y}, 3/4$; $1/2, 1/2 - y, 3/4$. On the basis of the data below, decide whether 0.05, 0.10, or 0.15 is the best value for y_U:

hkl	020	110
$|F_{hkl}|$	90	270
g_U	70	80

5.7. The following data determine atomic scattering factors by the formula $f = \sum_{j=1}^{4} a_j \exp(-b_j s^2) + c_j$, where $s = (\sin\theta)/\lambda$.[b] Calculate $|\mathbf{F}_{calc}|$ for the 111 and 222 reflections for sodium

---

[b]Brown PJ *et al.*, *International Tables for Crystallography*, Vol. C. International Union of Crystallography (2006).

chloride and potassium chloride. The unit-cell spacings were reported as $a_{NaCl} = 5.627$ Å and $a_{KCl} = 6.278$ Å.

	a_1	b_1	a_2	b_2	a_3	b_3	a_4	b_4	c
Na⁺	3.2565	2.6671	3.9362	6.1153	1.3998	0.2001	1.0032	14.0390	0.4040
K⁺	7.9578	12.6331	7.4917	0.7674	6.3590	−0.0020	1.1915	31.9128	−4.9978
Cl⁻	18.2915	0.0066	7.2084	1.1717	6.5337	19.5424	2.3386	60.4486	−16.378

Discuss the results in the light of the intensities of the following indexed early X-ray powder photograph; K^+ and Cl^- are isoelectronic:

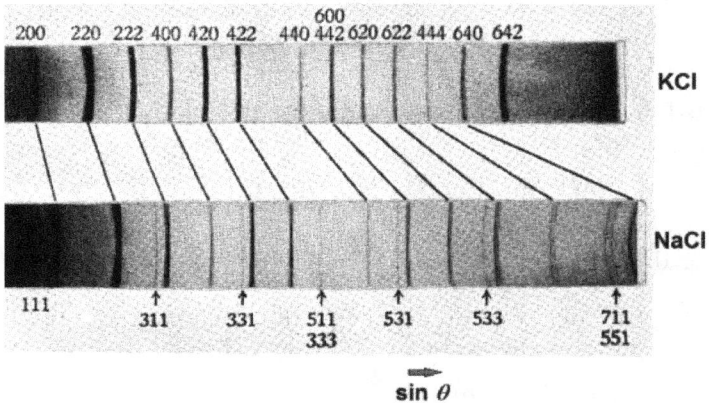

5.8. A crystal is found to have an overall isotropic temperature factor of 7 Å². (a) What is the fractional reduction at ambient temperature of the atomic scattering factor f_C for a carbon atom for a reflection at $\theta = 38.08°$, using Cu $K\alpha$ X-rays ($\lambda = 1.5418$ Å) compared to that for a carbon atom at rest and scattering under the same conditions? (b) What is the rms amplitude of vibration of the atom in a direction normal to the reflecting plane? The following data are available.

$(\sin\theta)/\lambda$	0.0	0.1	0.2	0.3	0.4	0.5
f_C	6	5.108	3.560	2.494	1.948	1.686

5.9. Compare the statistically distinguishable features of space groups $P2$, Pm and $P\frac{2}{m}$, none of which gives rise to systematic absences, and show how they may be distinguished.

5.10. (a) Derive the centric distribution of $|E|$-values. (b) Determine the average value $\langle \mathbf{E} \rangle$ for the centric distribution.

Answers to Self-Assessments

5.1. (a) $\lambda_{\min} = 12.398$ Å kV/20 kV $= 0.6199$ Å.

$$\text{Energy} = hc/\lambda = \frac{6.6261 \times 10^{-34}\, \text{J s} \times 2.9979 \times 10^{8}\, \text{m s}^{-1}}{0.6199 \times 10^{-10}\, \text{m}}$$

$$= 3.204 \times 10^{-15}\, \text{J} \equiv 20\, \text{keV}$$

(b) The ratio $I/I_0 = \exp(-0.023\, \text{cm} \times 0.74\, \text{cm}^{-1}) = 0.98$.

5.2. $\delta\lambda = \frac{h}{m_e c}(1 - \cos 2\theta) = 0.024263 \times (1 - 0.76604) = 0.005676$ Å.

5.3. $A' = 100\cos(0) + 50\cos(240°) = 75$; $B' = 100\sin(0) + 50\sin(240°) = -43.301$. Hence, $F = \sqrt{75^2 + (-43.301^2)} = 86.6$ and $\phi = \tan^{-1}(\frac{-43.301}{75}) = -30.0° \equiv 330°$.

5.4. There are several ways of showing the equivalence of the Bragg and Laue treatments, of which the following is one. In the diagram, $A_1, A_2, A_3, \ldots$ are regularly spaced scattering centres of spacing a in an infinite three-dimensional array. The X-ray beam makes an angle ϕ with the row of points and ψ is the angle between the row and the diffracted ray. The reflecting planes (hkl) make equal angles θ with both the incident and the reflected rays.

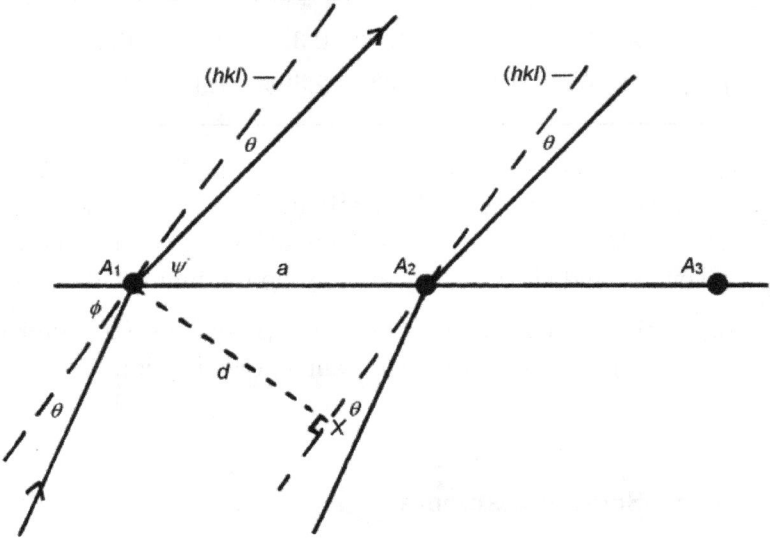

From the Laue equations, $a(\cos\psi - \cos\phi) = n\lambda$, where n is an integer. Expanding this equation: $-2a\frac{\sin(\psi+\phi)}{2}\frac{\sin(\psi-\phi)}{2} = n\lambda$. From the diagram, $\phi - \theta = \psi + \theta$. Then, $-2a\frac{\sin(\psi+\phi)}{2} = -a\sin(\psi+\theta) = -2d$, and $(\psi-\phi)/2 = (\psi/2)-[(\psi/2)+\theta] = -\theta$, so that $\phi/2 = (\psi/2) + \theta$. Additionally, $\frac{\sin(\psi-\phi)}{2} = \sin(-\theta) = -\sin\theta$. Thus, $2d\sin\theta = n\lambda$, which is the Bragg equation.

5.5.

$$A(hkl)/4 = \cos 2\pi\left(hx + lz + \frac{k+l}{4}\right)\cos 2\pi\left(ky - \frac{k+l}{4}\right)$$

$$= \left[\cos 2\pi(hx+lz)\cos 2\pi\left(\frac{k+l}{4}\right) - \sin 2\pi(hx+lz)\right.$$

$$\left. \times \sin 2\pi\left(\frac{k+l}{4}\right)\right]$$

$$\times \left[\cos 2\pi ky\cos 2\pi\left(\frac{k+l}{4}\right)\right.$$

$$\left. + \sin 2\pi ky\sin 2\pi\left(\frac{k+l}{4}\right)\right]$$

$$= \left[\cos 2\pi(hx + lz) \cos 2\pi ky \cos^2 2\pi \left(\frac{k+l}{4} \right) \right.$$

$$+ \cos 2\pi(hx + lz) \sin 2\pi ky \cos 2\pi \left(\frac{k+l}{4} \right)$$

$$\left. \times \sin 2\pi \left(\frac{k+l}{4} \right) \right]$$

$$- \left[\sin 2\pi(hx + lz) \cos 2\pi ky \cos 2\pi \left(\frac{k+l}{4} \right) \right.$$

$$\times \sin 2\pi \left(\frac{k+l}{4} \right)$$

$$\left. + \sin 2\pi(hx + lz) \sin 2\pi ky \sin^2 2\pi \left(\frac{k+l}{4} \right) \right]$$

Let the term $(k + l)/4 = m$. Then, $\cos 2\pi \frac{m}{4} \sin 2\pi \frac{m}{4} = \frac{1}{2} \sin 4\pi \frac{m}{4} = \frac{1}{2} \sin m\pi$, which is zero because $m = k + l$ which is integral. Thus,

$$A(hkl)/4 = \cos 2\pi \left(hx + lz + \frac{k+l}{4} \right) \cos 2\pi \left(ky - \frac{k+l}{4} \right)$$

$$= \cos 2\pi(hx + lz) \cos 2\pi ky \cos^2 2\pi \left(\frac{k+l}{4} \right)$$

$$- \sin 2\pi(hx + lz) \sin 2\pi ky \sin^2 2\pi \left(\frac{k+l}{4} \right)$$

Also, $\cos^2 2\pi(\frac{k+l}{4}) = 1$ for $h + k = 2n$ and zero for $h + k = 2n + 1$, whereas $\sin^2 2\pi(\frac{k+l}{4}) = 0$ for $h + k = 2n$ and unity for $h + k = 2n + 1$. Separating now, according to the parity of $h + k$:

$$\underline{h + k = 2n(\text{even})}$$

$$A(hkl) = 4 \cos 2\pi(hx + lz) \cos 2\pi ky$$

$$\underline{h + k = 2n + 1(\text{odd})}$$

$$A(hkl) = -4 \sin 2\pi(hx + lz) \sin 2\pi ky$$

5.6. (a) C unit cell, c-glide plane $\perp y$, a-glide plane $\perp y$, 2_1 axis $\parallel y$. Note, however, that the $h = 2n$ and $k = 2n$ conditions must arise with a C-centred unit cell: if $h + k = 2n$, then the conditions on h and on k follow. These conditions are redundant, or non-independent, and are listed in the International Tables [4] in parentheses as: $h0l : (h = 2n)$; $0k0: (k = 2n)$. The possible space groups are Cc and $C\frac{2}{c}$.

(b) P unit cell. The conditions $0k0: (k = 2n)$ and $h00: (h = 2n)$ are non-independent; they follow from the conditions on $h0l$ and $0kl$. The possible space groups are $Pba2$ and $Pbam$.

(c) The two coordinates in each set are related by the translation $(1/2, 1/2, 0)+$, so that they simulate C centring *for these positions*. Hence, the additional condition $h + k = 2n$ applies to them.

5.7. (a) $C2$, Cm, $C\frac{2}{m}$. (b) $Pnc2_1$, $Pmna$. (c) $Ima2$, $Imma$.

5.8. (a) From the given measurements, $b* = \frac{42.3\,\text{mm}/14}{60.0\,\text{mm}} \times 4.00 = 0.2014$, so that $b = 1.5418$ Å$/0.2014 = 7.66$ Å.

(b) The reflections on the photograph can be indexed by inspection. From the symmetry, only the upper right-hand quadrant need be considered. The reflections present are: 110, 130, 150, 170; 200, 220, 240, 260; 310, 330, 350, 370; 400, 420, 440, 460; 510, 530, 550; 600, 620, 640; 710, 730. These data present the following limiting conditions:

$hk0$: $h + k = 2n$, which indicates an n-glide plane $\perp$ to z;
$h00$: $(h = 2n)$ – redundant;
$0k0$: $(k = 2n)$ – redundant.

The partial diffraction symbol is $mmmP - n$, so that the space group (standard setting) is one of $Pnnn$, $Pban$, $Pccn$, $Pmmn$ or $Pbcn$.

5.9. From Section 5.5.1.1, $N = \frac{33.510}{(1.5418 \text{ Å})^3} \times 990.0$ Å$^3 \times \frac{(0.99619)^3}{(4 \times 2)} = 1118.56$, which means that 1118 is the nominal integral value for N_{max}. (The actual maximum would be less than 1118 because the A centring would exclude hkl reflections for $k + l = 2n + 1$.)

5.10. (a) $m = 4$.

(b) $d_{110} = a/\sqrt{h^2 + k^2} = 5/\sqrt{2}$; $\sin\theta = \lambda/2d = 1.5418$ Å$/(\sqrt{2} \times 5$ Å$) = 0.21804$, so that $\theta = 12.594°$. Thus, $L = 1/\sin 2\theta = 2.35$; $p = \frac{1}{2}(1 + \cos^2 2\theta) = 0.909$.

5.11. The systematic absences suggest space group $P2_1$ or $P\frac{2_1}{m}$. The value of $\langle |\mathbf{F}_{hkl}| \rangle^2 / \langle |\mathbf{F}_{hkl}|^2 \rangle = 77/99 = 0.778$. Taken together, the evidence indicates strongly the non-centrosymmetric space group.

References

[1] Röntgen W, *Sitzungsber. Physik.-medic. Gesellschaft Würzburg*, 137 (1895).

[2] James RW, *Optical Principles of the Diffraction of X-rays*. Kluwer Academic (1999).

[3] Woolfson MM, *An Introduction to X-ray Crystallography*, 2nd edn. Cambridge University Press (1977).

[4] Henry NFM and Lonsdale K (eds), *International Tables for X-ray Crystallography*, Vol. I, 3rd edn. International Union of Crystallography, Kynoch Press (1969). Online at ⟨https://archive.org/stream/InternationalTablesForX-ray CrystallographyVol1/HenryLonsdaleEds-internationalTablesForX-rayCrysta llographyVol1#page/n71/mode/2up⟩. *Note*: This source contains the earlier notation of $m3$ and $m3m$ in place of $m\bar{3}$ and $m\bar{3}m$ for those point groups, and for the space groups that derive from them.

[5] Aroyo MI (ed), *International Tables for Crystallography — Volume A: Space-Group Symmetry*, 6th edn., Wiley (2016); 2nd edn. Wiley (2014); 5th edn. Online at ⟨https://www.wiley.com/en-gb/International+Tables+for+Crys tallography,+6th+Edition,+Volume+A,+Space+Group+Symmetry-p-9780 470974230⟩.

[6] Hodgkin DMC, *Biogr. Mems. Fell. R. Soc.* **26**, 28 (1980).

[7] Ewald PP, *Z. Phys.* **14**, 465 (1913).

[8] Bruker GmbH, *D8 Venture*. Online at ⟨http://web.mit.edu/X-Ray/Ruf_talk_ 2014.pdf⟩.

[9] Rigaku Corporation. Online at ⟨https://www.labcompare.com/178-X -ray-Diffractometer-XRD-Instruments/42923-SuperNova-X-ray-Diffraction-System/⟩.

[10] Bruker GmbH, *D8 Advance*. Online at ⟨http://xray.chem.wisc.edu/Docu ments/Powder_XRD_operation_notes.pdf⟩.

[11] Zachariasen WH, *Acta Crystallogr.* **23**, 558 (1967).

[12] Wilson AJC, *Nature*, **150**, 152 (1942).

[13] Debye P, *Ann. Phys.* **348**, 49 (1921).

[14] Waller I, *Z. Phys.* **17**, 398 (1923).

[15] Ladd M and Palmer R, *Structure Determination by X-ray Crystallography*, 5th edn. Springer (2013).

[16] Woolfson MM, *Introduction to X-ray Crystallography*, 2nd edn. Cambridge University Press (1997).

[17] Rogers D, *Computing Methods in Crystallography*, Rollett JS (ed.) Pergamon (1965).

[18] Robertson JM and White JG, *J. Chem. Soc.* 358 (1947).

[19] Carrell CJ *et al.*, *Carcinogenesis*, **18**, 415 (1997).

[20] Rogers D and Wilson AJC, *Acta Crystallogr.* **6**, 439 (1953).

Chapter 6

Solving the Structure

A great advantage of X-ray analysis as a method of chemical structure analysis is its power to show some totally unexpected and surprising structure with, at the same time, complete certainty.
Dorothy Crowfoot Hodgkin

Key Topics

- Image formation
- Fourier series
- Electron density: functions and maps
- Patterson synthesis and interpretation
- Sharpened Patterson techniques
- Patterson search method
- Difference Fourier synthesis
- Direct methods of phasing
- Examples of complete structure analyses
- Isomorphous replacement
- Anomalous scattering

6.1. Introduction

There are many techniques available through which crystal structures have been and are being solved from data obtained by the diffraction of X-rays or of neutrons from crystals. Three of these methods have been responsible for solving the majority of crystal structures: they are the Patterson method, including molecular replacement (MR) the probability equations, usually called direct methods, and isomorphous replacement techniques. These are the procedures that will be discussed in this chapter. Discussions of other, less used methods, including theoretical techniques, can be found in other modern texts devoted to X-ray crystallography [1–3].

6.2.   Image Formation and the Phase Problem in X-ray Structure Analysis

A simple example of image formation occurs in the reading of written information with a hand lens. Another similar example resides in the projection of the slide of a crystal structure on to a screen, whereupon an image such as that in Fig. 6.1(a) would appear. If the lens is removed from the projector, the screen will appear like Fig. 6.1(b), although the object has not been moved. The information about the object is still present in the scattered light patch, as can be demonstrated with the aid of a hand lens and a white card inserted into the path of scattered radiation. The lens contains no information about the image; its purpose is to combine the scattered light such

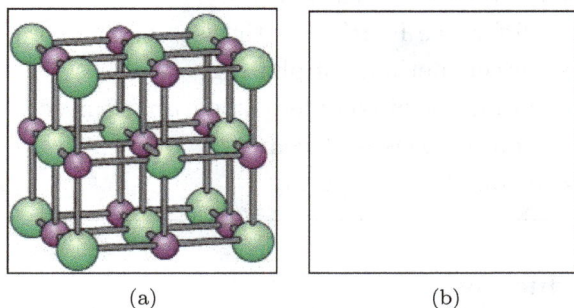

(a)   (b)

Fig. 6.1.   (a) Image of a colour slide of the unit cell and environs of the crystal structure of sodium chloride: Na^+ grey, Cl^- green; the unit cell contains four Na^+ and four Cl^- entities. (b) The unfocussed radiation from a projector lens scattered by the colour slide of the sodium chloride structure.

that the image is sharp, or 'in focus', and so reveal the nature of the object. Knowledge of the expected appearance of the object is presumed so that it will be realized when the image is in focus. Alternatively, the object slide could be marked with an ×, so that when the known × is sharply defined the image will be focussed correctly also. Image formation, then, involves the scattering of light by an object followed by the recombination of the scattered radiation in the correct relative phases to give an interpretable result.

The eye can resolve separations of $0.1 - 0.2\,\text{mm}$ in an object; a high-quality optical microscope can resolve to a limit of *ca.* $2000\,\text{Å}$, or to an even higher order. The resolving power R of a microscope is given by the Rayleigh formula:

$$R = \frac{0.61\lambda}{n\sin\theta} \tag{6.1}$$

where n is the refractive index of the medium between the object and the optical system and 2θ is the angle of scatter, and $n\sin\theta$ is known as the numerical aperture of the optical system. For visible light, $\langle\lambda\rangle = 5500\,\text{Å}$ and a microscope in oil immersion has a numerical aperture of *ca.* $1.46\,\text{Å}$, so that R would be approximately $2300\,\text{Å}$. Better resolution is available through electron microscopy. In particular, a resolution approaching $2\,\text{Å}$ has been claimed for high-resolution transmission electron microscopy (HRTEM), but only simple structures have been solved by this method. Nothing can yet compete with the quality of the structure analysis results that are obtained by X-ray and neutron diffraction methods.

A simple analogy to X-ray diffraction is the observation of a sodium lamp through a fine net, such as a handkerchief. When it is exposed to the monochromatic radiation, such as a sodium street lamp, a spot pattern is seen. This pattern is invariant with respect to translation of the net but it rotates as the net is rotated; these properties will discussed further shortly.

In order to form the object of a diffraction pattern image, the scattered radiation must be recombined in terms of both amplitude and phase. The lens system of an optical instrument, such as a microscope, carries out this process of recombination. With

X-rays or neutrons, however, the focussing cannot be carried out directly. In fact, phases are not recorded: the experimental procedure records the intensities $I(hkl)$ of reflections, and, while this can be converted to the ideal value $|\mathbf{F}(hkl)|^2$, the phase information is lost in the process. This can be envisaged from Fig. 5.12: $|\mathbf{F}(hkl)|^2 = A'(hkl)^2 + B'(hkl)^2 = |\mathbf{F}(hkl)|^2 \cos^2 \phi(hkl) + |\mathbf{F}(hkl)|^2 \sin^2 \phi(hkl) = |\mathbf{F}(hkl)|^2 = I(hkl)$, which is the collected data; the phase information is not disclosed by the process.

This loss of phase information constitutes the *phase problem* in X-ray crystallography, and crystal structure analysis is concerned first with recovering this phase information from the measured data. The periodic nature of the electron density of a crystal and, hence, of its diffraction pattern enables it to be represented by a Fourier series. Once the phases have been determined, the recombination of the scattered information, or diffraction pattern, can be achieved by calculation, and the object revealed in terms of its electron density; atomic positions are deemed to lie at the maxima in the electron density function.

6.3. Fourier Series

A portion of a periodic function $\psi(X)$ is shown in Fig. 6.2; it repeats exactly at intervals of 2π. According to Fourier's theorem, the series may be represented by a series of sine and cosine term:

$$\psi(X) = \sum_{h=-\infty}^{\infty} \{C(h) \cos 2\pi hx + S(h) \sin 2\pi hx\} \qquad (6.2)$$

where C and S are amplitudes and the index h is a wave number that defines the hth term of the series.

If Friedel's law is introduced for the coefficients C and S (Section 5.4.3.1), then in a one-dimensional case,

$$\psi(X) = C(0) + 2\sum_{h=1}^{\infty} \{C(h) \cos 2\pi hx + S(h) \sin 2\pi hx\} \qquad (6.3)$$

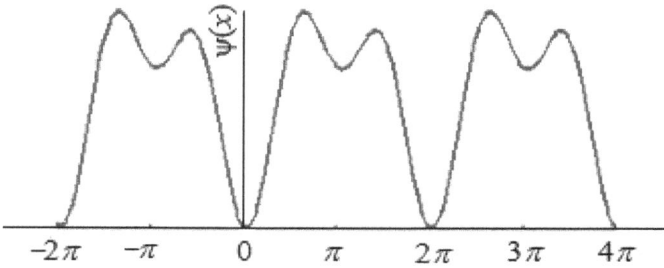

Fig. 6.2. Portion of a regularly repeating function $\psi(x)$ with a period of 2π.

It can be shown [2, 4] that, for a repeat period a,

$$C(h) = \frac{1}{a} \int_0^a \psi(X) \cos(2\pi h X/a)\mathrm{d}X \tag{6.4}$$

$$S(h) = \frac{1}{a} \int_0^a \psi(X) \sin(2\pi h X/a)\mathrm{d}X \tag{6.5}$$

Thus, if the function $\psi(X)$ is known, the coefficients C and S can be determined; the process is demonstrated next for a square wave.

The square wave function $\psi(X)$, in Fig. 6.3, has the following properties:

$$X < 0, \qquad \psi(X) = 0$$
$$0 < X < \pi, \quad \psi(X) = \pi$$

From Eqs. (6.4) and (6.5)

$$C(h) = \frac{1}{2\pi} \int_0^\pi \pi \cos(2\pi h X/2\pi)\mathrm{d}X = \frac{1}{2\pi} \int_0^\pi \pi \cos(hX)\mathrm{d}X$$

Integration shows that for $h = 0$, $C(h) = \frac{1}{2\pi}\int_0^\pi \pi\mathrm{d}X = \pi/2\pi$, whereas for $h \neq 0$, $C(0) = 0$. Similarly,

$$S(h) = \frac{1}{2\pi} \int_0^\pi \pi \sin(2\pi h X/2\pi)\mathrm{d}X$$

In this case, integration shows that for $h = 0$, $S(h) = 0$, whereas for $h \neq 0$, $S(h) = \frac{1}{2h}(1 - \cos \pi h)$. Applying these results to Eq. (6.3)

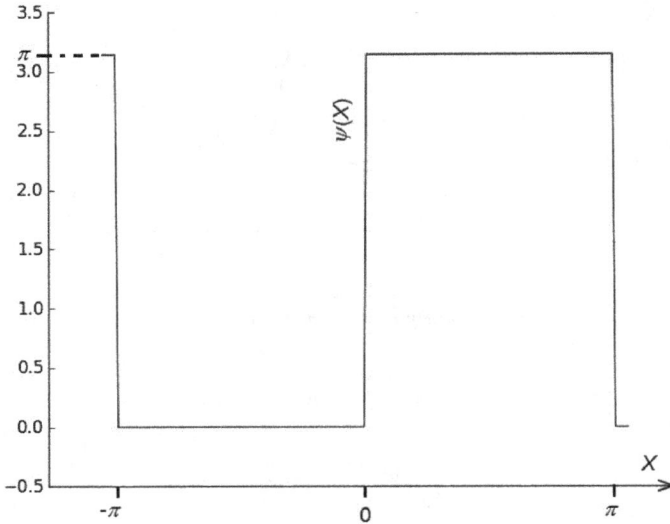

Fig. 6.3.   Square wave $\psi(X)$ of repeat period of 2π, shown over the range $-\pi$ to π.

gives

$$\psi(X) = \pi/2 + 2\sum_{h=1}^{\infty}\frac{1}{2h}(1 - \cos \pi h)\sin hX \qquad (6.6)$$

Since $(1 - \cos \pi h) = 0$, for $h = 2n$ and 2 for $h = 2n+1(n = 1, 2, 3, \ldots)$, it follows that, subject to $h = 2n + 1$, Eq. (6.6) becomes

$$\psi(X) = \pi/2 + 2\sum_{h=1}^{\infty}\frac{1}{h}\sin hx \qquad (6.7)$$

The variable X defines a sampling point m within the repeat period, for example, $m(2\pi/100)$, where $m = 0, 1, 2, 3, \ldots, m_{\text{max}})$. The index h, although ideally ranging from $-\infty$ to ∞ runs in practice from h_{min} to h_{max}. The larger the number of data in the summation, the better the representation of the function, as Fig. 6.4 shows. The series termination errors, or 'ripples', depend upon the number of terms included in the summation of the series. As more terms are added in to the series, the combination of waves tends to the ideal representation of the function. This effect is important in electron

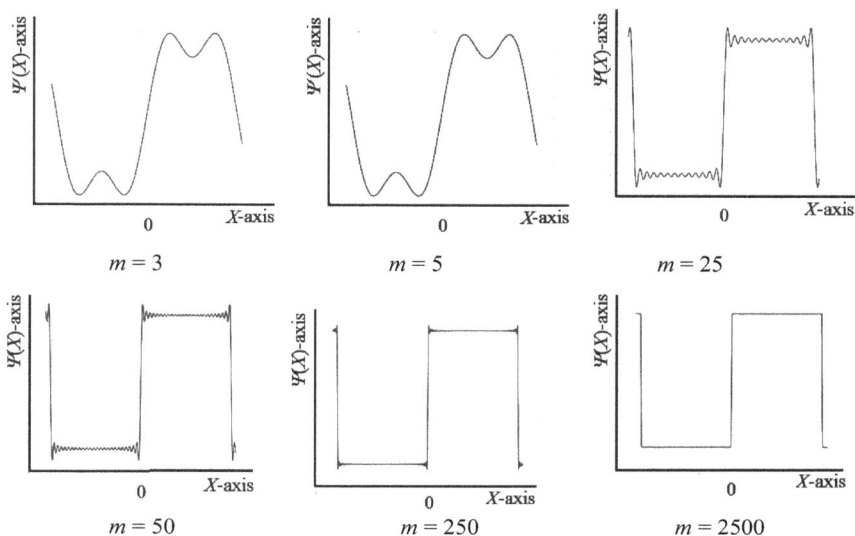

Fig. 6.4.   The square wave function with varying number of terms h. As h increases, the series termination errors (aka ripples) are decreased to zero by the combination of waves of increasingly higher frequencies.

density maps of crystal structures, particularly in the neighbourhood of heavy atoms that scatter X-rays strongly.

6.3.1.   Fourier series in exponential form

Let a function $\mathbf{G}(h)$ be defined on an Argand diagram, such that

$$\mathbf{G}(h) = C(h) + iS(h)$$
$$\mathbf{G}(\bar{h}) = C(h) - iS(h) \tag{6.8}$$

Multiplying throughout by $\exp(-2\pi hX/a)$ and forming the sum of the terms,

$$\mathbf{G}(h)\exp(-2\pi hX/a) + \mathbf{G}(\bar{h})\exp(-2\pi hX/a)$$
$$= [C(h) + iS(h)]\exp(-2\pi hX/a)$$
$$+[C(h) - iS(h)]\exp(-2\pi hX/a)$$
$$= 2C(h)\cos(2\pi hX/a) + 2S(h)\sin(2\pi hX/a)$$

By analogy with Eq. (6.2), using Euler's theorem $[\exp(\pm iX) = \cos X \pm i \sin X]$, $\psi(X)$ may written as

$$\psi(X) = \mathbf{G}(0) + \sum_{h=1}^{\infty} |\mathbf{G}(h)| \exp(-2\pi hX/a)$$
$$+ |\mathbf{G}(\bar{h})| \exp(-2\pi hX/a) \qquad (6.9)$$

or, introducing Friedel's law, as

$$\psi(X) = \sum_{h=-\infty}^{\infty} |\mathbf{G}(h)| \exp(-2\pi hX/a) \qquad (6.10)$$

Then, following the procedure that led to the deduction of Eqs. (6.4) and (6.5),

$$\mathbf{G}(h) = \frac{1}{a} \int_0^a \psi(X) \exp(2\pi hX/a)\mathrm{d}X \qquad (6.11)$$

Equations (6.10) and (6.11) are known as *Fourier transforms* of each other; *the differing signs* of the two exponential terms should be noted [4, 5]. Fourier transforms and some of their applications in crystal structure analysis are discussed in Appendix A10.

6.4.   Fourier Series Applied to X-ray Crystallography

The periodicity of a crystal structure is repeated in its election density which can, therefore, be represented by a Fourier series. This application can be considered for one-, two- and three-dimensional electron density distributions.

6.4.1.   One-dimensional electron density function $\rho(X)$

The plots shown in Figs. 4.2 and 6.2 may be regarded as one-dimensional electron density functions which are termed $\rho(X)$, where X is any sampling point within the repeat distance $0 - 2\pi$ in Fig. 6.2. In an infinitesimally small interval $\mathrm{d}X$, the electron density is constant and constitutes an amount of electrons. Its contribution to a structure factor $\mathbf{F}(h)$ follows from the Argand diagram as $\rho(X)\mathrm{d}X \exp(i2\pi hX)$, where the exponential term represents the phase of the contribution $\rho(X)dX$ with respect to the origin. Thus, the value of $\mathbf{F}(h)$ is given by

$$\mathbf{F}(h) = \int_0^a \rho(X)\exp(i2\pi hX/a)dX \qquad (6.12)$$

which is a generalized one-dimensional structure factor equation. Replacing $\rho(X)$ in Eq. (6.12) by Eq. (6.10) gives

$$\mathbf{F}(h) = \int_0^a \sum_{h=-\infty}^{\infty} |\mathbf{G}(h')|\exp(-i2\pi hX/a)\exp(i2\pi hX/a)dX \qquad (6.13)$$

where h' is another index not generally equal to h. Equation (6.13) can be written as

$$\mathbf{F}(h) = \sum_{h=-\infty}^{\infty} |\mathbf{G}(h')| \int_0^a \exp[-i2\pi(h-h')X/a]dX \qquad (6.14)$$

The value of the integral is $\frac{a}{(i2\pi h - h')}\exp[i2\pi(h-h')X/a]\big|_0^a$. Since h and h' are both integers, the integral is zero except for the case that $h = h'$. Then it becomes $\int_0^a dX$, which has the value a, whereupon it follows that $\mathbf{G}(h) = \mathbf{F}(h)/a$. Then, from Eq. (6.10) and using $\rho(X)$ in place of $\psi(X)$,

$$\rho(X) = \frac{1}{a}\sum_{h=-\infty}^{\infty} \mathbf{G}(h)\exp(-2\pi hX/a) \qquad (6.15)$$

which is the Fourier transform of Eq. (6.12).

From a consideration of Friedel's law and Fig. 5.12, Eq. (6.15) can be recast as a one-dimensional electron density function, now using the fractional coordinate x $(= X/a)$ and $h_{\max}$ as the upper practical limit of the index h:

$$\rho(x) = \frac{1}{a}\left\{ F(000) + 2\sum_{h=1}^{h_{\max}}[A'(h)\cos 2\pi hx + B'(h)\sin 2\pi hx]\right\} \qquad (6.16)$$

where $A'(h)$ and $B'(h)$ are the real and imaginary components of the structure factor $\mathbf{F}(h00)$. The term $F(000)$ is the number of electrons in the unit cell; it is a real number but this parameter and the factor

$1/a$ may be omitted in a calculation; the results can always be scaled as required (see Example 6.1). The full expression is important in three dimensions if electron counts of the atoms are to be extracted from the electron density function. Note that while in (6.15) $\rho(X)$ may be calculated at any value of X, in (6.16), it is dependent upon the integral reciprocal lattice points h.

Example 6.1. The following data were obtained for the centrosymmetric structure of hafnium disilicide. There are $4\mathrm{HfSi}_2$ entities in the unit cell in three sets of the special positions $\pm(0, y, 1/4; 1/2, 1/2 + y, 1/4)$. In this centrosymmetric structure, $A' = F_{\mathrm{obs}}$ and $B' = 0$.

$0k0$	020	040	060	080	010,0	012,0	014,0	016,0
$F_{\mathrm{obs}}(0k0)$	7	-14	-18	13	12	< 1	-20	< 1

Note that in a centrosymmetric structure with $\bar{1}$ at the origin, the structure factor $\mathbf{F}(hkl)$ may be written as $F(hkl)$ since it is a real number carrying a $+$ or $-$ sign (equivalent to a phase angle of 0 or π).

Omitting the two very weak reflections, the electron density function is obtained using the program FOUR (a) including the repeat distance a of $14.55\,\text{Å}$ and the $F(000)$ term of value 400, and (b) without these two parameters. The plots made with the program GRFN are shown in Figs. 6.5(a) and 6.5(b).

The four highest peaks are the sites of $4\,\mathrm{Hf}$, with $y = \pm(0.107, 0.607)$. Eight peaks are expected for silicon atoms but eight small peaks and two medium peaks can be seen. The most likely interpretation is four Si at $\pm(0.033, 0.533)$ and two Si at *ca.* ± 0.25. There remain four smaller peaks marked S in the neighbourhood of hafnium which must be regarded as spurious; they are examples of series termination errors. A calculation of structure factors confirmed the signs allocated to the reflection data, so that Fig. 6.5 is the best result with the given amount of data.

> **Example 6.1.** (*Continued*) The two plots in Fig. 6.5 are identical
> except for the scaling effect of $1/b$ and $F(000)$: for plot (a), $\rho(0) =$
> 24.74, and, for plot (b), $\rho(0) = -40.00$; as expected, $\frac{1}{b}[\rho(0)_b +$
> $F(000)] = \frac{360}{14.55}$ Å $= 24.74$ Å^{-1}.

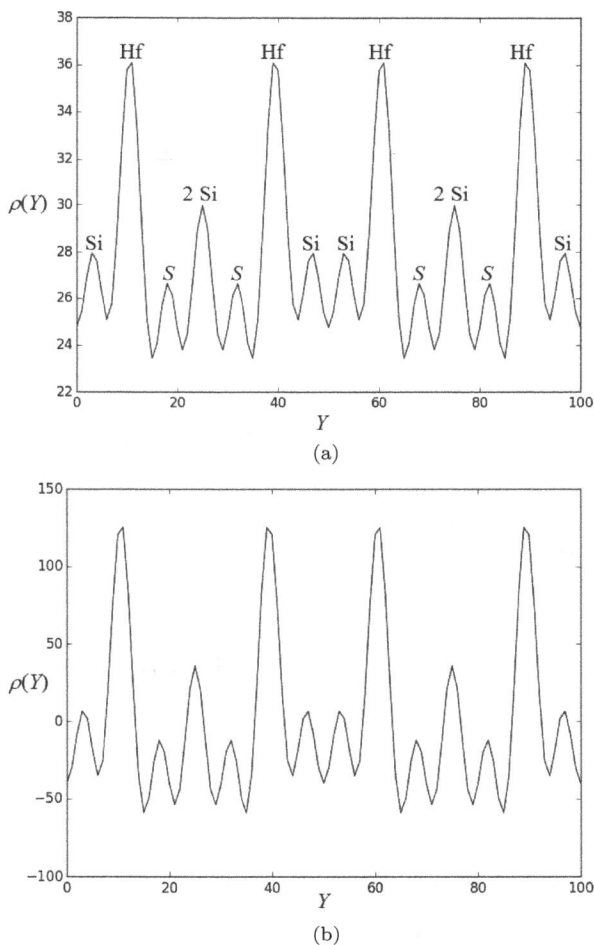

Fig. 6.5. One-dimensional electron density projection for hafnium disilicide HfSi$_2$ using
Eq. (6.8): (a) including the terms $F(000)$ and $1/b$; (b) excluding these two terms. The
difference is in the vertical scale only: Hf = hafnium, Si = silicon; 2 Si represents two
silicon atoms overlapping in projection at x approximately 0.25. The peaks marked S
are spurious maxima (ripples) in the neighbourhood of the heavy hafnium atom.

Self-assessment 6.1. Magnesium fluoride MgF_2 is tetragonal, with the centrosymmetric space group $P\frac{4_2}{m}nm$; $a = b = 4.625\,\text{Å}$, $c = 3.052\,\text{Å}$ and $Z = 2$. From space-group data [6,7], the probable atomic sites are: $2\,Mg$ at $0,0,0; 1/2,1/2,1/2$ and $4\,F$ at $\pm(x,x,0; 1/2+x, 1/2-x, 1/2)$. From the following $|\mathbf{F}_{obs}(h00)|$ data, calculate and plot $\rho(x)$, and find x_F.

$h00$	200	400	600	800	10,00	12,00
$|\mathbf{F}_{obs}(h00)|$	−2.7	12.0	7.2	0.1	3.2	0.1

In order to demonstrate the importance of correct phasing, reverse the sign of the 400 reflection and re-calculate and plot $\rho(x)$.[a]

6.5. Two- and Three-Dimensional Electron Density Equations

Expressions analogous to Eq. (6.16) can be written for two-dimensional and three-dimensional electron density functions. The program system XSYST in the Program Suite makes use of two-dimensional examples, for obvious reasons, and the electron density equation may be written as

$$\rho(xy) = \frac{1}{A}\left\{ F(000) + 2\sum_{h=1}^{\infty}\sum_{k=-\infty}^{\infty}[A'(hk)\cos 2\pi(hx+ky)\right.$$
$$\left. +B'_{(hk)}\sin 2\pi(hx+ky)]\right\} \tag{6.17}$$

where A is the area of the projection. From the practical point of view, this expression will be considered in the simpler form

$$\rho(xy) \propto \sum_{h}\sum_{k}[A'(hk)\cos 2\pi(hx+ky) + B'(hk)\sin 2\pi(hx+ky)] \tag{6.18}$$

---

[a]Answers to all Self-assessments are given at the end of the chapter.

as only peak positions on the electron density map will be of interest. The sums range over the values of h and k available in the data set. If the structure is centrosymmetric with the centre of symmetry at the origin, then Eq. (6.18) is simplified to

$$\rho(xy) \propto \sum_h \sum_k A'(hk) \cos 2\pi(hx + ky) \tag{6.19}$$

where $A' = \pm|\mathbf{F}_{\text{obs}}|$.

In three dimensions, the electron density equation is

$$\rho(xyz) = \frac{1}{V} \left\{ F(000) + 2\sum_{h=1}^{\infty} \sum_{k=-\infty}^{\infty} \sum_{l=-\infty}^{\infty} [A'(hkl) \right.$$

$$\left. \times \cos 2\pi(hx + ky + lz) + B'(hkl) \sin 2\pi(hx + ky + lz)] \right\} \tag{6.20}$$

or, from a practical point of view,

$$\rho(xyz) \propto \sum_h \sum_k \sum_l [A'(hkl) \cos 2\pi(hx + ky + lz)$$

$$+ B'(hkl) \sin 2\pi(hx + ky + lz)] \tag{6.21}$$

and for a centrosymmetric structure

$$\rho(xyz) \propto \sum_h \sum_k \sum_l A'(hkl) \cos 2\pi(hx + ky + lz) \tag{6.22}$$

and, again, $A' = \pm|\mathbf{F}_{\text{obs}}|$. An alternative expression for Eq. (6.21) arises from a consideration of Fig. 5.12 as

$$\rho(xyz) \propto \sum_h \sum_k \sum_l |\mathbf{F}_{\text{obs}}(hkl)| \cos[2\pi(hx + ky + lz) - \phi(hkl)] \tag{6.23}$$

This equation demonstrates clearly the dependence of the electron density on the phase angles: $|F(hkl)|$ is measured experimentally but $\phi(hkl)$ must be determined from the data set; this is the phase problem of which mention has been made earlier.

6.5.1.   Units of electron density

A consideration of the equations for electron density shows that its units are $\frac{1}{(\text{length})^n}$, where $n = 1, 2$ or 3 for one-, two- and three-dimensional syntheses, respectively. It is often quoted as $e\,\text{Å}^{-3}$ for the three-dimensional function; it is *number* (of electrons) *density*.

6.5.2.   Interpreting electron density maps

Atoms are revealed in an electron density map by the density contours that rise to a maximum at an atomic site position and fall to low values between the atoms. Figure 6.6 shows an electron density contour map of azidopurine monohydrate [8]. The ratio of two peak heights H_1/H_2 in an electron density map is approximately the ratio of the corresponding atomic numbers Z_1/Z_2. For the azidopurine structure, $H_N/H_C \approx 80/70 = 1.14$ and $Z_N/Z_C = 1.17$ which is a particularly good agreement. Hydrogen atoms, attached to carbon in this structure, are not revealed on the map because of their very small scattering power. Their sites may be located by a difference Fourier synthesis, as will be discussed shortly, or positioned through geometrical arguments. Current crystallographic program systems include routines that select the coordinates of electron density maxima, and also calculate hydrogen atom positions, according to their type, from standard geometry.

6.6.   Extracting Phase Information

In Chapter 4, the structural information obtainable from knowledge of the space group of a crystal together with its experimental density was examined by means of several examples. Subsequently, the measurement of unit-cell dimensions and the collection of X-ray intensity data, with the corrections to be applied to it, were studied. At this stage, then, the preliminary data available on the crystal structure may be summarized as follows:

- Crystal system
- Space group

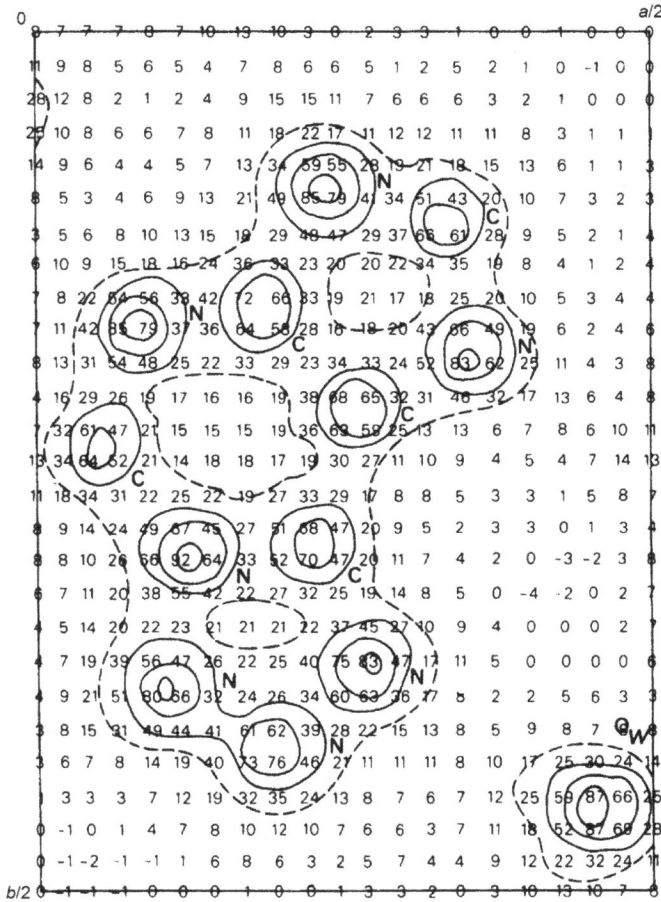

Fig. 6.6. Two-dimensional electron density projection $\rho(xy)$ for azidopurine monohydrate $C_5H_3N_7 \cdot H_2O$ contoured over a grid of field figures [8]. The peak O_W is the oxygen atom of the water molecule of crystallization; hydrogen atoms are resolved only very rarely in electron density maps.

- Unit-cell dimensions: $a, b, c, \alpha, \beta, \gamma$
- Unit-cell volume V_c
- Density D_x (experimental), D_c (calculated)
- Number Z_c of formula-entities per unit cell
- $|\mathbf{F}_{obs}(hkl)|$ observed structure factor amplitudes (data set) on a relative scale
- Overall temperature factor (Section 5.5.3.5)

6.6.1. Patterson synthesis

The Patterson function is a Fourier summation that uses the corrected (ideal) intensity data $|\mathbf{F}_{\mathrm{obs}}(hkl)|^2$. It is therefore a phase-free calculation [9] and so may be computed directly from the experimental data. The three-dimensional Patterson function may be written as

$$P(uvw) = (2/V) \sum_h \sum_k \sum_l |\mathbf{F}_{\mathrm{obs}}(hkl)|^2 \cos 2\pi(hu + kv + lw)$$

(6.24)

where the summations are taken over all recorded, unique values of h, k and l; in practice, the multiplier $(2/V)$ may be set to unity and the results brought to a convenient scale as desired. A two-dimensional form for a Patterson projection may be written as

$$P(uv) = (2/A) \sum_h \sum_k |\mathbf{F}_{\mathrm{obs}}(hk0)|^2 \cos 2\pi(hu + kv) \qquad (6.25)$$

A Patterson map is a contour map, similar to that of electron density, but wherein the peaks correspond to interatomic vectors in the structure. An analysis of the function leading to this result has been given elsewhere and need not be repeated here [2]. A maximum in the Patterson function at the coordinates (uvw) indicates a vector $\mathbf{r}(uvw)$ from the origin to that point. The Patterson unit cell has the same dimensions as the real unit cell, and its space group is centrosymmetric whether or not the crystal structure itself is centrosymmetric. In other words, it is the Laue symmetry plus unit cell centring, that is, it is a symmorphic space group, and it is shown for each space group on the first page of the space group entry in the International Tables for Crystallography, Vol. A; Figs. 4.17 and 4.24 are two examples.

The Patterson function may be thought of as a product (convolution) of an electron density function $\rho(\mathbf{r})$ with its inversion in the origin $\rho(-\mathbf{r})$. The formation of a Patterson function may be exemplified by the one-dimensional electron density function containing two peaks, 1 and 2 shown in Fig. 6.7. Consider sweeping a vector

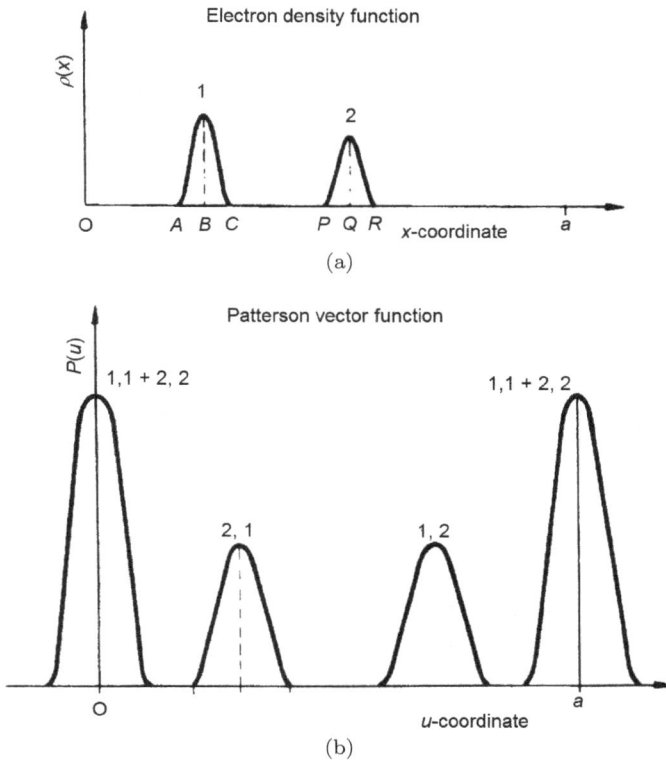

Fig. 6.7. (a) A one-dimensional electron density distribution $\rho(x)$ and (b) the corresponding Patterson function $P(u)$. In the electron density function, let the vector of length $u_{\min}$ be the distance CP and $u_{\max}$ the distance AR; u_{peak} is BQ. The maximum at $u = 0$ in the Patterson function represents the superposition of all self-vectors (1–1 and 2–2 in this example) in the unit cell on to the origin of $P(u)$. The vectors between B and C in $\rho(x)$ are the peaks 2–1 and 1–2 in $P(u)$.

of length $|\mathbf{u}|$ through the electron density function starting from a point approximately mid-way between O and A, calculating the electron density product at the two ends of the vector at small intervals along x. As long as u is less than $u_{\min}$, which is equal to the distance CP, the electron density product for that vector is zero. It is sufficient that one end of the vector $\mathbf{u}$ lies in a zero region of $\rho(x)$ for the corresponding value of $P(u)$ to be zero.

When $|\mathbf{u}|$ just exceeds $u_{\min}$, the electron density product $P(u)$ becomes positive. As the sweeping vector u is made larger, $P(u)$

shows an increased value, and when $|\mathbf{u}|$ is equal to the distance BQ, which is the interatomic distance $1-2$, $P(u)$ then has its maximum value for the peak under consideration; this position is peak marked 2, 1 on the Patterson function. As $|\mathbf{u}|$ is increased further, the peak height decreases and, eventually, it moves into a zero region of $\rho(x)$ whereupon the corresponding $P(u)$ falls to zero.

Since the electron density is a repeating function, the vector $\mathbf{u}$ will reach a value spanning peak 2 in $\rho(x)$ and peak 1 in the next unit cell, thus giving rise to the peak 1, 2. The centrosymmetric nature of the Patterson function follows from this analysis, as it is true for all atom pairs.

If the sweeping vector is allowed to decrease from the above starting value for any atom, then both ends will lie ultimately within one and the same electron density distribution for that atom. Thus, the largest peak in the Patterson function will exist always at the origin of the Patterson unit cell since the electron density products for all atoms are superimposed there.

The Patterson function may be summarized as a map of interatomic vectors all taken to a common origin, as illustrated in Fig. 6.8 for a two-dimensional system. Three atoms labelled i, j and k give

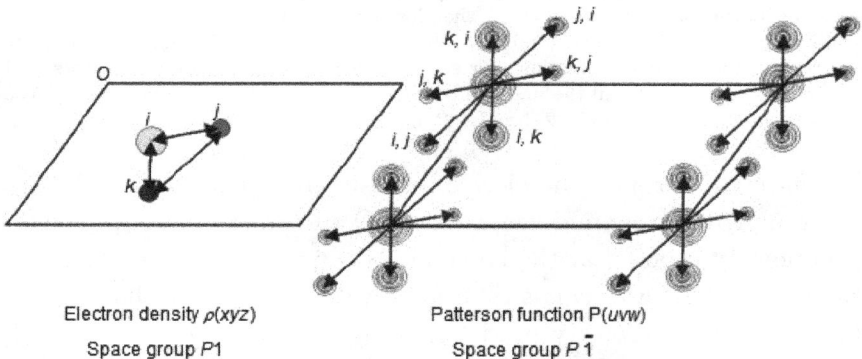

Electron density $\rho(xyz)$ Patterson function $P(uvw)$

Space group $P1$ Space group $P\bar{1}$

Fig. 6.8.   Three atoms i, j and k forming a non-centrosymmetric structure with electron density $\rho(xyz)$ and space group $P1$ give rises to 3^2 peaks in the Patterson function $P(uvw)$ with the centrosymmetric space group $P\bar{1}$; three of these nine peaks are self-vectors superimposed at the origin.

rise to nine vectors in Patterson space, three of which, the 'self-vectors', coincide at the origin. In general, a crystal structure with N atoms in the unit cell will give rise to $(N^2 - N)$ non-origin peaks in its Patterson function.

6.6.2. Patterson maxima

In order to show how the Patterson symmetry and the number of interatomic maxima arise for any given space group, it is necessary to determine all pairs of symmetry-related atoms in the unit cell; this exercise will be carried out for space group $P2_1/c$, shown in Fig. 4.17. The general equivalent positions and the vectors derived from them are tabulated hereunder; for convenience, a value of $-1/2$ has been changed to the crystallographically equivalent value of $1/2$:

General equivalent positions: (1) x, y, z (2) $\bar{x}, \bar{y}, \bar{z}$ (3) $\bar{x}, 1/2 + y, 1/2 - z$ (4) $x, 1/2 - y, 1/2 + z$

Vectors $u, v, w = f(x, y, z)$ between atoms i and j

i, j	1, 1	1,2	1,3	1,4
u, v, w	0,0,0	$2\bar{x}, 2\bar{y}, 2\bar{z}$	$2\bar{x}, 1/2, 1/2 - 2z$	$0, 1/2 - 2y, 1/2$
i, j	2,1	2,2	2,3	2,4
u, v, w	$2x, 2y, 2z$	0,0,0	$0, 1/2 + 2y, 1/2$	$2x, 1/2, 1/2 + 2z$
i, j	3,1	3,2	3,3	3,4
u, v, w	$2x, 1/2, 1/2 + 2z$	$0, 1/2 - 2y, 1/2$	0,0,0	$2x, 2\bar{y}, 2z$
i, j	4,1	4,2	4,3	4,4
u, v, w	$0, 1/2 + 2y, 1/2$	$2\bar{x}, 1/2, 1/2 - 2z$	$2\bar{x}, 2y, 2\bar{z}$	0,0,0

It is evident from this tabulation that certain peaks have twice the weight of others; they can be separated as follows:

Single weight: $\quad 2x, 2y, 2z; \qquad 2\bar{x}, 2\bar{y}, 2\bar{z}; \qquad 2x, 2\bar{y}, 2z; \qquad 2\bar{x}, 2y, 2\bar{z}$

Double weight: $2x, 1/2, 1/2 + 2z; \ 2\bar{x}, 1/2, 1/2 - 2z; \ 0, 1/2 + 2y, 1/2;$
$\qquad \qquad \quad 0, 1/2 - 2y, 1/2 \quad 0, 1/2 + 2y, 1/2; \quad 0, 1/2 - 2y, 1/2$

It should be noted also that, although the peaks of both weights refer to vectors between symmetry-related atoms, the double weight peaks occur in particular regions of space, namely, planes such as $2x, 1/2, 1/2 + 2z$ or lines such as $0, 1/2 + 2y, 1/2$, as was shown first by Harker [11]; they are termed *Harker sections* and *Harker lines*, respectively.

The height of the Patterson origin peak for a given crystal on the scale of the observed data follows from Eq. (6.24) as

$$P(000) = (2/V) \sum_h \sum_k \sum_l |\mathbf{F}_{\text{obs}}(hkl)|^2 \qquad (6.26)$$

This equation expresses the superposition of all products $\rho(\mathbf{r})\rho(-\mathbf{r})$ in the unit cell, and since electron density is proportional to atomic number, it follows that $P(000) \propto \sum_{j=1}^{N} Z_j^2$. If the height of the origin peak on a scaled Patterson map is H_0, then the height of a single weight peak between two atoms A and B is given approximately by

$$H_{A,B} \approx \frac{H_0 Z_A Z_B}{\sum_{j=1}^{N} Z_j^2} \qquad (6.27)$$

It is approximate because (a) the space occupied by N atoms in electron density space contains N^2 atoms in the Patterson space, and (b) in a molecule, several atoms pairs not related by symmetry could lie in approximately the same direction; so, Eq. (6.27) is only a guide.

6.6.3.   Sharpened Patterson function

A certain degree of compensation for the breadth of Patterson peaks is given by the use of *sharpened coefficients*. The $|\mathbf{F}_{\text{obs}}(hkl)|^2$ coefficients may be corrected for effects of thermal vibrations by approximating the atoms to point scattering species by using coefficients in the form

$$^s|\mathbf{F}_{\text{obs}}(hkl)|^2 = \frac{|\mathbf{F}_{\text{obs}}(hkl)|^2}{\left(\sum_j f_j^2\right) \exp[-2B(\sin^2\theta)\lambda^2]} \qquad (6.27a)$$

where $^s|\mathbf{F}_{\text{obs}}(hkl)|^2$ is a sharpened coefficient and B is an overall isotropic *temperature factor*, which modifies effectively the value of the atomic scattering factor; it will be discussed more fully in

Section 7.2.3. Other modifications of the sharpening expression have been reported. A useful alternative to Eq. (6.27a) employs the coefficients $|\mathbf{E}(hkl)|^2 - 1$ (Section 5.6.6), which has the additional property of eliminating the origin peak that often obscures small but important vectors. A sharpening process increases relatively the weight of the higher order intensities, which can lead to the introduction of spurious maxima by terminating the Fourier summation with moderately large coefficients; a compromise procedure includes the normalized structure factor $|\mathbf{E}|$, by using the coefficients $|\mathbf{E}(hkl)||\mathbf{F}_{obs}(hkl)|$ in place of $|\mathbf{F}_{obs}(hkl)|^2$ as coefficients in the Patterson summation.

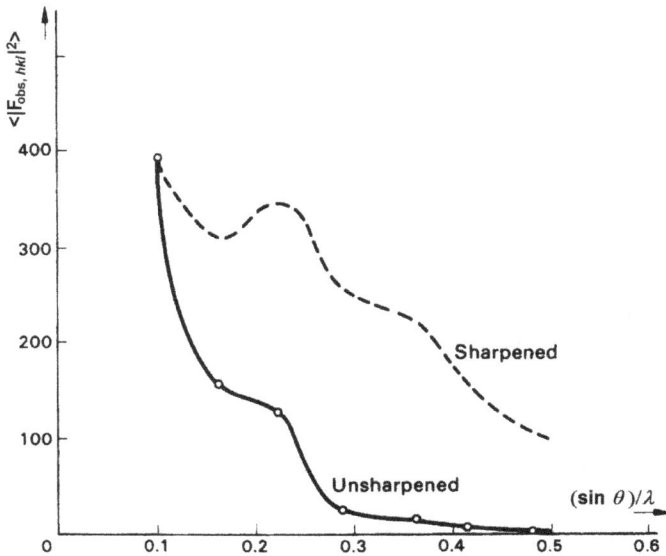

Fig. 6.9. The plot of $\langle|\mathbf{F}_{obs}(hkl)|^2\rangle$ vs. $(\sin\theta)/\lambda$ for euphenyl iodoacetate, showing enhancements in the average value of $|\mathbf{F}_{obs}(hkl)|^2$ that arise from a coincidence of interatomic vectors in the molecule. [*Structure Determination by X-ray Crystallography*, Mark Ladd and Rex Palmer, 5th edn. (2013). Reproduced by permission of Springer Science + Business Media, NY.]

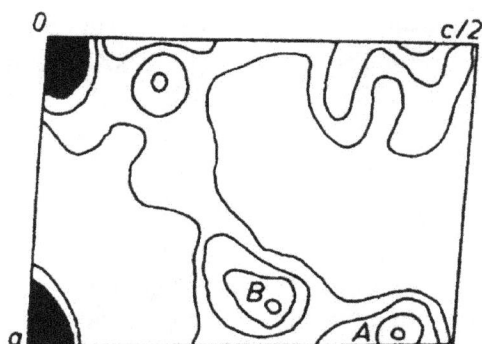

Fig. 6.10. Asymmetric unit of the Patterson projection $P(uw)$ for euphenyl iodoacetate. For space group $P2_1$, one heavy peak representing the I–I vector would be expected in the asymmetric unit, whereas two such peaks A and B were found. [*Structure Determination by X-ray Crystallography*, Mark Ladd and Rex Palmer, 5th edn. (2013). Reproduced by permission of Springer Science + Business Media, NY.]

Example 6.2. The curve $\langle|\mathbf{F}_{obs}(hkl)|^2\rangle$ vs. $(\sin\theta)\lambda$ for the crystal structure of euphenyl iodoacetate $C_{32}H_{53}O_2I$, space group $P2_1, Z_c = 2$

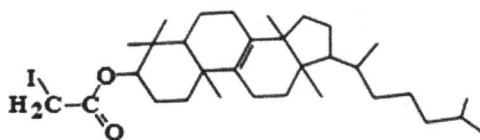

Euphenyl iodoacetate

is shown in Fig. 6.9. The enhancement shown by the unsharpened curve corresponds approximately to the width of a 6-membered ring and sharpening further increases this portion of the curve. Distances that are characteristic of a 6-membered ring are 2.4–2.88 Å and they give rise to an enhancement in the region 0.18–0.22 Å^{-1} in $(\sin\theta)\lambda$.

Fig. 6.11. The sharpened Patterson-Harker section $(u\, 1/2\, w)$ for euphenyl iodoacetate. Peaks A and B are still of similar weight, as judged by the number of contours. [*Structure Determination by X-ray Crystallography*, Mark Ladd and Rex Palmer, 5th edn. (2013). Reproduced by permission of Springer Science + Business Media, NY.]

Example 6.2. (*Continued*) In the Patterson projection $P(uw)$ in Fig. 6.10, two heavy peaks A and B of approximately equal weight arise where only one was expected; the large origin peak has not been contoured. The Harker section $(u\, 1/2\, w)$ shown in Fig. 6.11 has resolved more peaks, but leaves peaks A and B with approximately equal heights of contours.

The method of *Patterson selection* was devised [12] as a Patterson function, but with the exclusion of those $|\mathbf{F}_{\text{obs}}|^2$ data responsible for the local enhancement shown by the graph in Fig. 6.9. The result of this procedure is shown by the Harker section $(u, 1/2, w)$ in Fig. 6.12, which now reveals peak B as the I–I vector.

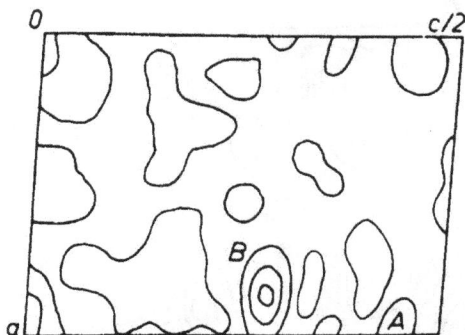

Fig. 6.12. The Harker section of Fig. 6.11, but computed without those data corresponding to the region of enhancement of $\langle|\mathbf{F}_{obs}(hkl)|^2\rangle$ shown in Fig. 6.9. Peak B is now revealed clearly as that of the I–I vector; peak A is much reduced in height. This result was confirmed by the structure analysis. [*Structure Determination by X-ray Crystallography*, Mark Ladd and Rex Palmer, 5th edn. (2013). Reproduced by permission of Springer Science + Business Media, NY.]

6.6.4.   Selected Patterson function

In a large molecule, it is to be expected that a number of interatomic vectors may lie in approximately the same direction, and they are additive on a Patterson map. Their presence may be revealed by a local enhancement of a curve of $\langle|\mathbf{F}_{obs}(hkl)|^2\rangle$ vs. $(\sin\theta)/\lambda$, as shown in Fig. 6.9, and illustrated by Example 6.2.

6.6.5.   Pseudosymmetry

The coordinates of the I–I peak in the Patterson function of Fig. 6.12 are $2x, \frac{1}{2}, 2z$, from which the x and z coordinates of the iodine atom follow immediately. In this space group, the origin along y is not determined uniquely, so that the y coordinate of the iodine atom can be given any value between 0 and 1. Since the trial structure based on just two iodine atom will exhibit centrosymmetry, it is convenient to choose the y coordinate of iodine as $\frac{1}{4}$. The coordinates of the iodine atoms can be used to generate a trial set of phases. However, in space group $P2_1$ for example, a structure containing two iodine atoms per unit cell related by inversion symmetry is compatible with the implied space group $P\frac{2_1}{m}$ so that the phases will be either 0 or π, given that the centre of symmetry is at the origin; the trial structure based on these phases leads to two molecules in the asymmetric unit. The correct choice may be made then by selecting the set of peaks that creates an acceptable chemical structural image together with other criteria.

6.6.5.1. *R-factor*

One criterion of correctness is based on the agreement between the observed and the corresponding calculated structure amplitudes. The conventional and weighted R-factors are given, respectively, by

$$R = \sum_{h,k,l} \frac{||\mathbf{F}_{obs}| - |\mathbf{F}_{calc}||}{|\mathbf{F}_{obs}|}$$

and

$$R_w = \sqrt{\left\{ \sum_{h,k,l} \frac{w(\mathbf{h})(|\mathbf{F}_{obs,h}|^2 - |\mathbf{F}_{calc,h}|^2)}{w(\mathbf{h})|\mathbf{F}_{obs}|^2} \right\}} \qquad (6.28)$$

where $w(\mathbf{h})$ is a weighting factor for the reflection $\mathbf{h}$ that is based on counting statistics. Once atoms other than iodine have been located, the pseudosymmetry is broken and the structure determination can be completed. The proposed structure is now subjected to refinement, a process that will be considered shortly.

Another source of pseudosymmetry, albeit inexact, can occur in the crystallography of natural product chemistry. For example, consider the crystal structure of an organic compound that crystallizes in space group $P2_1$ with $Z_c = 4$, that is, with two molecules in the asymmetric unit. These two molecules may pack in an arrangement that approximates to a two-fold symmetry relationship between them, which then has an effect of the observed intensity data although not in a systematic manner.

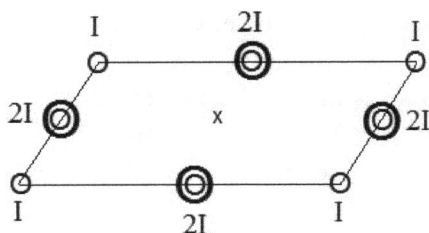

Fig. 6.13. Arrangement of Patterson vector peaks around the unit-cell origin (marked X) from a structure with two heavy atoms, not related by symmetry, in the asymmetric unit: single weight peaks are marked I and the heavier, double weight peaks are marked 2I. Note that the double weight peaks occur midway along the vectors joining the single weight peaks.

6.6.6.   Multiple heavy-atom structures

An example of a structure with more than one atom in the asymmetric unit occurred in the crystal structure analysis of diosgenin iodoacetate $C_{29}H_{43}O_4I$, space group $P2_1$, $Z_c = 4$ [13]. In this type of structure, the heavy atoms form a parallelogram of vectors around the origin of the unit cell and comprise 4 single- and 4 double-weight peaks, making up the expected $(4^2 - 4)$ non-origin peaks. The geometry is typified by the arrangement in Fig. 6.13, and is relatively easy to identify. Furthermore, phasing on two atoms unrelated by symmetry does not lead to pseudosymmetry in the electron density.

6.6.7.   Feasibility of a heavy-atom procedure

The expectation of a successful phasing by the heavy-atom method can be envisaged readily in the case of a centrosymmetric structure. Given a number of phases based on the heavy atom (H) that are either 0 or π, it may be assumed that *ca.* 50% of them are correct for the whole structure. The phase contribution from those reflections forming the rest of the data set (R) can be, again, 0 or π. For a given reflection, the phases from H and R may be expected to agree for half of them. Thus, it is reasonable to expect that a major proportion of the phases are correct. Then, subsequent cycles of structure factor and electron density calculations, with or without intermediate least-squares adjustments to the parameters, adding more and more atoms, would correct the phases that are in error and ultimately present a good trial structure for the final refinement process.

6.6.8. Patterson search technique

Another technique for solving the Patterson function involves choosing a known fragment of the molecule, built according to standard geometry, from which a set of x, y, z coordinates is obtained. These coordinates form a vector set which is moved through the unit cell of the Patterson function. Each position of the moving fragment is classified according to a measure-of-fit parameter, and a chosen number of the best result examined by structure factor and electron density calculations.

An example of this technique is provided by the crystal structure analysis of atropine $C_{17}H_{23}NO_3$:

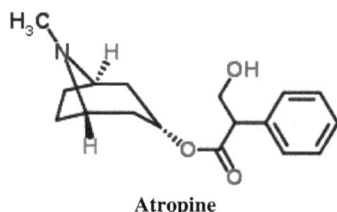

Atropine

The search entity comprised the phenyl ring and five other atoms, forming two rigid groups with a torsional linkage between them. The fragment was subjected to movement through the unit cell together with torsional adjustments between the two rigid groups. The analysis listed 20 sets of coordinates among which three sets indicated good fits. In fact, the best fit led to a successful analysis with a final R-factor of 4.5% [14].

6.6.9. Difference Fourier electron density summation

In cases where difficulty arises in locating all atoms correctly, the technique of difference Fourier synthesis can be helpful. At some stage in a crystal structure analysis, possible errors may exist owing to missing atoms, atoms incorrectly placed or, possibly, incorrect temperature factors. In such cases, a difference Fourier synthesis can be revealing. The technique uses coefficients $|\mathbf{F}_{obs}| - |\mathbf{F}_{calc}|$ in the Fourier summation. It is effectively a subtraction of the trial structure from a structure based on the known atoms, since the phases would normally be substantially correct at this stage; symbolically, it may be written as $\rho_{obs} - \rho_{calc}$.

Some of the important features of this synthesis are as follows:

- Incorrectly placed atoms correspond to high density in ρ_{calc} and low density in ρ_{obs}; $\Delta\rho$ is negative.
- Too small an atomic number or too high a temperature factor gives a positive $\Delta\rho$.
- A small error in position causes an atom to lie in a negative area near to positive unoccupied site.
- Light atoms, particularly hydrogen, may be located on a difference map provided that the measured $|\mathbf{F}_{obs}|$ data are of a sufficiently high quality.
- The difference map of a well-refined structure should appear almost featureless.

6.7. Example of a Complete Structure Analysis by the Heavy Atom-Method

Crystals of 2-bromobenzo[*b*]indeno[1,2-*e*]pyran, $M_r = 297.16$, crystallize in space group $P\frac{2_1}{c}$ with a unit cell of dimensions $a = 7.51$, $b = 5.96$, $c = 26.2$ Å, $\beta = 92.5°$; a drawing of this space group is given in Fig. 4.14. The density of $1.68\,\mathrm{g\,cm^{-3}}$ showed that $Z_c = 4$, which means that the molecules occupy general equivalent positions in the space group.

A convincing Wilson plot, shown in Fig. 6.14, gave 4.1 Å^2 for the overall isotropic temperature factor and 0.41 as the scale factor for $|\mathbf{F}_{obs}(hkl)|^2$.

The Harker section $(u\ 1/2\ w)$, shown in Fig. 6.15 for the asymmetric unit a_c, reveals a large peak which is the double weight vector

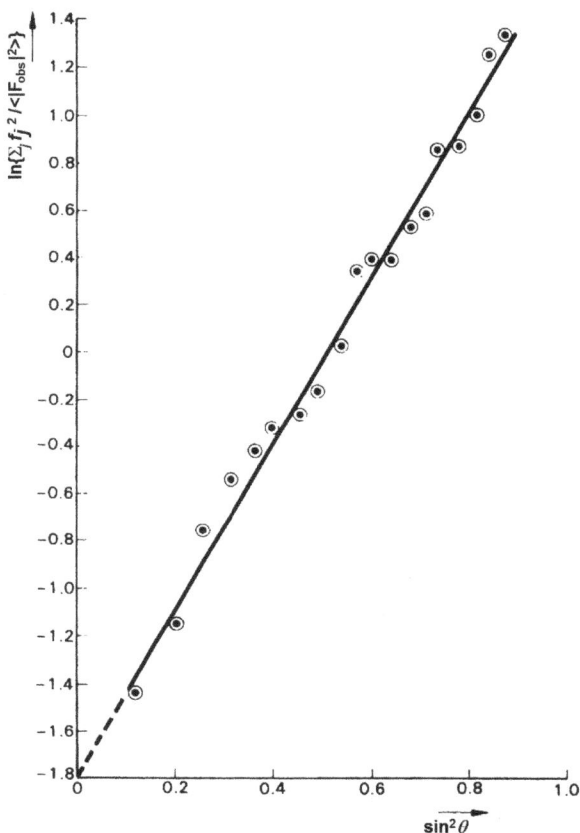

Fig. 6.14. The Wilson plot for 2-bromobenzo[b]indeno[1,2-e]pyran (BBIP) showing very satisfactory linearity; $B = 4.1\,\text{Å}^2$ and K (for $|\mathbf{F}_{\text{obs}}|^2$) = 0.41.

$(u\ ^1/_2\ w) = 2x, ^1/_2,\ 1 + 2z$, giving $x \approx 0.25$ and $z \approx 0.015$ for the bromine atom.

The Harker line in Fig. 6.16 shows the double weight vector $(0\ v\ ^1/_2) = 0\ ^1/_2 - 2y\ ^1/_2$ which leads to a y coordinate for bromine of *ca.* 0.19. The clustering of vector peaks parallel to the w-axis indicates that the molecules lie closely parallel to c in the unit cell.

Phases (signs) were determined based on the positions of the heavy bromine atoms and structure factors were calculated. An electron density map using the $|\mathbf{F}_{\text{obs}}|$ data together with the signs given

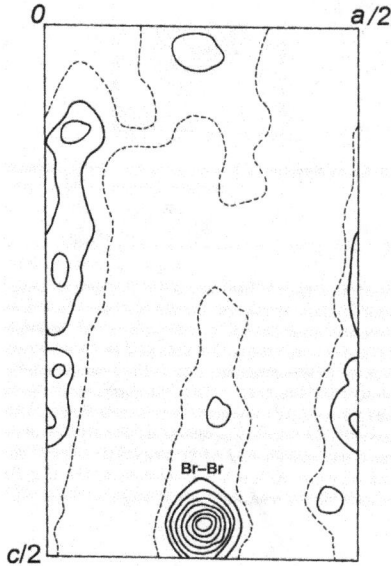

Fig. 6.15.   An asymmetric unit of the Harker section $(u\,1/2\,w)$ for BBIP; the double weight vector is clearly resolved and gives the values x_{Br} *ca.* 0.25, z_{Br} *ca.* 0.015. The final refined values are $x = 0.2398$, $z = 0.01520$.

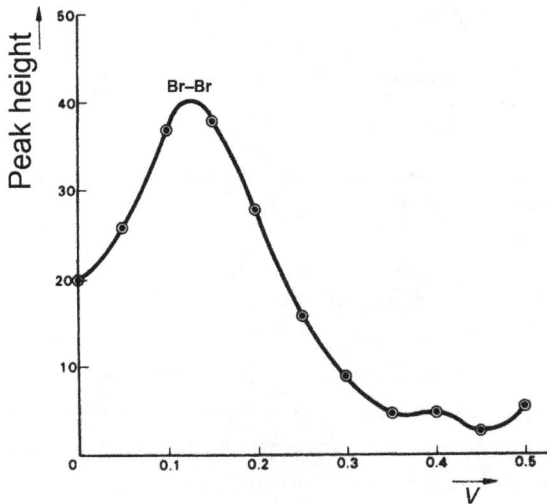

Fig. 6.16.   Asymmetric unit of the Harker line $(0\,1/2 - 2y\,1/2)$ for BBIP; the Br–Br vector gives y_{Br} *ca.* 0.19. The final refined value is $y = 0.1848$.

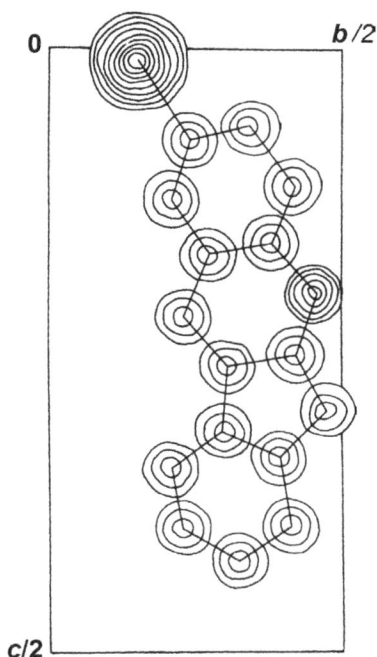

Fig. 6.17. Final electron density map for BBIP as seen along a; all atoms of the molecule, except hydrogen, are well resolved.

by the $|\mathbf{F}_{calc}|$ calculation enabled the remaining atoms of the molecule (16 carbon atoms and the oxygen atom) to be positioned in the structure. Refinement was carried out with anisotropic temperature factors. Hydrogen atom positions were determined from a difference electron density map. They were included in the structure factor calculations but were not refined. Figure 6.17 illustrates the final composite electron density map for the structure, excluding the hydrogen atoms, as seen along the direction of the short a unit-cell dimension.

The molecular structure is planar with a mean esd of 0.02 Å in departure from planarity. The structure analysis was refined to a final R-factor of 5.1%.

It is suggested that, at this stage, it could be beneficial for the reader to practice phase determination by the Patterson method, using the program XRAY in the Program Suite with the data sets NIOP, CL1P or CL2P, for example.

6.8.   Direct Methods of Phase Determination

The second of the three methods of solving crystal structures chosen here depends upon probability relationships between the observed structure amplitudes. Since the correct phases for X-ray reflections are ultimately extracted from the $|\mathbf{F}_{obs}|$ data, it is reasonable to assume that phase data information is encoded therein. This argument led to probability methods for determining phases independently of chemical information, the so-called 'direct methods' technique. The relationships involved in this method use normalized structure factors, the $|\mathbf{E}|$-values that are defined by Eq. (5.74) and include, therefore, scaling of the $|\mathbf{F}_{obs}|$ data and temperature correction of the atomic scattering factors. In particular, careful attention must be given to the correct choice of the ε-factor. The statistics of E-values have been discussed in Section 5.6.6.

6.8.1.   Fixing the origin: Structure invariants
and structure seminvariants

Direct phasing assumes some reflections with known phases, and a centrosymmetric structure will be used first as an example. It has been shown that it is advantageous to choose the origin of the unit cell at a centre of symmetry. In a centrosymmetric structure, there are eight possible positions for the centre of symmetry, and they may be classified according to the parity of h, k and l, as shown in Table 6.1.

In a centrosymmetric structure with $\bar{1}$ at the origin, the phase is either 0 or π. It is customary to refer to the *sign* s of a centric reflection as either $+$ or $-$, whereupon $s(hkl) = \frac{F(hkl)}{|F(hkl)|}$. Thus, it is usual, in a centrosymmetric structure, to write $F(hkl)$ for $\mathbf{F}(hkl)$ with phase $0, s(hkl) = 1$, and $-F(hkl)$ for $\mathbf{F}(hkl)$ with phase π, $s(hkl) = -1$.

Table 6.1.   Effect of a change in origin on the sign of a centrosymmetric structure factor.

	Parity group							
	1	2	3	4	5	6	7	8
	h even	h odd	h even	h even	h even	h odd	h odd	h odd
	k even	k even	k odd	k even	k odd	k even	k odd	k odd
	l even	l even	l even	l odd	l odd	l odd	l even	l odd
Origin ($\bar{1}$) coordinates	Structure factor sign							
0, 0, 0	+	+	+	+	+	+	+	+
$1/2$, 0, 0	+	−	+	+	+	−	−	−
0, $1/2$, 0	+	+	−	+	−	+	−	−
0, 0, $1/2$	+	+	+	−	−	−	+	−
0, $1/2$, $1/2$	+	+	−	−	+	−	−	+
$1/2$, 0, $1/2$	+	−	+	−	−	+	−	+
$1/2$, $1/2$, 0	+	−	−	+	−	−	+	+
$1/2$, $1/2$, $1/2$	+	−	−	−	+	+	+	−

In a primitive space group, signs may be allocated to three reflections in order to fix the origin, subject to certain rules. These signs form a *starting set* for phasing the reflections, and from which additional signs can be determined.

From Section 5.4.4,

$$\mathbf{F}(hkl)_{000} = \sum_j g_j \cos 2\pi(hx_j + ky_j + lz_j) \qquad (6.29)$$

where the subscript 000 indicates $\bar{1}$ at the point 0, 0, 0. If the centre of symmetry is moved to another point, say, $1/2$, 0, $1/2$, then by the change of origin procedure (Section 4.6),

$$\mathbf{F}(hkl)_{1/2\,0\,1/2} = \sum_j g_j \cos\left[2\pi(hx_j + ky_j + lz_j) - \frac{(h+l)}{2}\right] \qquad (6.30)$$

Expanding, remembering that $\sin[\pi(h+l)] = 0$ and $\cos[\pi(h+l)] = (-1)^{(h+l)}$,

$$\mathbf{F}(hkl)_{1/2\,0\,1/2} = (-1)^{h+l}\mathbf{F}(hkl)_{000} \qquad (6.31)$$

Thus, the amplitude $|\mathbf{F}(hkl)|$ is, not surprisingly, invariant under change of origin, and is known as a *structure invariant*. The corresponding sign $s(hkl)$ may change according to the parity of h, k and l, as shown by Table 6.1.

The origin position as 0, 0, 0 will be chosen, and a working set of $|\mathbf{E}|$-values with $|\mathbf{E}|$ in the range 1.5–4.0 is selected. In choosing the starting set, reflections of large $|\mathbf{E}|$-values are required as they will provide continuing phase development with the best probability of correctness but, in addition, the parity must be correct. The relationships of parity, using 'e' for even indices and 'o' for odd indices, are given by

$$e + e = o + o = e \tag{6.32}$$

$$e + o = o \tag{6.33}$$

Example 6.3. How would a starting set be chosen from among the following $|\mathbf{E}|$-values?

hkl	$6\bar{1}\bar{7}$	426	203	$8\bar{1}\,\bar{4}$	705
$|\mathbf{E}|$	3.1	2.7	2.4	2.3	2.2

The three reflections used to fix the origin in a structure with a primitive space group must be linearly independent, that is, the indices must not sum to give e e e. Reflections that are e e e do not change phase with change of the permitted origin; they are termed *structure seminvariants* and cannot be used to fix the origin, as Table 6.1 shows.

Reflection $6\bar{1}\bar{7}$ is a good choice (large $|\mathbf{E}|$ value) and is parity group 5, e o o. By giving it a positive sign, the origin is restricted to one of $0, 0, 0$; $1/2, 0, 0$; $0, 1/2, 1/2$; $1/2, 1/2, 1/2$. The next largest $|\mathbf{E}|$ is 426 but as it is a structure seminvariant, it cannot be used to restrict the origin.

Example 6.3. (*Continued*) Reflection 203 of parity e e o is a satisfactory second choice, and the origin is now restricted to 0, 0, 0 and $1/2$, 0, 0. Reflection $8\bar{1}\bar{4}$, parity e o e, cannot be used as the third reflection as it is not linearly independent of the first two: e o o + e e o = e o e.

Reflection 705, parity o e o, is satisfactory, and by choosing these three reflections with + signs, the origin is specified as 0, 0, 0.

6.9. Sayre's Equation

A general formula for equal-atom centrosymmetric structures was derived by Sayre in 1952 in the form [16]

$$s(hkl)s(h'k'l')s(h - h', k - k', l - l') \approx +1 \qquad (6.34)$$

where the sign $\approx$ means 'is probably equal to'. In Eq. (6.22) for a centrosymmetric structure, $A'(hkl) = \pm F_{obs}(hkl)$. A positive sign for $F_{obs}(hkl)$ implies that the (hkl) plane passes through the origin, like the full lines shown in Fig. 6.18, whereas a negative sign corresponds to the interleaving dashed lines in that figure. The diagram shows that triple intersections occur for three full lines (+ + +) or one full line and two dashed line.

The three planes in this *triple-product relationship* (TPR) will be written more simply as the vectors **h**, **k** and **h** − **k**; an example of these vectors is shown in Fig. 6.19 on a portion of the $h0l$ reciprocal lattice.

6.9.1. An early example of direct phase determination

An early, apparently inadvertent, application of direct phasing occurred in the determination of the crystal structure of hexamethyl benzene, space group $P\bar{1}$, with one molecule in the unit cell lying on a centre of symmetry.

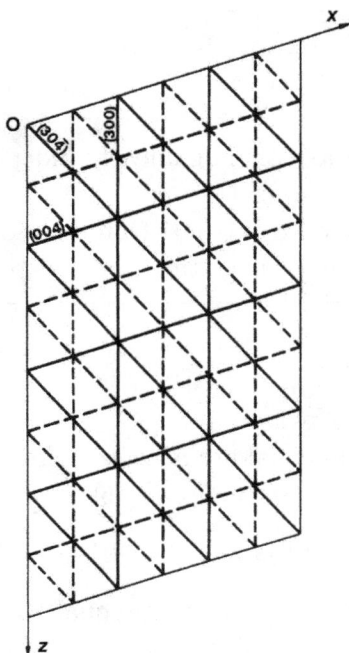

Fig. 6.18.   Physical implication of the triple-product relationship given by Eq. (6.34). Three families of planes are shown: {300}, {004} and {30$\bar{4}$}. The relationship $s(300)\, s(004)\, s(30\bar{4}) = +1$ is an expression of the Sayre equation.

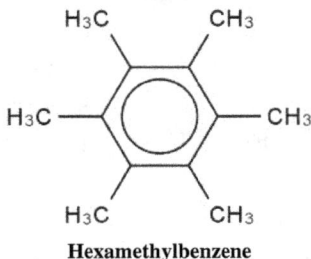

Hexamethylbenzene

The unit-cell dimensions indicated that the molecule would be well resolved in the x, y projection. The structure of hexamethyl-benzene was determined in 1929 from the construction of a figure such as Fig. 6.19 with the three strong, high-order reflections $7\bar{3}0(\mathbf{h})$, $340(\mathbf{k})$ and $4\bar{7}0(\mathbf{h} - \mathbf{k})$. Three carbon atoms were placed at the intersections of these planes, while the remaining atoms were

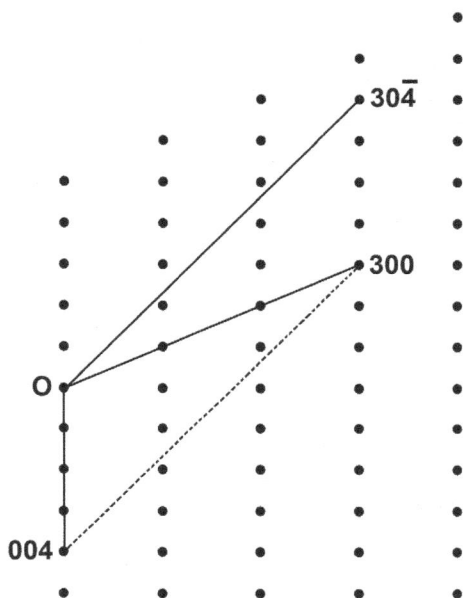

Fig. 6.19. The three vectors **300**, **004** and **30$\overline{4}$** from Fig. 6.18 shown on a portion of the *h0l* reciprocal lattice.

located by symmetry. The three chosen reflections can be seen now to have satisfied the Sayre equation. The structure and geometry of the benzene ring were thus confirmed by this structure analysis [17].

6.10. The Σ_2 Formula of Hauptman and Karle

An extended form of Eq. (6.34) is given in terms of |**E**|-values, for which the effect of the decrease in atomic scattering factor with increasing θ is reduced; |**E**|-values simulate structure factors for point atoms. The relation is given in the **h**, **k** notation by [18]

$$s(\mathbf{h}) \approx s \left\{ \sum_{\mathbf{k}} |\mathbf{E}(\mathbf{k})||\mathbf{E}(\mathbf{h-k})| \right\} \tag{6.35}$$

which is the Hauptman and Karle Σ_2 formula for centrosymmetric crystals. The development of this equation carried with it a measure of the probability that $s(\mathbf{h})$ is positive, given in the form

$$P(hkl)_+ = \tfrac{1}{2} + \tfrac{1}{2}\ \tanh[(\sigma_3/\sigma_2^{3/2})\alpha'] \tag{6.36}$$

where α' is given by

$$\alpha' = |\mathbf{E}(\mathbf{h})| \sum_{\mathbf{k}} |\mathbf{E}(\mathbf{k})||\mathbf{E}(\mathbf{h\text{-}k})| \tag{6.37}$$

and σ_n by

$$\sigma_n = \sum_j Z_j^n \tag{6.38}$$

where Z_j is the atomic number of the jth atom in the structure. If the atoms are identical, or nearly so, then

$$\sigma_3/\sigma_2^{3/2} = \frac{1}{\sqrt{N}} \tag{6.39}$$

Example 6.4. Pursuing Eq. (6.36), consider the following data relating to reflection $\mathbf{h} = 400$, with $|\mathbf{E}(\mathbf{h})| = 2.13$, in a hypothetical centrosymmetric structure in which the signs of 4 reflections are known. There are 64 equal atoms of atomic number 7 in the unit cell, so that $(\sigma_3/\sigma_2^{3/2}) = (N)^{-1/2}$.

$\mathbf{k}$	$|\mathbf{E}(\mathbf{k})|$	$\mathbf{h-k}$	$|\mathbf{E}(\mathbf{h-k})|$	$|\mathbf{E}(\mathbf{h})|\,\mathbf{E}(\mathbf{h-k})|$	$s(\mathbf{h})$
$10\bar{4}$	2.05	304	-1.85	-3.793	-1
004	-2.03	$40\bar{4}$	1.92	-3.898	-1
106	-1.85	$30\bar{6}$	1.65	-3.053	-1
$20\bar{8}$	1.10	208	0.94	$+1.034$	$+1$

$\alpha' = |\mathbf{E}(\mathbf{h})| \times \sum |\mathbf{E}(\mathbf{k})||\mathbf{E}(\mathbf{h-k})| = 2.13 \times (-9.710) = -20.682$ and $P(\mathbf{h})_+ = \tfrac{1}{2} + \tfrac{1}{2}\tanh[0.125 \times (-20.682)] = 5.65 \times 10^{-3}$ or 0.56%, which means that $s_{400} = -1$ with a probability of 99.4%.

The probability $P(hkl)_+$ varies with the number of atoms in the unit cell; the larger the value of N, the smaller the probability

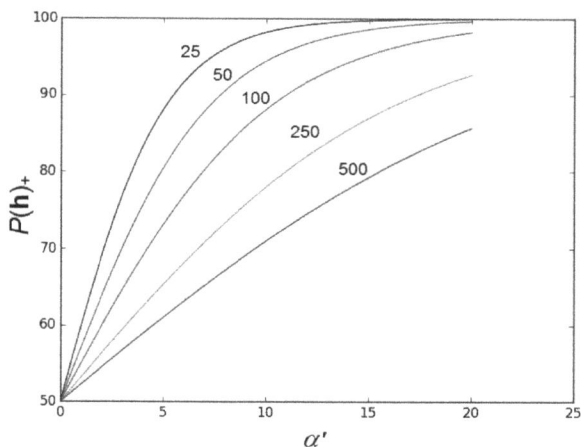

Fig. 6.20. Variation of the probability $P(\mathbf{h})_+$ with α' for varying numbers N of atoms in the unit cell: $N = 25, 50, 100, 250$ and 500.

becomes, as Fig. 6.20 shows. The graph shows the importance of using the large values of $|\mathbf{E}(\mathbf{h})|$, particularly in the early stages of sign allocation, in order to maximize the value of α'.

6.11. Using the Space Group Symmetry

The study of Problem 5.1 draws attention to relationships between the amplitudes of structure factors for differing combinations of signs of h, k and l, and similar equalities exist among their associated phases. Continuing with the examination of space group $P\frac{2_1}{c}$, relationships are found between signs that are important in direct methods of phasing. Thus, for the two parity classes $k + l$ even and odd, the structure factor equation for this space group leads to the following relationships:

$$k + l = 2n :$$
$$s(hkl) = s(\bar{h}\,\bar{k}\,\bar{l}) = s(h\bar{k}l) \neq s(\bar{h}kl); \ s(\bar{h}kl) = s(hk\bar{l})$$

$$(6.40)$$

$$k + l = 2n + 1 :$$

$$s(hkl) = s(\overline{h}\,\overline{k}\,\overline{l}) = -s(h\overline{k}l) \neq s(\overline{h}kl); \ s(\overline{h}kl) = -s(hk\overline{l})$$
$$(6.41)$$

These equations enable the introduction of negative signs at an early stage; the starting set comprises three positive signs, as has been shown.

In non-centrosymmetric space groups, the changes in magnitude and sign must be considered for both the A' and the B' components. From a study of Problem 5.2, the following geometrical structure factors were deduced for space group $Pma2$.

$h = 2n$

$$A(hkl) = 4\cos 2\pi hx \cos 2\pi ky \cos 2\pi lz \qquad (6.42)$$

$$B(hkl) = 4\cos 2\pi hx \cos 2\pi ky \cos 2\pi lz \qquad (6.43)$$

$h = 2n + 1$

$$A(hkl) = -4\sin 2\pi hx \sin 2\pi ky \sin 2\pi lz \qquad (6.44)$$

$$B(hkl) = -4\sin 2\pi hx \sin 2\pi ky \sin 2\pi lz \qquad (6.45)$$

The amplitude relationships are independent of parity:

$$|\mathbf{F}(hkl)| = |\mathbf{F}(\overline{h}\,\overline{k}\,\overline{l})| = |\mathbf{F}(\overline{h}kl)| = |\mathbf{F}(h\overline{k}l)| = |\mathbf{F}(hk\overline{l})| \qquad (6.46)$$

but the pattern of the phase changes varies with the parity:

$h = 2n$

$$\phi(hkl) = -\phi(\overline{h}\,\overline{k}\,\overline{l}) = \phi(\overline{h}kl) = \phi(h\overline{k}l) = -\phi(hk\overline{l}) \qquad (6.47)$$

$h = 2n + 1$

$$\phi(hkl) = -\phi(\overline{h}\,\overline{k}\,\overline{l}) = \pi + \phi(\overline{h}kl) = \pi + \phi(h\overline{k}l) = -\phi(hk\overline{l}) \qquad (6.48)$$

The symmetry relationships of the amplitudes $|\mathbf{F}(hkl)|$ and of the phases $\phi(hkl)$ can be found for all space groups in the International Tables [6, 7].

6.12. Direct Phasing in Centrosymmetric Crystals

The first step in direct phasing is the conversion of $|\mathbf{F}_{obs}|$ data to the corresponding $|\mathbf{E}|$-values, according to Eq. (5.74), and selecting a subset with $|\mathbf{E}|$ greater than a certain value, often chosen as 1.5, and associated with high probability. A Σ_2 listing is prepared, such as that shown for $|\mathbf{E}(\mathbf{h})|$ in Example 6.4, for all data in the subset.

In the absence of translational symmetry, a TPR will generate only positive signs, and in order to propagate the sign determination, one procedure involves allocating sign symbols (a, b, c) to reflections. Then, further data can be phased in terms of these symbols. Generally, 3–5 such symbols suffice, and usually the completed data set will show relationships between the signs. For example, the number of unknown signs may decrease finally to just 2, a and b. At this stage, electron density maps are calculated with the signed $|\mathbf{E}|$-values. The use of $|\mathbf{E}|$-values as Fourier coefficients introduces a useful degree of sharpness into the synthesis, but may lead also to some small spurious peaks which can be eliminated often on chemical grounds.

For two sign symbols, there would be four possible $|\mathbf{E}|$-maps. However, the synthesis with $a = b = +1$ would be rejected since it would imply a large peak at the origin. If this situation were true, then the structure could probably have been solved from the Patterson function. Thus, there will be three maps to compute and examine: $a = 1$, $c = -1$; $a = -1$, $c = 1$; $a = c = -1$. From these three maps, it is highly likely that a chemically feasible structure can be postulated and an electron density map calculated with the complete $|\mathbf{F}_{obs}|$ data set. Then, missing atoms (if any) can be located and the structure refined. In the presence of translational symmetry, the process is more straightforward because negative signs arise in the early stage of the sign determination process.

6.13. Direct Phasing in Non-centrosymmetric Crystals

Phase determination for non-centrosymmetric structures is not as straightforward as the centrosymmetric case just studied. While three phases may be chosen as $0°$ in order to specify the origin in a primitive unit cell, other phases may take any value between $0°$ and $360°$. An equation which is a more generalized form of Eq. (6.35) is used in most techniques for phasing non-centrosymmetric crystals. The preliminary stages of forming $|\mathbf{E}|$-values, fixing an origin and building a Σ_2 listing, are carried out as before.

Given vector triplets $\mathbf{h}$, $\mathbf{k}$ and $\mathbf{h} - \mathbf{k}$, it has been shown [19] that a structure factor $\mathbf{F}(\mathbf{h})$ may be considered to lie on a circle of radius $\mathbf{F}(\mathbf{h})$ on an Argand diagram of centre O within a small circle of radius $|\mathbf{r}|$ and centre $\delta(\mathbf{h}, \mathbf{k})$ from O, as shown in Fig. 6.21. The quantity $\delta(\mathbf{h}, \mathbf{k})$ is given by $\mathbf{F}(\mathbf{k})\mathbf{F}(\mathbf{h} - \mathbf{k})/F(000)$. The product $\mathbf{F}(\mathbf{k})\mathbf{F}(\mathbf{h} - \mathbf{k})$ may be defined from the Argand diagram as

$$|\mathbf{F}(\mathbf{k})||\mathbf{F}(\mathbf{h} - \mathbf{k})| \exp\{i[\phi(\mathbf{k}) + \phi(\mathbf{h} - \mathbf{k})]\} \qquad (6.49)$$

It has been shown that the larger the value of $|\mathbf{F}(\mathbf{k})|^2$ and $|\mathbf{F}(\mathbf{h} - \mathbf{k})|^2$, the closer $|\mathbf{F}(\mathbf{h})|$ approaches $|\delta(\mathbf{h}, \mathbf{k})|$. For a given $\mathbf{h}$, as $\mathbf{k}$ is varied, $\mathbf{F}(\mathbf{h})$ is given by

$$\mathbf{F}(\mathbf{h}) \propto \langle \mathbf{F}(\mathbf{k})\mathbf{F}(\mathbf{h} - \mathbf{k}) \rangle_\mathbf{k} \qquad (6.50)$$

where the constant of proportionality is $F(000)$ and $\langle \cdots \rangle_\mathbf{k}$ implies an average over a number of values of $\mathbf{k}$ in a TPR; this equation is a generalized form of Eq. (6.35). Since $\mathbf{F}(\mathbf{h}) = |\mathbf{F}(\mathbf{h})| \exp i\phi(\mathbf{h})$, it follows from Eq. (6.49) that the phase addition formula is given as $\phi(\mathbf{h}) \approx \phi(\mathbf{k}) + \phi(\mathbf{h} - \mathbf{k})$, which, when several triplets are involved, becomes

$$\phi(\mathbf{h}) \approx \langle \phi(\mathbf{k}) + \phi(\mathbf{h} - \mathbf{k}) \rangle_\mathbf{k} \qquad (6.51)$$

On an Argand diagram, and introducing $|\mathbf{E}|$-values, each vector $\mathbf{h}$ is linked through its TPR to the product $|\mathbf{E}(\mathbf{k})||\mathbf{E}(\mathbf{h} - \mathbf{k})$ which may be represented by real and imaginary components A and B such that

$$A_\mathbf{h} = |\mathbf{E}(\mathbf{k})||\mathbf{E}(\mathbf{h} - \mathbf{k})| \cos[\phi(\mathbf{k}) + \phi(\mathbf{h} - \mathbf{k})] \qquad (6.52)$$

and

$$B_\mathbf{h} = |\mathbf{E}(\mathbf{k})||\mathbf{E}(\mathbf{h} - \mathbf{k})| \sin[\phi(\mathbf{k}) + \phi(\mathbf{h} - \mathbf{k})] \qquad (6.53)$$

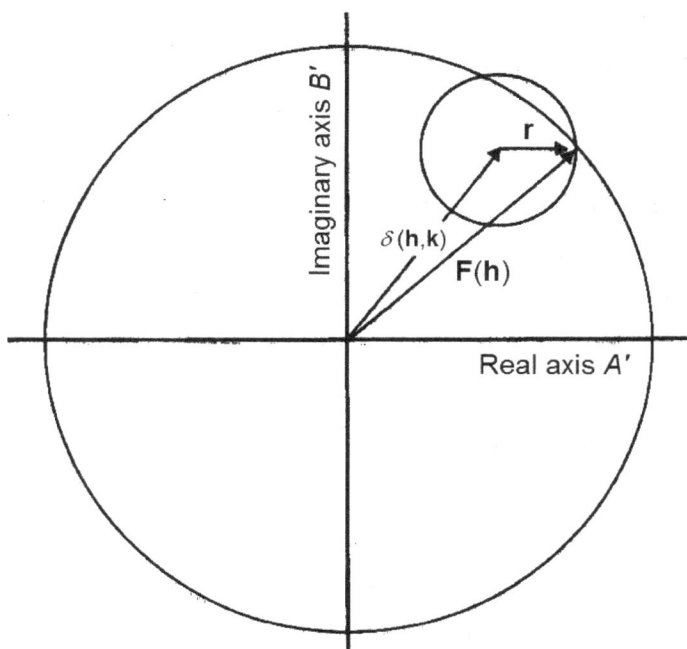

Fig. 6.21. Representation of the relationship $|\mathbf{F}(\mathbf{h})| - \delta(\mathbf{h}, \mathbf{k}) \leq |\mathbf{r}|$ for a non-centrosymmetric crystal.

The phase angle follows from the equation

$$\tan \phi(\mathbf{h}) = \frac{\mathbf{B}}{\mathbf{A}} = \frac{\sum_{\mathbf{h}} w_{\mathbf{h}} |\mathbf{E}(\mathbf{k})| |\mathbf{E}(\mathbf{h} - \mathbf{k})| \sin[\phi(\mathbf{k}) + \phi(\mathbf{h} - \mathbf{k})]}{\sum_{\mathbf{h}} w_{\mathbf{h}} |\mathbf{E}(\mathbf{k})| |\mathbf{E}(\mathbf{h} - \mathbf{k})| \cos[\phi(\mathbf{k}) + \phi(\mathbf{h} - \mathbf{k})]}$$

(6.54)

which is the *tangent formula*, and $w_{\mathbf{h}}$ is a weighting function that can be designed to downgrade the effect of the less-well determined phases. Many program systems for phasing the reflections of non-centrosymmetric structures are based on a formula similar to Eq. (6.54).

After a sufficient number of phases have been determined, ideally *ca.* eight per atom in the asymmetric unit, $|\mathbf{E}|$-maps are calculated in order to locate atoms of the structure. When this process is complete, the structure is then ready for refinement. In some cases, a partial refinement at an earlier stage may lead to the placement of missing atoms.

6.13.1. Enantiomorph definition in non-centrosymmetric crystals

Non-centrosymmetric space groups that contain no form of inversion symmetry ($\overline{R}$, including m) can exhibit enantiomorphous forms, and two arrangements can be defined that lead to the same values of $|\mathbf{F}(hkl)|$. Two such space groups that occur frequently are $P2_12_12_1$ and $P2_1$, particularly among natural and synthetic organic compounds. The crystal structure of euphenyl iodoacetate, illustrated under Example 6.2, crystallizes in the latter space group with two molecules in the unit cell. It can have two enantiomorphic arrangements related by inversion in the origin. For these two configurations, a structure factor for an hkl reflection can be represented by the structure (S) or by its inverse I. From Section 5.4.3.1,

$$\mathbf{F}(\mathbf{h})_S = A(\mathbf{h})_S + B(\mathbf{h})_S$$
$$\mathbf{F}(\mathbf{h})_I = A(\mathbf{h})_I + B(\mathbf{h})_I \tag{6.55}$$

Furthermore,

$$A(\mathbf{h})_S + A(\mathbf{h})_I$$
$$B(\mathbf{h})_S - B(\mathbf{h})_I \tag{6.56}$$

Thus, for the structure (or its inverse), B'_{h} may be chosen to be a positive quantity, whereupon the phase angle is defined by $0 \leq \phi_{\mathrm{h}} \leq \pi$. This procedure was followed in the structure determination of tubercidin, the structure determination of which has been described elsewhere [2, 25], where the enantiomorph was defined by allocating a phase of $3\pi/4$ to a general reflection.

Self-assessment 6.2. What are the relationships between (1) structure amplitudes $|\mathbf{F}(hkl)|$ and (2) the phase angles $\phi(hkl)$ in space group $P2_1$?. The following data apply to this space group.

$$k = 2n : \quad A = 2\cos 2\pi(hx + lz)\cos 2\pi ky$$
$$B = 2\cos 2\pi(hx + lz)\sin 2\pi ky = 0 \text{ if } k = 0$$
$$k = 2n + 1 : \quad A = -2\sin 2\pi(hx + lz)\sin 2\pi ky$$
$$B = -2\sin 2\pi(hx + lz)\cos 2\pi ky$$
$$A = B = 0 \text{ if } h = l = 0$$

Other powerful phase determining formulae exist, such as the positive and negative quartets (NQRs). For example, in the case of four $|\mathbf{E}|$-values $|\mathbf{E}_{\mathbf{h}_1}|$, $|\mathbf{E}_{\mathbf{h}_2}|$, $|\mathbf{E}_{\mathbf{h}_3}|$ and $|\mathbf{E}_{\mathbf{h}_4}|$ for which the expression $(2/N)|\mathbf{E}_{\mathbf{h}_1}||\mathbf{E}_{\mathbf{h}_2}||\mathbf{E}_{\mathbf{h}_3}||\mathbf{E}_{\mathbf{h}_4}|$, where N is the number of atoms, is a sufficiently large number, and also that the values of $|\mathbf{E}_{\mathbf{h}_1+\mathbf{h}_2}|$, $|\mathbf{E}_{\mathbf{h}_1+\mathbf{h}_3}|$ and $|\mathbf{E}_{\mathbf{h}_1+\mathbf{h}_4}|$ are large values, then the following equation holds: $\cos(\phi_1 + \phi_1 + \phi_1 + \phi_1) \approx +1$, whereas if $|\mathbf{E}_{\mathbf{h}_1+\mathbf{h}_2}|$, $|\mathbf{E}_{\mathbf{h}_1+\mathbf{h}_3}|$ and $|\mathbf{E}_{\mathbf{h}_1+\mathbf{h}_4}|$ are all small in magnitude, then $\cos(\phi_1 + \phi_1 + \phi_1 + \phi_1) \approx -1$. These formulae feature in the current structure determination software systems. Detailed discussions and examples of phasing in non-centrosymmetric structures are given in the literature [20–24].

6.14. Example of a Complete Structure Analysis by Direct Methods

An example of an analysis by the direct phasing of a centrosymmetric structure is afforded by the solving of the structure of potassium 2-hydroxy-3,4-dioxocyclobut-1-ene-1-olate monohydrate $C_4HO_4^-K^+ \cdot H_2O$, known also by the trivial name of potassium squarate (KHSQ).

The crystals are monoclinic, space group $P2_1/c$ with the unit-cell dimensions $a = 8.641(1)$, $b = 10.909(1)$, $c = 6.563(2)$ Å, $\beta = 99.81(1)°$, $Z_c = 4$ where the figures in parentheses refer to the estimated standard deviations to be applied to the last significant

Table 6.2.   Statistics of $|\mathbf{E}|$-values distributions for KHSQ, and the acentric and centric values.

	KHSQ	Acentric	Centric
$\langle|\mathbf{E}|^2\rangle$	1.00	1.00	1.00
$\langle|\mathbf{E}|\rangle$	0.81	0.89	0.80
$\langle|\mathbf{E}|^2 - 1\rangle$	0.95	0.74	0.97
$\% \geq 1.00$	33.9	36.8	31.7
$\% \geq 1.50$	14.6	10.5	13.4
$\% \geq 1.75$	8.4	4.7	8.0
$\% \geq 2.00$	4.9	1.8	4.6
$\% \geq 3.00$	1.1	0.2	1.2

figure. This space group is determined from systematic absences, and agreement is provided by the good statistics listed in Table 6.2 (see also Fig. 6.22).

A crystal of dimensions *ca.* 0.3, 0.3, 0.5 mm was mounted on a goniometer head (Fig. 5.21) and a total of 946 unique reflections were collected with a kappa X-ray diffractometer. The data were corrected as described in Section 5.5.3, and a Wilson plot allowed the determination of an approximate scale factor and an overall isotropic temperature factor for the data set.

A set of 142 data with $|\mathbf{E}| \geq 1.5$ were selected, and a Σ_2 listing prepared; a section of the Σ_2 listing is shown in Table 6.3. Symmetry-related reflections are of significance in generating TPRs: for example, on account of centrosymmetry, $\mathbf{h}$, $\mathbf{k}$, $\mathbf{h} - \mathbf{k}$ and $\mathbf{h}$, $\mathbf{k}$, $\mathbf{h} + \mathbf{k}$ are both valid TPRs. In addition, with space group $P\frac{2_1}{c}$, the phase symmetry is given by Eqs. (6.40) and (6.41). Thus, Table 6.3 takes all symmetry relations into consideration in determining TPRs. The products $|\mathbf{E}(\mathbf{h})||\mathbf{E}(\mathbf{k})||\mathbf{E}(\mathbf{h} - \mathbf{k})$ are large values, thus ensuring reliable sign determination.

Three reflections from those listed in Table 6.3 were used to specify the origin, and from them several additional signs were developed, as shown by Table 6.4. In this way, despite a large number of TPRs (1278), only 26 signs were developed. Progress was made by allocating letter symbols to reflection, for example, $112 = a$, $332 = b$, $010, 4 = c$; when symmetry relationships are applied, there are effectively 12 symbolic signs.

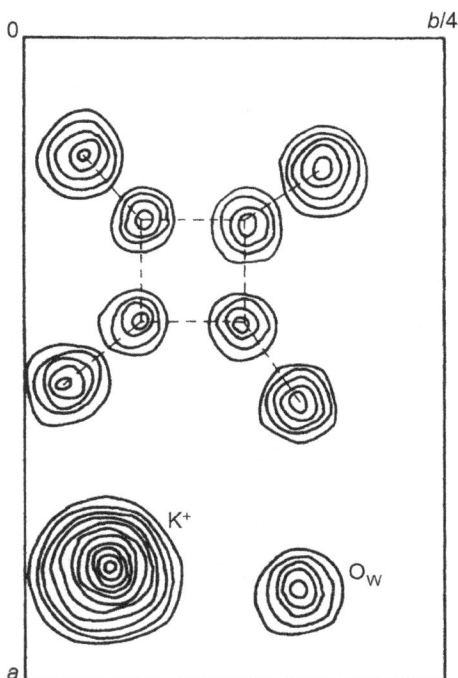

Fig. 6.22. Final electron density map of the KHSQ structure, as seen along c.

Sign determination proceeded in terms of signs and symbols, and, in the process, relationships were determined between the symbols. When the process was complete, the relationship $a = ac = b = -1$, which shows that $c = +1$. Then, substituting the values of the symbols a, b and c, a total of 142 signed $|\mathbf{E}|$ values were used to obtain a first Fourier map. All atoms other than the oxygen atoms of the water molecule were revealed together with three peaks that turned out to be spurious. A further Fourier synthesis, based on the located atoms, revealed the remaining atom, and a composite electron density map of the structure is shown in Fig. 6.23.

The three hydrogen atoms in the asymmetric unit are not resolved on the electron density map. However, a difference Fourier map showed their positions clearly. The remainder of the difference map is almost flat, as is the case for a well-refined structure. A few very

Table 6.3.   Partial Σ_2 listing for KHSQ.

h	$|E_h|$	k	$|E_k|$	h − k	$|E_{h-k}|$	$|E|_{TPR}$
531	2.6	010,4[a]	2.8	573	2.6	18.93
(37)[b]		041	2.2	572	3.3	18.88
		041	2.0	570	2.7	14.04
		114	2.3	625	1.7	10.17
		032	1.7	563	2.0	8.84
114	2.3	572	3.3	482	1.9	14.42
(45)		664	1.8	570	2.7	11.18
		681	1.5	573	2.6	8.97
		563	2.0	451	1.5	6.90
		454	1.6	540	1.5	5.52
032	1.7	531	2.6	563	2.0	8.84
(54)		572	3.3	540	1.5	8.42
		482	1.9	454	1.6	5.17
		451	1.5	481	2.0	5.10
112	2.5	572	3.3	664	1.8	14.85
(39)		482	1.9	570	2.7	12.83
		114	2.3	002	1.9	10.93
		571	1.7	681	1.5	6.38
010,4	2.8	332	2.2	372	1.9	11.70
(35)		625	1.7	681	1.5	7.14
332	2.2	114	2.3	242	1.7	8.60
(46)		313	1.8	041	2.0	7.92
		625	1.7	313	1.8	6.73
002	1.9	114	2.3	116	1.5	6.56
(25)		681	1.5	681	1.6	4.56

[a] A two-digit Miller index (or zone symbol) is followed by a comma unless it is the third digit.
[b] The numbers in parentheses are the total numbers of TPRs for the given **h**.

small, spurious maxima are present, most noticeably in the region of the potassium ion, as shown in Fig. 6.24. A possible reason for the residual maxima is that, in an ionic structure, the tabulated atomic scattering factors will be slightly incorrect owing to charge displacements in the molecule. An *ab initio* calculation of electron populations is shown in Table 6.5. The value of Σp (Table 6.5) is 86, which is $F(000)/4$, the number of electrons in a formula-entity. The withdrawal of electron density from hydrogen atoms by the more electronegative species is the reason that bond lengths to hydrogen

Table 6.4. Origin-fixing reflections and the development of signs.

Origin-fixing reflections and their symmetry equivalents			
h	$s_\mathbf{h}$	**h**	$s_\mathbf{h}$
531	+	$5\bar{3}1$	+
$\bar{5}\bar{3}1$	+	$\bar{5}3\bar{1}$	+
032	+	$0\bar{3}2$	−
$0\bar{3}\bar{2}$	+	$03\bar{1}$	−
114	+	$1\bar{1}4$	−
$\bar{1}\bar{1}4$	+	$\bar{1}\bar{1}4$	−

Developed signs					
h	$s_\mathbf{h}$	**k**	$s_\mathbf{k}$	$\mathbf{h \pm k}$	$s_{\mathbf{h}\pm\mathbf{k}}$
531	+	$1\bar{1}4$	−	625	−
531	+	032	+	563	+
032	+	$1\bar{1}4$	−	126	−
$5\bar{6}3$	−	032	+	$5\bar{9}1$	−
563	+	114	+	$45\bar{1}$	+
563	+	$1\bar{1}4$	−	$47\bar{1}$	−

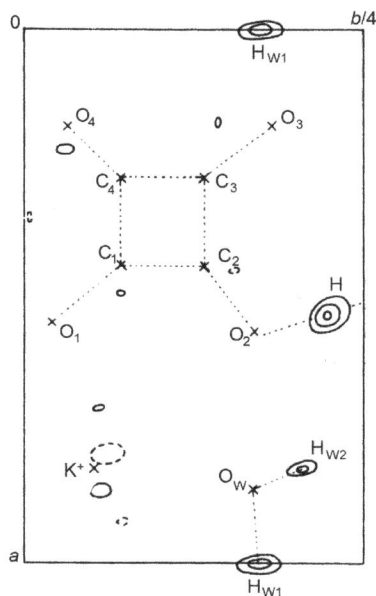

Fig. 6.23. Difference electron density map for KHSQ; the three hydrogen atoms of the structure are revealed clearly. A few very small spurious maxima exist, probably on account of electron drift towards the more electronegative species present (see Table 6.5).

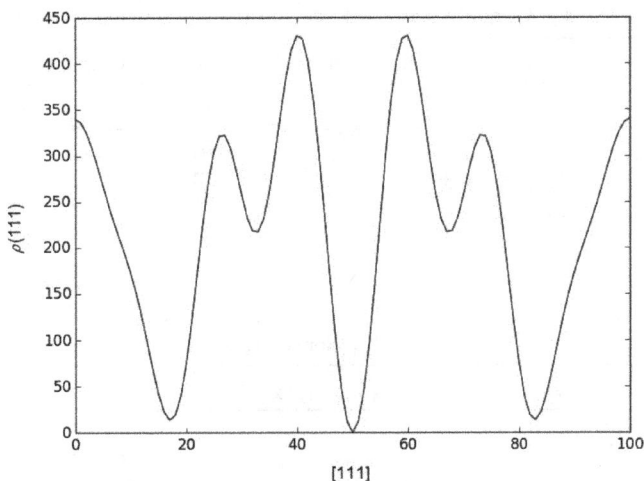

Fig. 6.24.   One-dimensional electron density $\rho(111)$ plot along the direction [111] of the potassium aluminium sulphate structure $KAl(SO_4)_2.12H_2O$.

Table 6.5.   Electron population parameters p in the KHSQ entity.[a]

Atom	p	Atom	p	Atom	p	Atom	p
K	18.111	O_1	8.808	O_2	8.760	O_3	8.626
O_4	8.544	C_1	5.581	C_2	5.833	C_3	5.605
C_4	5.599	O_W	8.855	H	0.564	H_{W1}	0.560
H_{W2}	0.554						

[a] $\sum p = 86 = M_r$.

atoms are measured shorter by X-ray diffraction than by neutron diffraction or by spectroscopy.

The crystal data set for nitroguanidine (NO2G) in the Program Suite is an example that is well suited to a direct methods solution.

6.15.   Isomorphous Replacement

The isomorphous replacement technique is applicable where two or more structures that differ in only one or more heavy atoms crystallize in almost identical unit-cell dimensions and symmetry patterns. It can be applied to both centrosymmetric and non-centrosymmetric structures.

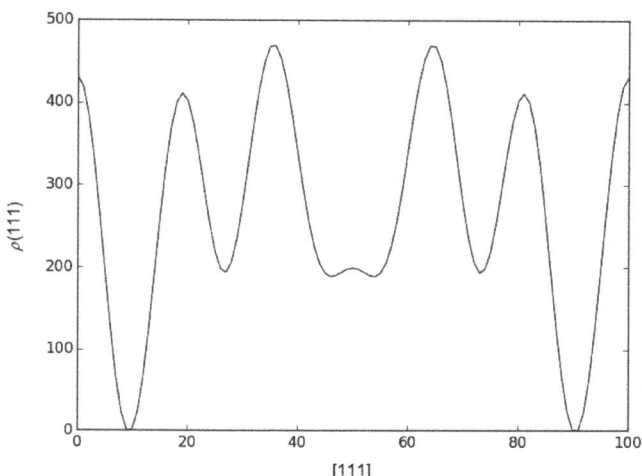

Fig. 6.25. One-dimensional electron density $\rho(111)$ plot along the direction [111] of the potassium aluminium selenate structure $KAl(SeO_4)_2.12H_2O$. The enhancement of the peak at $x = ca.\,0.19$, compared with Fig. 6.24, shows that this site is selenium (and sulphur).

6.15.1. Isomorphous replacement in centrosymmetric crystals

An early example of the technique involved the structures of the alums, which have the general formula $MAl(SO_4)_2 \cdot 12H_2O$, where M is NH_4^+, K^+, Rb^+ or Tl^+. A study of the unit cell and space group data was given in Section 4.12.2, where it was shown that the sulphur atoms must lie on the diagonals of the cubic unit cell at positions such as x, x, x. A one-dimensional Fourier electron density summation along the [111] direction, assuming that the heavy atoms (Rb or Tl) govern the phases as zero ($s = +1$), is shown in Fig. 6.24. The large peak at the origin arises from the potassium and aluminium atoms in superposition in this projection. The next highest peak at $x = ca.\,0.35$ was taken to represent the sulphur atom. However, a satisfactory structure could not be obtained on this model. The isomorphous derivative with selenium in the place of sulphur was used for a second analysis; its electron density plot is shown in Fig. 6.25. The enhancement of the peak at $x = ca.\ 0.19$ means that the selenium (or sulphur) atom occupies this site; the peak at 0.35 was shown later to arise from a superposition of oxygen atoms.

Isomorphous replacement in centrosymmetric crystals is straightforward as the difference in the intensities of a given reflection **h** for two derivatives is related to the difference in scattering power of the replaceable atoms at the value of $\theta(\mathbf{h})$ since the other atoms are approximately constant in their positions in the unit cells of the isomorphs.

Self-assessment 6.3. The following $|\mathbf{F}_{obs}(hhh)|$ data have been determined for a series of isomorphous alums. By considering the changes in nominal scattering power, allocate signs to the reflections as far as possible. Thallium is sufficiently heavy for all its $|\mathbf{F}_{obs}(hhh)|$ signs in the table to be $+1$.

hhh	NH_4^+ (10 e)	K^+ (18 e)	Rb^+ (36 e)	Tl^+ (80 e)
111	86	38	19	113
222	< 1	19	79	195
333	111	125	158	236
444	25	6	55	125
555	24	49	64	131
666	86	86	122	164
777	53	34	< 1	18
888	< 1	16	22	56

6.15.2. Isomorphous replacement in non-centrosymmetric crystals

In non-centrosymmetric crystals, the problem is a little more complicated since the phase may take any value from 0 to 2π. As for the centrosymmetric crystal, the basic equation may be written as

$$\mathbf{F}_{PH} = \mathbf{F}_P + \mathbf{F}_H \tag{6.57}$$

where the replaceable (heavy) atom is labelled H and the structure without the heavy atom is labelled P; this method is most important for the structure analysis of proteins and other macromolecules. Assuming that data are available for two derivatives, P and PH, the *single isomorphous replacement* (SIR) equation (6.57) may be re-cast as

$$|\mathbf{F}_P| \exp(\mathrm{i}\phi_P) = \mathbf{F}_{PH} + \mathbf{F}_H \qquad (6.58)$$

Some proteins have endogenous metals in their structures; ions of metals such as potassium, calcium, manganese and nickel can occur in the macromolecule, and they provide built-in heavy-atom sources. In other cases, metals can be introduced which combine with structural moieties of the macromolecule, such as cysteine, methionine and histamine; the metals introduced may be, for example, mercury in the form of mercury(II) chloride, or platinum in the form of potassium hexachloroplatinate. Another method involves the replacement of an atom by a chemically similar but heavier atom, such the replacement of sulphur by selenium.

Given the value of $\mathbf{F}_H$ from a Patterson analysis of a heavy atom derivative PH, as discussed in Section 6.6.1, Fig. 6.26 shows that there are two solutions for the phase ϕ_P. One of the two circles shown, that for the protein P, has its centre at the origin O of an Argand diagram and its radius is $|\mathbf{F}_P|$, while the second circle, for the heavy-atom derivative PH, has its centre at the end of the vector $-\mathbf{F}_H$ and a radius of $|\mathbf{F}_{PH}|$. Two points of intersection X and Y of the two circles show an ambiguity for the phase ϕ_P, namely, that one of the two angles made by OX and OY with the real axis approximates to the true value of ϕ_P.

At this stage, the following procedure may be used to establish which of the two phases is the preferred value. In the *multiple isomorphous replacement* technique (MIR), one or more additional derivatives are acquired. In Fig. 6.27, a second derivative PQ is introduced, using a *different* (Why?) replaceable site in the molecule. The intersections X and Y again appear, as expected. In addition however, the second derivative PQ produces two intersections with the

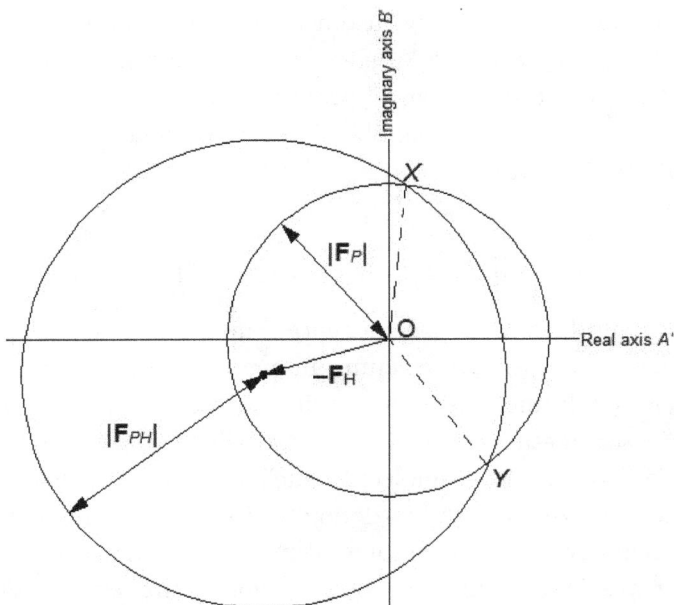

Fig. 6.26. Single isomorphous replacement (SIR) function; the derivatives are P and PH, where P represents a protein or other macromolecule and PH is the macromolecule with an attached heavy atom H. The SIR procedure leaves a phase ambiguity the angles $X-O-A'$ and $Y-O-A'$.

$\mathbf{F}_P$ circle also, namely, X and Z, which shows that the required intersection is now X and that ϕ_P is the angle $X-O-A'$.

The exact coincidence that is suggested in Fig. 6.27 is not a general experience. There is usually a 'circle of error' within which the correct phase angle lies. However, if the phases so obtained are sufficiently close to the true values such that interpretable Fourier and difference Fourier electron density maps can be extracted, then refinement of the structure will lead normally to a satisfactory structure analysis. Generally, a 'best' phase estimate is calculated from a probability function which provides also a weight for each reflection.

Whatever be the technique used, crystals are examined by X-ray diffraction as discussed in Section 5.5.2 so as to produce as complete a diffraction record as possible. In this context, the use of synchrotron radiation is of great importance since its very high intensity means

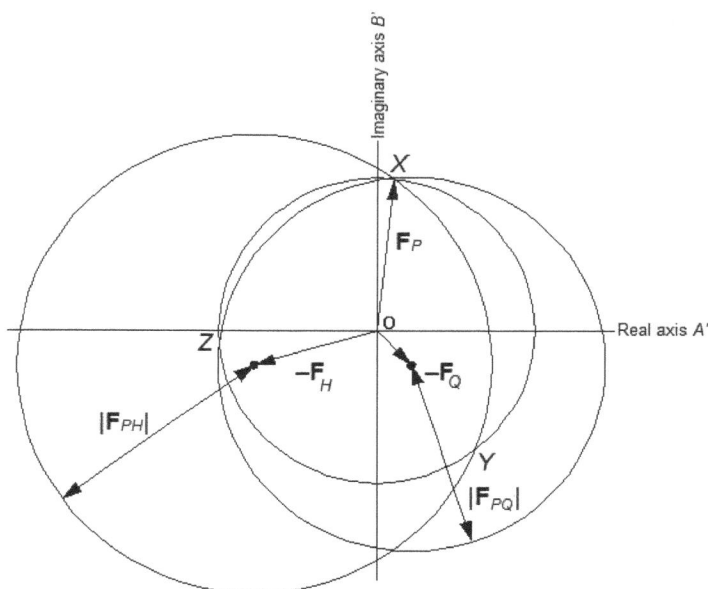

Fig. 6.27. Multiple isomorphous replacement (MIR) function; two derivatives are shown in this figure, *PH* and *PQ*, where *PQ* represents the macromolecule *P* with the second derivative *PQ* using a different replaceable site in the molecule. The ambiguity is resolved, with ϕ_P as the angle between *OX* and the real axis.

that the data set can be obtained in a very short period of time, even with very small crystals. Protein crystals and other macromolecular structures are usually subject to radiation damage, and more than one crystal may be necessary to obtain sufficient data for a structure analysis. Even more important is the X-ray free electron laser (XFEL) at Hamburg. This source emits 40 fs pulses of X-rays that allow collection of data before damage to the crystal occurs, and envisages also the use of crystals of nanosize in X-ray diffraction.

Self-assessment 6.4. In a single isomorphous replacement, a given $|\mathbf{F}_{\mathrm{obs}}|$ for a protein structure was 63.0, and, for the corresponding heavy-atom derivative, the $|\mathbf{F}_{\mathrm{obs}}|$ was 84.5; the heavy atom contribution $\mathbf{F}_H$ is 29 with a phase angle of 240°. What are the two possible indications for the phase of *P* for this reflection?

6.16. Anomalous Scattering

The atomic scattering factor is a complex quantity that may be represented by the equation

$$f = f_0 + \Delta f' + i\Delta f'' = f' + i\Delta f'' \qquad (6.59)$$

where the component f_0 is independent of wavelength and $f' = f_0 + \Delta f'$. The existence of Friedel's law, which was discussed in Section 5.4.3.1, implies that in many circumstances the two correcting terms $\Delta f'$ and $i\Delta f''$ are sufficiently small to be neglected, which is generally the case provided that the atoms undergoing X-radiation do not have absorption levels in the neighbourhood of the wavelength of the radiation. However, as the atomic number increases the magnitudes of the two corrections may become significant in terms of anomalous scattering, whereupon $|\mathbf{F}(hkl)|$ is no longer equal to $|\mathbf{F}(\bar{h}\bar{k}\bar{l})|$ and $\phi(hkl)$ is no longer equal to $-\phi(\bar{h}\bar{k}\bar{l})$ and The normal situation for the structure factor of a reflection $\mathbf{h}$ from a crystalline species comprising a heavy atom H and the rest R of the structure may be written as

$$\mathbf{F}(\mathbf{h}) = \mathbf{R}(\mathbf{h}) + \mathbf{f}(\mathbf{h})_H \qquad (6.60)$$

and is shown diagrammatically on an Argand diagram in Fig. 6.28. Friedel's law applies, so that $|\mathbf{F}(\mathbf{h})| = |\mathbf{F}(\bar{h})|$ and $\phi(\mathbf{h}) = -\phi(\bar{h})$, where $\mathbf{h}$ represents hkl.

In the presence of anomalous scattering, of the two corrections arising from the heavy atom, $\Delta f'$, which has the same phase as f_0, is usually negative, whereas the correction $\Delta f''$ is positive and with a phase advance of $90°$ for both $\mathbf{h}$ and $\bar{\mathbf{h}}$. The result is that $|\mathbf{F}(\mathbf{h})| \neq |\mathbf{F}(\bar{\mathbf{h}})|$ and $\phi(\mathbf{h}) \neq -\phi(\bar{\mathbf{h}})$, as shown in Fig. 6.29.

Two important features arise in connection with anomalous scattering, namely, phasing of reflections and determination of an absolute configuration.

Under Friedel's law, an X-ray diffraction pattern is a three-dimensional weighted reciprocal lattice of symmetry corresponding to point group of the crystal with the addition of a centre

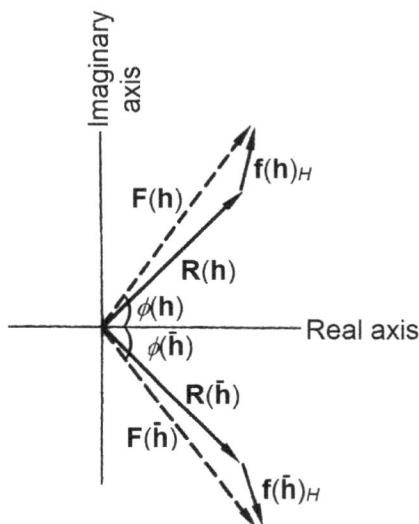

Fig. 6.28. A structure factor $\mathbf{F}(\mathbf{h})$ (and its conjugate $\mathbf{F}(\bar{h})$), comprising the contribution $\mathbf{f}(\mathbf{h})_H$ from the heavy atom H; that from the rest of the structure is $\mathbf{R}(\mathbf{h})$. In the absence of significant anomalous scattering, $|\mathbf{F}(\mathbf{h})| = |\mathbf{F}(\bar{h})|$ and $\phi(\mathbf{h}) = -\phi(\bar{h})$.

of symmetry, that is, to one of the eleven Laue classes shown by bold type in Table 2.5. When anomalous scattering is significant, the centre of symmetry is destroyed, as indicated in Fig. 6.29, and the symmetry is then that of the true point group of the crystal. In general, the *anomalous difference* $\Delta|\mathbf{F}_{\text{anom}}|$ is given by

$$\Delta|\mathbf{F}_{\text{anom}}| = |\mathbf{F}(hkl)| - |\mathbf{F}(\bar{h}\bar{k}\bar{l})| \tag{6.61}$$

In the case of space group $P2_1$, for example, $\Delta|\mathbf{F}_{\text{anom}}|$ may be given also by a pair such as

$$\Delta|\mathbf{F}_{\text{anom}}| = |\mathbf{F}(hkl)| - |\mathbf{F}(h\bar{k}l)| \tag{6.62}$$

The term $\Delta|\mathbf{F}_{\text{anom}}|$ is known as a *Bijvoet difference*, and the two reflections form a *Bijvoet pair* [28].

While the phase ambiguity shown in Fig. 6.26 can be resolved by a further heavy atom derivative but with a different replaceable site, as

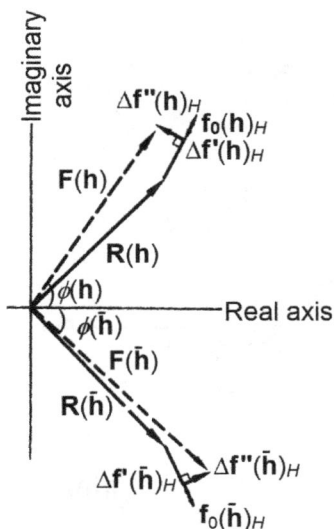

Fig. 6.29. A structure factor $\mathbf{F(h)}$ (and its conjugate $\mathbf{F(\bar{h})}$), comprising the real contribution $\mathbf{f_o(h)}_H$ from the heavy atom H, its real and imaginary components are $\Delta\mathbf{f'(h)}_H$ and $\Delta\mathbf{f''(h)}_H$; that from the rest of the structure is $\mathbf{R(h)}$. The imaginary component has a phase advance of $90°$ for *both* $\mathbf{h}$ and $\mathbf{\bar{h}}$. In the presence of anomalous scattering, $|\mathbf{F(h)}| \neq |\mathbf{F(\bar{h})}|$ and $\phi(\mathbf{h}) \neq -\phi(\mathbf{\bar{h}})$, and the phase ambiguity can be resolved.

discussed through Fig. 6.27, the required additional derivative may be furnished by the end position of the vector $\Delta f''_H$; the ambiguity may be then resolved, within the limits of experimental error. Further discussions on phasing by anomalous scattering may be found in the literature [2, 3].

In a non-centrosymmetric crystal, a structure and its inverse give rise to the same set of $|\mathbf{F}_{calc}|$ data. One configuration corresponds to the true or absolute configuration of the chiral chemical species in the structure. An early test for configuration given by the Hamilton ratio was based on the ratio of R factors for the derived and refined structure with atomic coordinates (x, y, z) and its inverse $(\bar{x}, \bar{y}, \bar{z})$ [27].

A more powerful method uses the Flack parameter, which indicates clearly the absolute configuration of the structure. The

Flack parameter x is calculated during the refinement, using the equation

$$|\mathbf{F}_{\mathrm{obs}}(hkl)|^2 = (1-x)|\mathbf{F}_{\mathrm{calc}}(hkl)|^2 + x|\mathbf{F}_{\mathrm{calc}}(\overline{h}\,\overline{k}\,\overline{l})|^2 \qquad (6.63)$$

By determining x for all data, x is usually found to lie between 0 and 1. If the value is near 0, with a small uncertainty, the absolute structure given by the refinement is probably correct, whereas if the value is near 1, then the inverted structure is the more probable. A value close to 0.5 indicates the possibility that the crystal is a racemic mixture or a twin crystal.

Example 6.5. The compound $C_{31}H_{44}O_2S_2$ crystallizes in space group $P2_1$ with $Z_c = 2$; it was solved by Patterson methods and its absolute configuration determined [29]. It is another example of a structure with two heavy atoms in the asymmetric unit. The structure analysis was refined to $R = 3.4\%$ and $R_w = 4.7\%$. The structure was refined a second time, but with the signs of the imaginary components $\Delta f_H''$ reversed, whereupon $R' = 3.6\%$ and $R_w' = 4.9\%$. The ratio $R_w'/R_w = 1.04$ is significant at the 0.01 level [27]. Thus, the absolute configuration corresponds to the coordinates as determined. This result was confirmed subsequently by the Flack parameter.

Problems

6.1. Uranium monosilicide crystallizes in the orthorhombic system with unit-cell dimensions $a = 5.7$, $b = 7.7$ and $c = 3.9\,\text{Å}$. The structure is centrosymmetric with the uranium atoms occupying the special equivalent positions $\pm(x, y, 1/4; 1/2 - x, 1/2 + y, 1/4)$; the geometrical structure factor is given by $A = 4\cos 2\pi \left[ky + \left(\frac{h+k+l}{4}\right)\right] \cos 2\pi \left[hx - \left(\frac{h+k}{4}\right)\right]$. Use the data below to determine approximate values for the x and y coordinates of uranium.

hkl	200	111	210	231	040	101	021	310
$|I_{obs}(hkl)|$	< 1	240	250	200	< 1	170	180	< 1

(**Hint:** Determine an approximate value for x from 200; then, find y from 111, 231 and 040.)

6.2. What values exist for the integral $\int_{-\pi}^{\pi} \cos(2\pi mx) \cos(2\pi nx) \, dx$? (Appendix A7 may be helpful.)

6.3. The crystalline complex of methylamine and boron trifluoride is monoclinic space group $P2_1/m$ and $Z_c = 2$. (a) Draw a diagram to show the general equivalent positions and symmetry elements of the asymmetric unit of space group $P2_1/m$; origin on $\bar{1}$. (b) What can be determined about the crystal structure?

6.4. A crystalline compound with space group $P2_1/n$ contains four heavy atoms H related by the space group symmetry. The Harker sections $(u, 1/2, w)$ and $(u, 0.1, w)$ are shown in Fig. P6.1; the heavy atom H has its maximum peak value on the $(u, 0.1, w)$ section. The general equivalent positions for the space group $P2_1/n$ are $\pm(x, y, z; 1/2 + x, 1/2 - y, 1/2 + z)$. (a) What are the Patterson coordinates and weights of the heavy atom vectors? (b) By measurement on the diagrams given, determine approximate values for the coordinates on the heavy atoms.

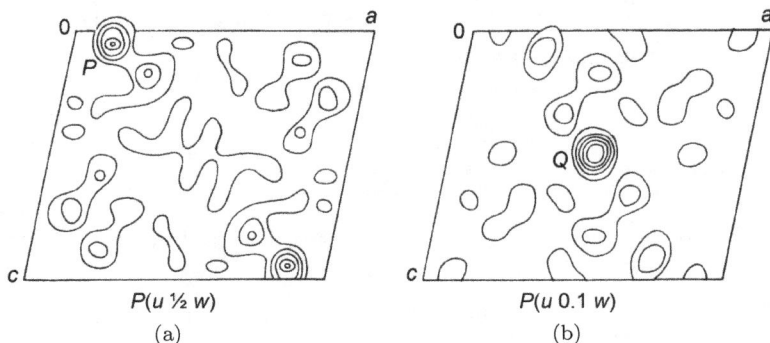

Fig. P6.1.   Harker sections for Problem 6.4.

6.5. Sodium hydride NaH has the sodium chloride structure type (Fig. 4.1(c)). The scattering factors f for X-rays and the scattering cross-sections σ for neutrons are listed below. (a) Formulate a structure factor equation for this structure type. (b) Calculate $|\mathbf{F}_{111}|$ and $|\mathbf{F}_{220}|$ for NaH and NaD for X-rays and neutrons.

hkl	f_{Na}^+	f_{H^-/D^-}	σ_{Na}^+	σ_{H^-}	σ_{D^-}
111	8.05	0.381	0.348	−0.372	0.673
220	6.72	0.208	0.348	−0.372	0.673

6.6. Euphenyl iodoacetate $C_{32}H_{53}O_2I$ crystallizes in space group $P2_1$ with $a = 7.26$, $b = 11.55$, $c = 19.22$ Å, $\beta = 94.07°$, $Z_c = 2$. The sharpened and selected Harker section $(u, 1/2, w)$ is shown below:

Harker section $(u, \frac{1}{2}, w)$ for euphenyl iodoacetate

$(\sin\theta)/\lambda$	0	0.05	0.10	0.15	0.20	0.25	0.30	0.35	0.40
f_I		53	51.6	48.1	44.0	40.0	36.4	33.0	29.8	26.8

Determine probable signs for the following reflections:

hkl	001	0014	300	106
$|\mathbf{F}_{obs}|$	40	37	35	33

(a) Determine the x, y and z coordinates for the iodine atoms in the unit cell.

(b) Calculate the length of the shortest I–I vector in the crystal.

(c) The following data are the temperature-corrected scattering factors for iodine.

6.7. A crystal contains six atoms in the unit cell. Five of these atoms scatter in accordance with Friedel's law and contribute an amount $90.0\exp(i\phi)$ to the 002 reflection. The sixth atom has fractional coordinates 0.050, 0.100, 0.200, and the components of its scattering factor are $f_0 = 55.5, \Delta f' = -3.3$ and $\Delta f'' = 9.0$. If the phase angle ϕ is 30°, calculate $|\mathbf{F}(002)|, \phi(002), |\mathbf{F}(00\bar{2})|$ and $\phi(00\bar{2})$.

6.8. The following $|\mathbf{E}|$-values were calculated for the $0kl$ zone of a monoclinic crystal which was allocated the space group $P\frac{2_1}{a}$. Determine a suitable origin and assign + and − signs to as many of the data as possible. Extend the sign determination with symbols as appropriate. In the $0kl$ zone of $P\frac{2_1}{a}$, $F(0kl) = F(0\bar{k}\bar{l}) = (-1)^k F(0\bar{k}l)$.

$0kl$	$|\mathbf{E}|$	$0kl$	$|\mathbf{E}|$
024	2.8	035	1.8
0018	2.4	0817	1.8
081	2.2	011,7	1.3
011,9	2.2	011	1.0
038	2.1	026	0.3
0310	1.9	0312	0.1
059	1.9	021	0.1

6.9. Crystals of 3-β-acetoxy-6,7-epidithio-19-norlanosta-5,7,9,11-tetraene $C_{31}H_{44}O_2S_2$ are monoclinic: $a = 20.1896\,\text{Å}$, $b = 11.0709\,\text{Å}$, $c = 6.4953\,\text{Å}$, $\beta = 90.578°$, space group $P2_1$ and $Z = 2$.

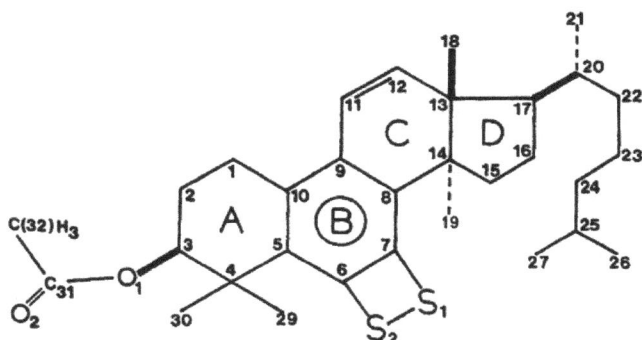

3-β-Acetoxy-6,7-epidithio-19-norlanosta-5,7,9,11-tetraene

The coordinates of the two sulphur atoms in the asymmetric unit were found to be $S(1)\, 0.2092, 3/4, -0.1522$; $S(2)\, 0.1095, 0.6933, -0.1216$. (a) Make a sketch of the relative positions and the weights of the heavy-atom vectors in projection on to (001) from 0 to $\pm 1/2$ along a and b. (b) Calculate the length of the $S(1)$–$S(2)$ bond.

6.10. In a crystal of space group $P2_12_12_1$, the following indications were obtained for the $|E(\mathbf{h})|$ reflection, $\mathbf{h} = 771$. Determine $\phi(771)$ from (a) Eq. (6.51) and (b) Eq. (6.54); equal weights of unity may be assumed for each observation.

$|\mathbf{E}(\mathbf{k})|$	$\phi(\mathbf{k})/\mathrm{deg}$	$|\mathbf{E}(\mathbf{h}-\mathbf{k})|$	$\phi(\mathbf{h}-\mathbf{k})/\mathrm{deg}$
2.1	0	2.2	-37
1.7	177	2.0	-180
1.8	90	1.7	-144
2.0	90	1.8	-90
1.7	102	2.4	-64
1.9	-79	2.3	92

6.11. The following data were taken from the crystal structure analysis of potassium squarate monohydrate. Set an origin, draw up a $\sum_2$ listing and allocate signs to the reflections as far as possible.

The space group of this substance is $P\frac{2_1}{a}$, for which $s(hkl) = s(\bar{h}\,\bar{k}\,\bar{l}) = (-1)^{k+1}s(h\bar{k}l)$

TPR	h	k	h − k	$\lvert\mathbf{E}_h\rvert\lvert\mathbf{E}_k\rvert\lvert\mathbf{E}_{h-k}\rvert$
1	800	670	$2\bar{7}0$	10.095
2		340	$5\bar{4}0$	7.672
3		411,0	$4\bar{1}\bar{1},0$	4.860
4		040	$8\bar{4}0$	4.176
5	300	040	$3\bar{4}0$	3.497
6		840	$\bar{5}\bar{4}0$	6.014
7		570	$\bar{2}\bar{7}0$	10.046
8	730	$0\bar{4}0$	770	3.069
9		$5\bar{4}0$	270	6.924
10	700	570	$2\bar{7}0$	12.974
11	340	$7\bar{7}0$	$\bar{4}11,0$	4.051

6.12. (a) The amplitudes of the structure factors of a protein P and its derivative PH with a heavy atom H are as follows: $\lvert\mathbf{F}_{PH}\rvert = 75.15$, $\lvert\mathbf{F}_P\rvert = 53.84$ and $\mathbf{F}_H = 41.23\exp i(200.00°)$. By construction, determine the possible phase angles for the structure factor $\lvert\mathbf{F}_P\rvert$.

(b) The ambiguity may be resolved, within experimental error, by using a second heavy-atom derivative. To the diagram already drawn in answer to (a), a second heavy-atom derivative at a different replaceable site in the protein is employed. Using a derivative PK, with $\mathbf{F}_K = 16.40\exp i(-52.43°)$, decide on the correct phase for $\lvert\mathbf{F}_P\rvert$.

6.13. An organic compound, believed to be $C_{12}H_8N_2$, was examined by X-ray diffraction and a structure obtained from a ρ_{obs} electron density map. A molecular formula was fitted to this map as shown below:

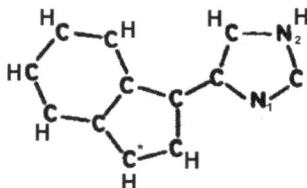

The compound C_{12} H_8 N_2

However, the structure would not refine as well as was expected from the quality of the $|F_{obs}|$ data. In a further investigation, a $\rho_{obs} - \rho_{calc}$ difference density map was calculated and is shown below. What might be another interpretation of the results?

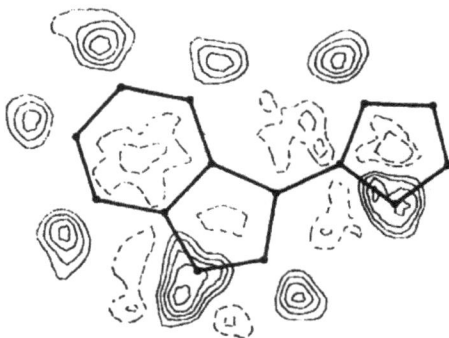

Difference Fourier map of 'C_{12} H_8 N_2'

Answers to Self-Assessments

6.1. From the first map hereunder, $x_F = ca.0.203$. With the erroneous sign, an interpretation of the second map is very difficult; not even Mg shows up as the highest peak at the expected values of $x = 0$ and $1/2$.

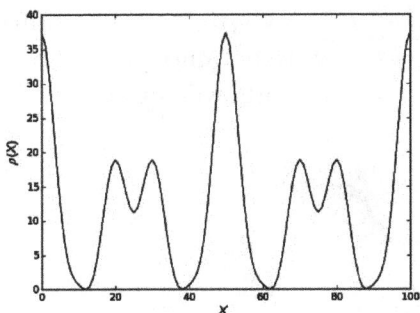

ρ (x) excluding the 1/*a* and F(000) terms

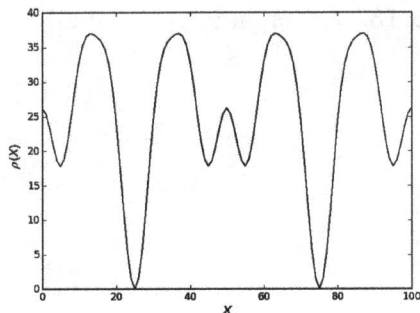

ρ (x) as before but with 400 sign change

6.2. $|F(hkl)| = |F(\bar{h}\,\bar{k}\,\bar{l})| = |F(\bar{h}kl)| \neq |F(\bar{h}kl)|;\ |F(\bar{h}kl)| = |F(hk\bar{l})$

$k = 2n:\quad \phi(hkl) = -\phi(\bar{h}\,\bar{k}\,\bar{l}) = -\phi(h\bar{k}l) \neq \phi(\bar{h}kl);$

$\phi(\bar{h}kl) = -\phi(hk\bar{l})$

$k = 2n + 1:\quad \phi(hkl) = -\phi(\bar{h}\,\bar{k}\,\bar{l}) = \pi - \phi(h\bar{k}l) \neq \phi(\bar{h}kl);$

$\phi(\bar{h}kl) = \pi + \phi(hk\bar{l})$

6.3. Using $|\mathbf{F}(111)|_{M_1} = |\mathbf{F}(111)|_{M_2} + 4|f(111)|_{M_1} - f(111)|_{M_1}| \approx |\mathbf{F}(111)|_{M2} + 4|Z_{M1} - Z_{M1}|$, the following assignment can be made; an * implies that $|\mathbf{F}_{\text{obs}}|$ is too small to make a reasonable allocation of sign:

111	NH_4^+ (10 e)	K^+ (18 e)	Rb^+ (36 e)	Tl^+ (80 e)
111	−86	−38	+19	+113
222	*	+19	+79	+195
333	+111	+125	+158	+236
444	−25	*	+55	+125
555	+24	+49	+64	+131
666	−86	+86	+122	+164
777	−53	−34	*	+18
888	*	+16	+22	+56

The reader may wish now to check the assignments made by calculating electron density maps along [111] for each derivative, using the programs FOUR and then plotting the results with the program GRFN.

6.4. A circle of radius 63.0 units is drawn with its centre at the origin O of an Argand diagram. The vector $-\mathbf{F}$ is plotted from the same origin. A circle of radius 84.5 units is drawn centred on the end of the vector $-\mathbf{F}$. The two intersections give the possible values $A'OP$ and $A'OQ$ for the phase angle with respect to the real axis A' as shown by the diagram hereunder: by measurement, the two possible values for ϕ are *ca.* 12.5° and 108.0°.

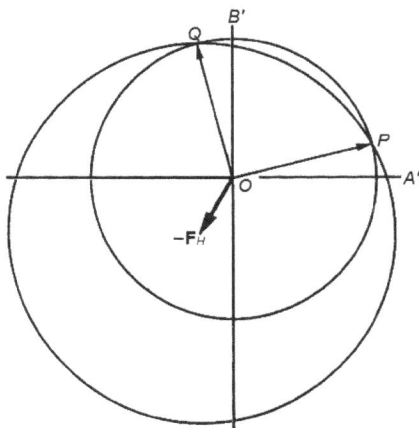

Diagram for a single isomorphous replacement (SIR)

References

[1] Blake AJ *et al.*, *Crystal Structure Analysis*: *Principles and Practice*, 2nd edn. International Union of Crystallography/Oxford University Press (2007).

[2] Ladd M and Palmer R, *Structure Determination by X-ray Crystallography*, 5th edn. Springer (2013).

[3] Luger P, *Modern X-ray Analysis on Single Crystals*, 2nd edn. De Gruyter (2014).

[4] Woolfson MM, *Introduction to X-ray Crystallography*, 2nd edn. Cambridge University Press (2010).

[5] Lipson H and Taylor CA. *Fourier Transforms and X-ray Diffraction*. Bell and Sons (1958).

[6] Henry NFM and Lonsdale K. (eds), *International Tables for X-ray Crystallography*, Vol. I, 3rd edn., International Union of Crystallography. Kynoch Press (1969). Online at ⟨https://archive.org/stream/International TablesForX-rayCrystallographyVol1/HenryLonsdaleEds-internationalTables ForX-rayCrystallographyVol1{#}page/n71/mode/2up⟩. *Note*: This source

contains the earlier notation of $m3$ and $m3m$ in place of $m\bar{3}$ and $m\bar{3}m$ for those point groups, and for the space groups that derive from them.

[7] Aroyo MI (ed), *International Tables for Crystallography*, Vol. A, *Space-Group Symmetry*, 6th edn., International Union of Crystallography, Wiley (2016); 2nd edn., Wiley (2014); 5th edn.

[8] Glusker JP, *Acta Crystallogr. B***24**, 359 (1968).

[9] Patterson AL, *Phys. Rev.* **46**, 372 (1934).

[10] Patterson AL, *Z. Kristallogr.* **90**, 517 (1935).

[11] Harker D, *J. Chem. Phys.* **4**, 381 (1936).

[12] Ladd MFC, *Z. Kristallogr.* **124**, 64 (1967).

[13] O'Donnell EA and Ladd MFC, *Acta Crystallogr.* **23**, 460 (1967).

[14] Tanczos AC *et al.*, *Comput. Biol. Chem.* **28**, 375 (2004).

[15] Ladd MFC and Povey DC, *J. Cryst. Mol. Struct.* **2**, 243 (1972).

[16] Sayre D, *Acta Crystallogr.* **5**, 60 (1952).

[17] Lonsdale K, *Proc. Roy. Soc.A* **123**, 494 (1928).

[18] Hauptman H and Karle J, *Solution of the Phase Problem, 1.* American Crystallographic Association/Polycrystal Book Service NY (1952).

[19] Karle J and Hauptman H, *ActaCrystallogr.* **3**, 181 (1950).

[20] Giacovazzo C, *Direct Phasing in Crystallography.* International Union of Crystallography/Oxford Science Publications (1999).

[21] Ladd MFC and Palmer RA (eds), *Theory and Practice of Direct Methods in Crystallography.* Plenum Press (1989).

[22] Woolfson MM and Hai-Fu F, *Physical and Non-physical Methods of Solving Crystal Structures.* Cambridge University Press (1995).

[23] Hauptman H, *Crystal Structure Determination: The Role of Cosine Seminvariants.* Plenum (1972).

[24] Hauptman H, *ActaCrystallogr.* **B28**, 2337 (1972).

[25] Stroud R, *Acta Crystallogr. B* **29**, 690 (1973); see also reference [2, p. 367ff].

[26] Peerdeman AF and Bijvoet JM, *Acta Crystallogr.* **9**, 1012 (1956).

[27] Hamilton WC, *Acta Crystallogr.* **18**, 502 (1965).

[28] Flack HD, *Acta Crystallogr. A* **39**, 876 (1983).

[29] Ladd MFC and Povey DC, *Acta Crystallogr. B* **32**, 1311 (1976).

Chapter 7

Refinement and Molecular Geometry

Never is there either work without reward,
or, reward without work being expended.
Titus Livius (Livi)

Key Topics

- Least squares in structure analysis
- Temperature and scale factors
- Weighting schemes and precision
- Refinement and measures of correctness
- Molecular geometry
- Significance testing
- Typical presentation of results
- Databases

7.1. Introduction

While it is possible to refine reasonably well the result of a crystal structure analysis by Fourier and difference Fourier techniques, it is a normal practice to complete a structure determination by a least-squares process of refinement. In Appendix A11, a short discussion of the least-squares procedure is given, albeit in the straightforward case of fitting a straight to a set of data points, and of determining the estimated standard deviations (esd) in the derived parameters.

The refinement of a crystal structure analysis is a more complicated process. However, the central problem is the same, namely, that of finding the best values of the determined parameters that reproduce an observed data set to within acceptable limits.

The unit-cell dimensions are refined normally at the initial stage of the intensity data collection. When the atomic positions, excluding hydrogen atoms, have been determined sufficiently well, the parameters that are then refined are atomic x, y and z coordinates, atom occupancy factors (less than unity for atoms in special equivalent positions) if appropriate, temperature factors, scale factor for the $|\mathbf{F}_{obs}|$ unique data set, extinction and absorption corrections where necessary and, if an absolute configuration is sought, also the Flack parameter x is included. If it is not feasible to refine the positional parameters of the hydrogen atoms, they can be fixed in geometrically acceptable positions according to their type, and the refinement can then include their positions and isotropic temperature factors U.

7.2. Least Squares Applied to Crystal Structure Analysis

The parameters of the structure determination are adjusted so as to obtain the best agreement between the observed $|\mathbf{F}_{obs}|$ data set and the corresponding $|\mathbf{F}_{calc}|$ values that are calculated from the structural parameters. The usual procedure is the minimization of a function that includes all desired structural parameters. Such a function Φ may be of the form

$$\Phi_F = \sum_{|F|>n\sigma} w(\mathbf{h})(|\mathbf{F}_{obs}| - G|\mathbf{F}_{calc}|)^2 \qquad (7.1)$$

where all data with $|\mathbf{F}_{obs}|$ greater than $n\sigma(|\mathbf{F}_{obs}|)$ are used, n is a number usually between 2 and 4, σ is the esd of the intensity data, G is a scale factor for $|\mathbf{F}_{obs}|$ applied inversely to $|\mathbf{F}_{calc}|$, and $w(\mathbf{h})$ is the weighting function for a reflection $\mathbf{h}$ based on counting statistics.

A superior technique is a refinement based on $|\mathbf{F}|^2$:

$$\Phi_{F^2} = \sum_{|F|} w(\mathbf{h})(|\mathbf{F}_{obs}|^2 - G^2|\mathbf{F}_{calc}|^2)^2 \qquad (7.2)$$

The R-factor from Eq. (7.1) is generally less than that from Eq. (7.2). However, Φ_{F^2} allows all reflections to be included, even those that are recorded as zero, or as slightly negative on account of experimental error: reflections of low intensity are also a feature of the structure, but their higher values of σ, acting through the weighting function w, will influence the R-factors.

As with other crystallographic calculations, least-squares refinement is available with well-tested and documented programs, such as the SHELX, WinGX and CRYSTALS systems, together with CCP4 for proteins and other macromolecules [1]. These program systems solve and refine a structure, and *inter alia* identify peaks of the electron density function and calculate molecular geometry. Knowledge of the processes taking place in these programs enables the user to apply them most advantageously. Detailed user manuals accompany the program systems. The parameters that usually enter the refinement process will be discussed briefly.

7.2.1. Unit-cell dimensions

The measurement of unit-cell dimensions from photographic data was discussed in Section 5.5.1, and more precise measurements are obtained with an X-ray diffractometer. The measurements are best made at very high values of the Bragg angle θ, since the variation of $\sin\theta$ with θ is less in that region, as the following diagram shows:

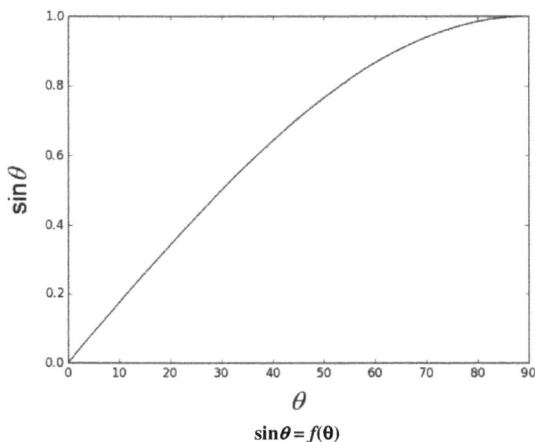

$$\sin\theta = f(\theta)$$

Diffractometer software includes a unit-cell refinement procedure. Consider the example of a monoclinic crystal for which the θ-values of 25 or more high-order reflections have been measured to a precision of $0.01°$. In the monoclinic system,

$$4\sin^2\theta = h^2 a^{*2} + k^2 b^{*2} + l^2 c^{*2} + 2lh c^* a^* \cos\beta^* \qquad (7.3)$$

In order to obtain the best-fit values of the parameters a^*, b^*, c^* and β^* with respect to the measured i values of θ, the expression

$$\sum_i (h^2 a_i^{*2} + k^2 b_i^{*2} + l^2 c_i^{*2} + 2lh c_i^* a_i^* \cos\beta_i^* - 4\sin^2\theta_i)^2 \qquad (7.4)$$

is minimized with respect to a^*, b^*, c^* and $\cos\beta^*$ over all i measurements of θ by the least-squares procedure. The software delivers the best-fit values of the unit-cell parameters together with their esds.

7.2.2.   Coordinate parameters

Once a set of approximate x, y and z coordinates has been derived for the atoms, the refinement of the parameters of the structure can be carried out. The necessary parameters always include atomic coordinates, temperature factors, and a scale factor which is applied inversely to the calculated values of $|\mathbf{F}(\mathbf{h})|$. Following Eq. (7.1), refinement on $|\mathbf{F}(\mathbf{h})|$ minimizes the function

$$\Phi = \sum_{\mathbf{h}} w(\mathbf{h})(|\mathbf{F}_{\text{obs}}(\mathbf{h})| - G|\mathbf{F}_{\text{calc}},(\mathbf{h})|)^2 \qquad (7.5)$$

where the sum is over all $\mathbf{h}$ terms ($\mathbf{h} \equiv hkl$) of weight $w(\mathbf{h})$. The variables in $|\mathbf{F}_{\text{calc}}(\mathbf{h})|$ are $p_j(j = 1 - n)$, and for each variable p the least-squares technique requires $\partial\Phi/\partial p_j = 0$, or

$$\sum_{\mathbf{h}} w(\mathbf{h})\Delta\frac{\partial|F_{\text{calc}}(\mathbf{h})|}{\partial p_j} = 0 \qquad (7.6)$$

where $\Delta = |\mathbf{F}_{\text{obs}}(\mathbf{h})| - |\mathbf{F}_{\text{calc}}(\mathbf{h})|$. As the corrections to the parameters are usually small values at this stage, Eq. (7.5) may be expressed as a Taylor series truncated after the first term:

$$\Delta_{\mathbf{p},\delta} = \Delta_{\mathbf{p}} - \sum_{i=1}^{n} w(\mathbf{h})\delta_i \frac{\partial |\mathbf{F}_{\text{calc}}(\mathbf{h})|}{\partial p_j} \tag{7.7}$$

where δ_i is the shift applied to parameter p_i, $\mathbf{p}$ and $\boldsymbol{\delta}$ represent the complete sets of variables and shifts. Substitution of Eq. (7.7) in Eq. (7.6) produces the normal equations

$$\sum_{i=1}^{n} \left[\sum_{\mathbf{h}} w(\mathbf{h}) \frac{\partial |\mathbf{F}_{\text{calc}}(\mathbf{h})|}{\partial p_i} \frac{\partial |\mathbf{F}_{\text{calc}}(\mathbf{h})|}{\partial p_j} \right] \delta_i = \sum_{\mathbf{h}} w(\mathbf{h})\Delta \frac{\partial |\mathbf{F}_{\text{calc}}(\mathbf{h})|}{\partial p_j} \tag{7.8}$$

or concisely as

$$\sum_i a_{ij}\delta_i = b_j \tag{7.9}$$

where $a_{ij} = \sum_{\mathbf{h}} w(\mathbf{h}) \frac{\partial |\mathbf{F}_{\text{calc}}(\mathbf{h})|}{\partial p_i} \frac{\partial |\mathbf{F}_{\text{calc}}(\mathbf{h})|}{\partial p_j}$ and $b_j = \sum_{\mathbf{h}} w(\mathbf{h})\Delta \frac{\partial |F_{\text{calc}}(\mathbf{h})|}{\partial p_j}$, which must be solved for the shifts in the p_i parameters.

The simplest solution of the least-squares equations uses the terms of the diagonal of the a_{ij} matrix, that is, its off-diagonal terms are ignored. It is a relatively rapid calculation, and it is the procedure that is used in the XRAY system in the Program Suite for use with the crystal structure exercises. A better approximation uses the diagonal terms and blocks on each side of it; however, this method underestimates esd by *ca.* 25%.

The preferred and generally used technique is that of full matrix refinement, which is the method that is used in the current crystallographic program systems. Crystallographic computing procedures are described in detail in the manuals of the program systems.[a]

---

[a]See, for example, ⟨http://www.unics.uni-hannover.de/nhccurla/chemie/programme/shelx/SHELX-97%20Manual.pdf⟩ and ⟨http://shelx.uni-ac.gwdg.de/SHELX/⟩.

7.2.3.   Displacement factors (aka temperature factors)

The subject of temperature factors was introduced in Chapter 6 in discussing their effect on calculations involving the atomic scattering factor.

The displacement of an atom from its ideal position and shape results mainly from the effect of temperature-dependent vibrations of atoms or atomic groups, but also from disorder in a crystal structure and from intermolecular attractions. The illustration below shows the electron density map of the iodine atom and part of the ring structure of euphenyl iodoacetate $C_{32}H_{53}O_2I$. The contours around the (heavy) iodine atom in the molecule show significant anisotropy arising from enhanced vibrations at the end of the acetyl side-chain carrying the iodine atom; hydrogen atoms are not shown in this diagram.

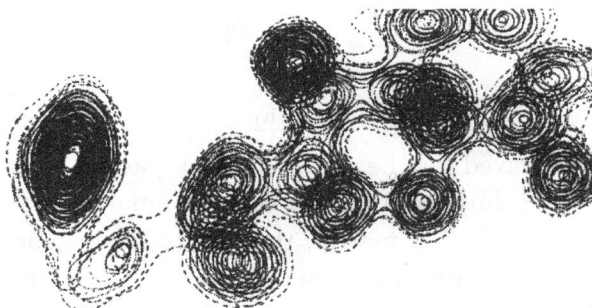

Partial electron density contour map of euphenyl iodoacetate

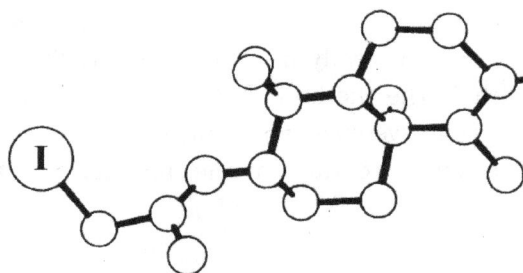

Idealized diagram of the partial structure of euphenyl iodoacetate shown above

The term *temperature factor* is used traditionally to embrace all atomic displacements. For the early stages of a structure determination, approximate overall temperature factor and scale factor are determined usually by the Wilson method described in Section 5.5.3.5. At this stage, the temperature factor is isotropic and is applied equally to all atoms. During refinement, the overall temperature factor B can be refined separately to an isotropic factor for each atom. However, atomic vibrations are generally not isotropic, particularly with structures of organic molecules. In the general case, the vibration is represented by a triaxial ellipsoid, specified by three principal vibration directions x, y and z:

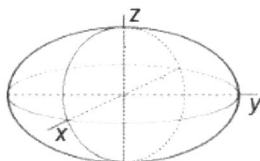

Thermal ellipsoid: x, y and z are principal vibration directions

Thus, there exist three parameters that may be used to represent a temperature correction parameter T to an atomic scattering factor

- B: Overall, isotropic temperature factor; $T = \exp[-B(\sin^2\theta)/\lambda^2]$
- B_j: Individual isotropic temperature for the jth atom; $T_j = \exp[-B_j(\sin^2\theta)/\lambda^2]$
- U: Individual anisotropic symmetrical tensor of six unique components $(U_{ij} = U_{ji})$; $T_j = \exp[-2\pi^2(U_{11}h^2a^{*2} + U_{22}k^2b^{*2} + U_{33}l^2c^{*2} + 2U_{12}hka^*b^* + 2U_{23}klb^*c^* + 2U_{31}lhc^*a^*)]$

The parameter B is related to atomic displacement by the equation

$$B = 8\pi^2\langle U^2\rangle = \frac{8\pi^2}{3}\text{trace}(\mathbf{U}) \tag{7.10}$$

where $\langle U^2\rangle$ is the projection of the mean square displacement of an atom in the direction of a given vector $\mathbf{h}$. In the case of a B_j parameter, $\langle U^2\rangle$ is replaced by $\langle U_j^2\rangle$ for the jth atom; $\langle U^2\rangle$ is sometimes given by the symbol U. The 3×3 tensor $\mathbf{U}$ is symmetrical with six unique components in the general case, namely, three diagonal

Table 7.1. Values of B, $\langle U^2 \rangle$ and rms displacement $\langle U^2 \rangle$.

$B/\text{Å}^2$	$\langle U^2 \rangle / \text{Å}^2$	$\langle U^2 \rangle / \text{Å}$
1.0	0.0127	0.113
3.0	0.0380	0.195
6.0	0.0760	0.276
10.0	0.127	0.356

terms and three off-diagonal terms. The presence of symmetry modifies the number of non-zero components. Thus, in the orthorhombic system, for example, only the U_{jj} terms are non-zero. Numerical relationships between B and U are listed in Table 7.1 for typical values of B. Large values of temperature factors should be examined carefully as they could represent inaccuracies in atomic positions or even wrong atom types. An example representation of anisotropic displacements of atoms, excluding hydrogen atoms, is shown hereunder for the molecule of pentamethylcyclopentadiene. It can be seen that the carbon atoms of the terminal methyl groups exhibit more vibration than do those bound into the ring structure.

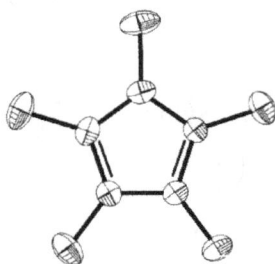

Molecule of pentamethylcyclopentadiene $C_{10}H_{16}$ (omitting hydrogen atoms) showing the thermal ellipsoids

Example 7.1. The surface of vibration of an orthorhombic crystal is a triaxial ellipsoid, and the mean square vibrational amplitude of the ith in the direction of a unit vector $\mathbf{V}(V_1, V_2, V_3)$ in reciprocal space is given by $\langle U_i^2 \rangle =$

Example 7.1. (*Continued*) $\sum_{i=1}^{3}\sum_{j=1}^{3}U_{ij}V_iV_j$. The component of $\mathbf{U}$ for atom i with $\mathbf{V}[100]$ parallel to a^*, is $\langle U_i^2 \rangle = U_{11}$. Let the vector $\mathbf{V}$ in the crystal lie in a direction $60°$ from a^* and in the $a^*\_b^*$ plane, and it has the components $[1/2, \sqrt{3}/2, 0]$. If $U_{11} = 0.01$ Å, $U_{22} = 0.03$ Å and $U_{33} = 0.02$ Å, then the component of $\mathbf{U}$ in that direction is given by $\langle U_j^2 \rangle = [(1/2)^2 U_{11}^2 + (\sqrt{3}/2)^2 U_{22}^2]^{1/2} = 0.026$ Å.

7.2.4. Scale factor

The scale factor in the early stages of an analysis can be modified by the expression $K = \sum_\mathbf{h} |\mathbf{F}_{\text{obs}}(\mathbf{h})| / \sum_\mathbf{h} |\mathbf{F}_{\text{calc}}(\mathbf{h})|$, but its final adjustment takes place in the least-squares refinement. Since the experimental data must not be modified in a least-squares process, the scale factor is applied as an inverse parameter on $|\mathbf{F}_{\text{calc}}(\mathbf{h})|$, namely, the term G in Eq. (7.1).

7.2.5. Weighting schemes

An esd $\sigma(\mathbf{h})$ for each observed reflection $\mathbf{h}$ may be obtained from X-ray diffractometer counting statistics as $\sigma_{|\mathbf{F}_{\text{obs}}(\mathbf{h})|} = \sqrt{N(\mathbf{h})}$, where $N(\mathbf{h})$ is determined by the peak and background counts for the reflection $\mathbf{h}$. A simple weighting scheme in the early stages of refinement is

$$w(\mathbf{h})^{-1} = \sigma_{|\mathbf{F}|_{\text{obs}}(\mathbf{h})}^2 \tag{7.11}$$

A more general weighting scheme is given by

$$w(\mathbf{h})^{-1} = \sigma_{|\mathbf{F}_{\text{obs}}(\mathbf{h})|}^2 + (aP)^2 + (bP) \tag{7.12}$$

where P is given by $P = 2|\mathbf{F}_{\text{calc}}(\mathbf{h})|^2 + 1/3 \max(|\mathbf{F}_{\text{obs}}(\mathbf{h})|^2, 0)$ [2]. A good weighting scheme leads to a constant or nearly constant value of $w(\mathbf{h})\Delta^2$ over local ranges of $|\mathbf{F}_{\text{obs}}(\mathbf{h})|$, where $\Delta = |\mathbf{F}_{\text{obs}}(\mathbf{h})| - |\mathbf{F}_{\text{calc}}(\mathbf{h})|$.

Self-assessment 7.1. An atom in a crystal was found to have an overall isotropic temperature factor of $3.5\,\text{Å}^2$. What is the mean displacement of this atom?[b]

7.2.6. Special conditions

It has been noted in Sections 4.3.1 and 4.11.2 that in certain space groups, the origin is not uniquely defined until the first atom is positioned in the unit cell. In space group $P2$, for example, the origin is taken on the two-fold axis, the line $[0, y, 0]$. Suppose that the unit cell contains one molecule of formula HR, where the atom H is heavier than those of the rest R of the molecule. The origin is specified by choosing a y coordinate for the atom H, say, $y_H = 0$, and it remains unchanged during the analysis and refinement. For an atom on the two-fold axis, the positions x, y, z and $\bar{x}$, y, $\bar{z}$ coincide then for $x = z = 0$. In refinement, the atom H would be given an occupancy factor of $1/2$ with respect to unity for the R atoms. In addition, for this atom, the anisotropic temperature factors are given as U_{11}, U_{22}, U_{33}, $\widetilde{U_{12}}$, $\widetilde{U_{23}}$, U_{31}, where the tilde is used here to indicate that U_{12} and U_{23} are both zero and invariant.

Limitations such as these are referred to as *constraints*. A constraint is an exact condition applied to an atom or a parameter that eliminates it from refinement while still contributing to a crystallographic procedure, such as structure factor calculation or least-squares refinement. Other examples of constraints are rigid bodies such as a phenyl ring regarded as a regular hexagon of carbon atoms with a C–C bond length of 1.4 Å and bond angle of 120°, or a hydrogen atom in *riding mode*, that is, in a position that is governed by the geometry of its attached entity. A related topic in refinement is a *restraint*, which is an item of additional information such as a carbon–carbon single bond distance restrained to be 1.54 Å within

---

[b]Answers to all self-assessments are given at the end of the chapter.

±0.02 Å, or the degree of planarity of a ring entity. Current crystallographic software provides facilities for incorporating these conditions in the refinement procedure.

7.2.7. Precision of the results

A satisfactory weighting scheme leads to parameters of lowest variances. The variance of a parameter p_j is related to the element a_{jj} of the matrix that is the inverse of matrix a_{jj} in Eq. (7.9). Then, $\sigma^2_{p_j}$ is given by

$$\sigma^2_{p_j} = \frac{1}{a_{jj}} \frac{\sum_{\mathbf{h}} w(\mathbf{h}) \Delta^2}{(N_r - N_p)} \tag{7.13}$$

where N_r and N_p are the numbers of reflections and the numbers of parameters, respectively. A value $N_r/N_p > 8$ is recommended for non-centrosymmetric crystals, while $N_r/N_p > 10$ leads generally to results of good precision centrosymmetric crystals.

7.2.8. Comments on the least-squares procedure

Normally, a crystal structure model for least-squares refinement should have all, or nearly all, atoms of the molecule located in approximately correct sites. The success of the least-squares procedure depends largely on the fact that the problem is overdetermined; the recommended data/variables ratio tends to compensate for discrepancies in individual intensity measurements. In some cases, the ratio can be increased by treating a group of atoms as a rigid body; a phenyl ring is one example of such a moiety.

A least-squares calculation operates only on the data supplied to it; it cannot find any atoms that are not present within the model. However, it can be advantageous in some cases to refine a lesser portion of the structure. This procedure may help with the example data set NIOP in the XRAY system of the Program Suite; this molecule contains nickel and sulphur atoms as heavy species in relation to the rest of the structure. Their coordinates, found from a Patterson synthesis, may be refined at that stage of the analysis.

The experimental data are subject to errors, which may be random or systematic in nature.Random errors are unavoidable and arise from the quantum nature of diffraction. The error is described through a Poisson distribution of counting statistics, with the error σ being $1/\sqrt{N}$, that is, the square root of the expected number of photons, as discussed earlier, and applied to both peak and background counts. Current quantum detectors allow random-error magnitude to be assessed, so that the data-collection strategy can be formalized. Reflections of low intensity and consequently high variance can be re-measured for a longer time, which is a straightforward matter in X-ray diffractometry. Systematic errors may be instrumental and/or operational and arise from many sources, some of which depend upon the correction for absorption, estimation of peak shape, and background count. They can be minimized by careful attention to the experimental procedure. Often, a data set can be improved by measurement at a low temperature, a procedure that increases the scattering through a decrease in thermal vibration. Diffractometer software is geared to handling these and many other data collection problems so as to deliver a data set of high quality. It is usual to measure more than just the asymmetric unit of intensity data. In such a case, the weighted mean intensity is given by

$$\langle I(\mathbf{h}) \rangle = \frac{\sum_j w_j(\mathbf{h}) I(\mathbf{h})}{\sum_j w_j(\mathbf{h})} \qquad (7.14)$$

where $w_j(\mathbf{h}) = 1/\sigma_j(\mathbf{h})^2$ and the sum is taken over all j symmetry-equivalent values of a given reflection $\mathbf{h}$. The quality of the merging procedure may be judged by R_{int} (aka R_{merge}), given as

$$R_{\text{int}} = \frac{\sum_{\mathbf{h}} \left[\sum_j |I_j(\mathbf{h}) - \langle I_j(\mathbf{h}) \rangle| \right]}{\sum_{\mathbf{h}} \left[\sum I_j(\mathbf{h}) \right]} \qquad (7.15)$$

A value of R_{int} in the region of 5% indicates a satisfactory data merging for refinement.

7.2.9. $|\mathbf{F}_{\text{obs}}|^2$ refinement

An alternative and, in many cases, preferred least-squares refinement procedure is based on $|\mathbf{F}_{\text{obs}}(\mathbf{h})|^2$ rather than on $|\mathbf{F}_{\text{obs}}(\mathbf{h})|$. The procedure is otherwise similar to that described above, and its progress can be tracked by the weighted R-factor, R_{w}, given by Eq. (6.28). One important advantage is the use of the complete data set rather than only those data that are limited by a cut-off value of $n\,\sigma(n = 2 - 4)$. Refinement based on $|\mathbf{F}_{\text{obs}}(\mathbf{h})|$ may lead to problems with very weak reflections and with reflections that measure negatively by virtue of the experiment method. These features introduce uncertainty in the evaluations of the esds of the refined parameters. The use of $|\mathbf{F}_{\text{obs}}(\mathbf{h})|^2$ generally avoids these problems, and is a feature of the widely used SHELX program system. The R-factor from an $|\mathbf{F}_{\text{obs}}(\mathbf{h})|^2$ refinement is always somewhat larger than that based on $|\mathbf{F}_{\text{obs}}(\mathbf{h})|$.

7.3. Measures of Correctness of an Analysis

The R-factor (Section 6.6.5.1) is one criterion of correctness because it compares the experimental structure amplitudes with those calculated from the model. An expectation value depends upon the size of the structure as well as on the analysis itself; for example, the R-factor for a refined protein structure is usually larger than that for a smaller molecule. This situation arises partially from the use of isotropic temperature factors; the change to anisotropic temperature correction decreases the data/variable ratio N_r/N_p significantly; a well-refined protein structure has an R-factor in the region of 0.1.

For smaller molecules, an R-factor can be 0.03 or even lower. The following histogram, derived from published data, shows that the majority of well-refined acceptable structures have R-factors within the range 0.03–0.07; R-factors greater than 0.11 total just 6%.

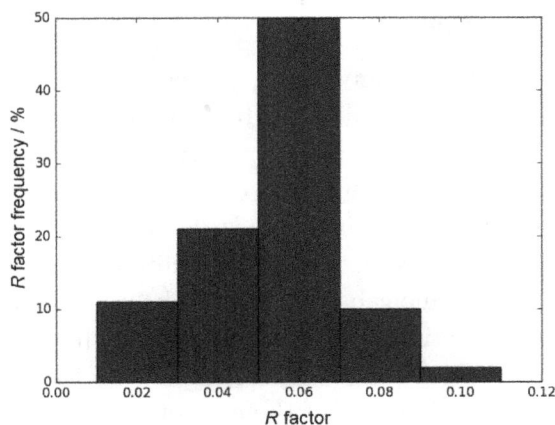

Frequency of published *R*-factors

In general, the *R*-factor, as an indicator of correctness, should always be treated with caution; a good structure determination must satisfy several criteria other than that of a low *R*-value:

- Bond lengths and bond angles in the model must be chemically sensible in terms of the published data. For example, a carbon–carbon single bond would be expected to be in the region of 1.54 Å. Standard bond lengths and bond angles are listed in Tables 7.2 and 7.3. Larger values than standard are known, but they must be supported by good structural evidence.

Table 7.2.   Selected standard bond lengths/Å.

Bond	Exemplar	Bond length	Bond	Exemplar	Bond length
C—H	R_2CH_2	1.07	C—O	Ester	1.36
C—H	Aromatic	1.08	C=O	Amide	1.24
C—H	RCH_3	1.10	C=C	Aldehyde	1.22
C—C	Hydrocarbon	1.54	C—S	R_2S	1.82
C—C	Aromatic	1.40	N—H	Amide	0.99
C=C	Ethene	1.33	O—H	Alcohol	0.97
C≡C	Ethyne	1.20	O=O	Peroxide	1.21
C—O	HCOOR	1.47	P—O	Ester	1.56
C=O	O=C—N	1.34	S—H	Thiol	1.33
C—O	Alcohol	1.43	S—S	Disulphide	2.05
Ph[a]—C	in Aromatics	1.52	Ph[a]—N	in Aromatics	1.42

[a]Phenyl ring, C_6H_5-.

An interesting example of an unusual bond length is the C–C bond of 1.704 Å in the diamondoid compound 2-(1-diamantyl) [121]tetramantane. The structure was solved and refined to $R = 0.045$ [3]:

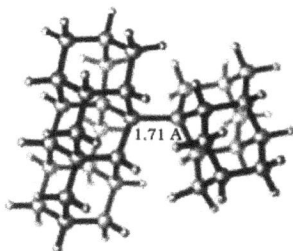

2-(1-Diamantyl)[121]tetramantane

- No temperature factors should be found to be non-positive definite. In such a case, the thermal ellipsoid should be checked for regularity in shape as an atom may be misplaced in position or type.
- The convergence of the refinement should have a shift/error ratio of *ca.* 0.05 or lower; all non-hydrogen anisotropic thermal parameters should be refined provided there are sufficient observed reflections for the number of variables. The Friedel pairs or equivalent data should be collected and merged if the structure is non-centrosymmetric.

Table 7.3.   Standard bond angles/deg.[a]

Atom type	Geometry	Angle
C4	Tetrahedral	109
C3	Planar	120
C2	Bent	109
C2	Linear	180
N4	Tetrahedral	109
N3	Pyramidal	109
N3	Planar	120
N2	Bent	109
N2	Linear	180
O3	Pyramidal	109
O2	Bent	109

[a]The 'Atom type' notation indicates the connectivity.

- The final difference Fourier map should be almost featureless. Any fluctuations arising from random errors should be less than the standard deviation of the electron density σ_ρ, given by

$$\sigma_\rho = \frac{1}{V} \left[\sum_{\mathbf{h}} (|\mathbf{F}_{obs}(\mathbf{h})| - |\mathbf{F}_{calc}(\mathbf{h})|)^2 \right]^{1/2} \tag{7.16}$$

- If appropriate, the absolute configuration of a non-centrosymmetric structure should be determined.

7.4.   Molecular Geometry

A significant part of the process of accepting a structure analysis as being reliable involves a determination of the geometry of the molecule, and a comparison of the results against standard chemical data such as bond lengths and bond angles. A judgement takes into account the esds of the parameters, together with the data of Tables 7.2 and 7.3. Appropriate data from Appendix A1 figure in these calculations.

7.4.1.   Bond lengths and bond angles

Three atoms, labelled 1, 2 and 3 in Fig. 7.1, form two bonds of lengths r_{12} and r_{32} and one bond angle ϕ_{123} with respect to the x-, y- and z-axes, which need not be orthogonal. Vectors $\mathbf{r}_j$ measured from the origin O may be written as

$$\mathbf{r}_j = x_j \mathbf{a} + y_j \mathbf{b} + z_j \mathbf{c} \tag{7.17}$$

The vector $\mathbf{r}_{12}$ between the two atoms 1 and 2 is then

$$\mathbf{r}_{12} = \mathbf{r}_2 - \mathbf{r}_1 = (x_2 - x_1)\mathbf{a} + (y_2 - y_1)\mathbf{b} + (z_2 - z_1)\mathbf{c} \tag{7.18}$$

The dot product of each side of Eq. (7.18) results in the general equation

$$r_{12}^2 = (x_2 - x_1)^2 \mathbf{a}^2 + (y_2 - y_1)^2 \mathbf{b}^2 + (z_2 - z_1)^2 \mathbf{c}^2$$
$$+ 2(y_2 - y_1)(z_2 - z_1)bc \cos\alpha + 2(z_2 - z_1)(x_2 - x_1)ca \cos\beta$$
$$+ 2(x_2 - x_1)(y_2 - y_1)ab \cos\gamma \tag{7.19}$$

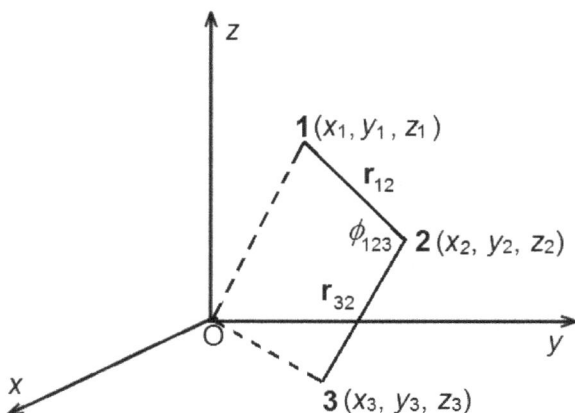

Fig. 7.1. Atoms 1, 2 and 3 forming bonds $\mathbf{r}_{12}(= \mathbf{x}_2 - \mathbf{x}_1)$ and $\mathbf{r}_{23}(= \mathbf{x}_3 - \mathbf{x}_2)$ and angle $\phi(= \angle 1 - 2 - 3)$, with respect to the axes x, y and z.

where the parameters have the meanings as before. The equation is simplified according to the crystal symmetry. For example, in a tetragonal crystal,

$$r_{12}^2 = [(x_2 - x_1)^2 + (y_2 - y_1)^2]\mathbf{a}^2 + (z_2 - z_1)^2\mathbf{c}^2 \qquad (7.20)$$

For the bond angle, since $\mathbf{r}_{12} \cdot \mathbf{r}_{32} = r_{12}r_{32}\cos\angle\mathbf{r}_{12}\mathbf{r}_{32}$, the bond angle ϕ_{123} in the diagram is given from the general case by

$$\cos\phi_{123} = \frac{\mathbf{r}_{12} \cdot \mathbf{r}_{32}}{r_{12}r_{32}} \qquad (7.21)$$

where the four parameters required are given by Eqs. (7.17)–(7.19). Again, symmetry can simplify the calculation.

Self-assessment 7.2. A monoclinic crystal has the unit-cell dimensions $a = 7.508$ Å, $b = 5.959$ Å, $c = 26.1720$ Å, $\beta = 92.55°$. Four atoms have the following coordinates: (1) 0.7820, 0.4187, 0.3789; (2) 0.7310, 0.4951, 0.4252; (3) 0.6524, 0.7075, 0.4297; (4) 0.6214, 0.8413, 0.3871. Using Eqs. (7.19)–(7.20), calculate the bond lengths and the bond angles in the aromatic molecular fragment 1–2–3–4.

7.4.2.   Torsion angles

Torsion angles χ are important in comparing related molecules or different possible conformations of one and the same molecule; the torsion angle range is $-180° < \chi \le 180°$. The value of a torsion angle in a situation of nominal freely-moving bonds may be a clue to steric factors in the solid state that inhibit any degree of rotation. In Fig. 7.2(a), four linked atoms 1–2–3–4 are shown; the torsion angle is χ_{1234}. By convention, the torsion angle is deemed *positive* if, when looking along the direction 2→3, a *clockwise* twist of the bond 2–3, carrying atom 1 with it, brings the bond 1–2 into an eclipsed position with respect to the bond 3–4. Figure 7.2(b) is the aspect of the four atoms as seen along the 2→3 direction; χ_{1234} is positive. In order to calculate the torsion angle, the two vector products in Eq. (7.22) are evaluated:

$$\mathbf{p}_1 = \mathbf{r}_{23} \times (-\mathbf{r}_{12})$$
$$\mathbf{p}_2 = \mathbf{r}_{23} \times \mathbf{r}_{24}$$

(7.22)

Then, the value of the torsion angle χ_{1234} is given by

$$\cos \chi_{1234} = \left(\frac{\mathbf{p}_1 \cdot \mathbf{p}_2}{p_1 p_2} \right)$$

(7.23)

However, if the following expression is used,

$$\chi_{1234} = \text{ATAN2}\{ |\mathbf{r}_{23}|\mathbf{r}_{12} \cdot (\mathbf{r}_{23} \times \mathbf{r}_{34}), (\mathbf{r}_{12} \times \mathbf{r}_{23}).(\mathbf{r}_{23} \times \mathbf{r}_{34})\}$$

(7.24)

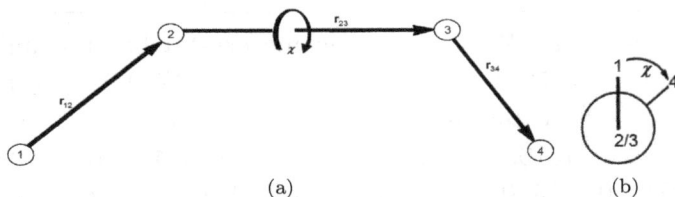

(a)   (b)

Fig. 7.2.   Torsion angles: (a) A sequence 1–2–3–4 of four linked atoms forming the torsion angle χ. (b) The same sequence as in (a), but with the view along the direction $2 \rightarrow 3$; atom 3 lies below atom 2 with respect to the line of observation. The torsion angle χ is positive when a clockwise twist of the bond 2–3, carrying atom 1 with it, brings the bond 1–2 into the eclipsed position with respect to the bond 3–4.

where ATAN2 is a function (as written here in Fortran) for $\tan^{-1}$ with two arguments, the torsion angle is obtained with its correct sign. This calculation is somewhat lengthy, and is included in the program MOLGOM in the Program Suite.

> **Self-assessment 7.3.** Use the data from Self-assessment 7.2 to determine the torsion angle χ_{1234} using MOLGOM, and comment on the result. The program will also return the values of the bond lengths and bond angles from Self-assessment 7.2.

7.4.3. Best-fit plane

The extent of planarity of a ring or ring system is required frequently to be determined in a structural investigation. A least-squares best-fit plane calculation is straightforward, and its application to a group of atoms is discussed in Section 8.3.8. The program PLANE in the Program Suite determines the best-fit plane and the perpendicular deviations of the data points from it.

7.4.4. Precision of molecular geometry

Desirably, geometrical parameters of a crystal structure determination require accompanying expressions of their precision. Normally, a least-squares refinement process reports esds for these parameters; thus, an atomic coordinate will be expressed, for example, as 0.1234(5), where the number in parentheses is to be applied to the final digit of the parameter.

In general, errors depend upon more than one variable, and Appendix A11 describes the propagation of errors. From Appendix A11.3, Eq. (A11.14) may be cast in the form

$$\sigma_d^2 = \sum_{j=1}^{n} \left(\frac{\partial d}{\partial p_j} \right)^2 \sigma_{p_j}^2 \tag{7.25}$$

where d is a bond length, or other parameter, dependent upon n variables p_j ($j = 1, n$), and σ^2 is the variance of a parameter and its esd is $\sqrt{\sigma^2}$.

Self-assessment 7.4. Consider a bond formed by two atoms lying along the c edge of a tetragonal unit cell of dimensions $a = 5.513(1)$ Å and $c = 10.060(2)$. The fractional coordinates of the atoms are $A = 0, 0, 0.5418$ and $B = 0, 0, 0.3712$. Calculate the bond length d_{AB} and its esd.

Similar calculations can be set up for all distance and angle parameters that emerge from a crystal structure analysis.

7.4.5. Significance testing

In addition to the calculation of R-factors for a structure determination and the esds in its derived parameters, both of which have been discussed already, another statistic that examines the quality of a refinement is the goodness-of-fit parameter S (aka *Goof*), defined by

$$S = \left\{ \frac{\sum_{\mathbf{h}} [w(\mathbf{h})(|\mathbf{F}_{\text{obs}}(\mathbf{h})|^2 - |\mathbf{F}_{\text{calc}}(\mathbf{h})|^2)^2}{N_r - N_p} \right\}^{1/2}$$

For a well-refined structure, S should tend to unity. In principle, $-1 \leq S \leq 1$, where -1 indicates total negative correlation, 0 implies no correlation and $+1$ represents complete positive correlation.

A further possible disparity may arise from two results that are expected to be equal within experimental error. A crystal structure containing two or more molecules in the asymmetric unit and unrelated by symmetry can provide such a situation. It can be studied by a form of chi-square test and is examined here for bond lengths in two molecules in the crystal structure of the organic species (I):

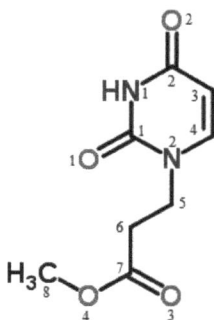

Molecule (I); the numbering relates to the atoms listed in Tables 7.4 and 7.5

It involves the calculation of the function $\sum_{j=1}^{n}(\Delta_j/\sigma_{\Delta_j})^2$ distributed with n degrees of freedom, where Δ_j is the difference in the lengths of bond j in the molecules 1 and 2; σ_{Δ_j} is the esd of Δ_j, and is determined as

$$\sigma_{\Delta_j} = \sqrt{\sigma_{d_{j1}}^2 + \sigma_{d_{j2}}^2}$$

The bond lengths in the molecule are examined in two sets (excluding bonds to the hydrogen atoms): the ring structure ($n = 8$ pairs of measurements), and the side-chain ($n = 6$ pairs of measurements); the two sets of measurements are assumed to be uncorrelated.

The null hypothesis is that all differences in the pairs of bond lengths arise from random experimental errors. Then, standard statistical tables are used to determine the significance level of the test, which is the probability p of incorrectly rejecting the hypothesis. The test is deemed not significant unless p is less than 0.05, the level that is chosen generally. The bond length and chi-square data are set out in Tables 7.4 and 7.5.

For the ring structure, the sum of the $(\Delta_j/\sigma_{\Delta_j})^2$ values, or χ^2, from Table 7.4 is 9.98 and for the side-chain from Table 7.5, it is 12.82. Turning now to standard statistical tables [4], part of which is listed in Table 7.6, for the ring of eight items ($n = 8$ degrees of freedom) the probability lies between 0.2 and 0.3 and so is not significant, that is, the discrepancies arise from random errors. In the case of the side-chain of six items ($n = 6$ degrees of freedom),

Table 7.4. Bond lengths for portion 1 of a molecule of formula $C_8H_{10}N_2O_4$.

Bond	Length/Å Mol. 1	Length/Å Mol. 2	Δ_j/Å	σ_{Δ_j}/Å	$\Delta_j/\sigma_{\Delta_j}$	$(\Delta_j/\sigma_{\Delta_j})^2$
C_1—N_1	1.476(2)	1.479(2)	0.003	0.00283	1.060	1.124
C_2—N_1	1.469(2)	1.474(5)	0.005	0.00283	1.767	3.122
C_2—C_3	1.512(3)	1.508(3)	0.004	0.00424	0.943	0.889
C_3—C_4	1.511(2)	1.513(2)	0.002	0.00283	0.707	0.499
C_4—N_2	1.474(2)	1.472(2)	0.002	0.00283	0.707	0.499
C_1—N_2	1.480(3)	1.478(3)	0.002	0.00424	0.472	0.222
C_1—O_1	1.217(2)	1.215(2)	0.002	0.00283	0.707	0.499
C_2—O_2	1.214(2)	1.209(2)	0.005	0.00283	1.767	3.122

Table 7.5. Bond lengths for portion 2 of a molecule of formula $C_8H_{10}N_2O_4$.

Bond	Length/Å Mol. 1	Length/Å Mol. 2	Δ_j/Å	σ_{Δ_j}/Å	$\Delta_j/\sigma_{\Delta_j}$	$(\Delta_j/\sigma_{\Delta_j})^2$
C_5—N_2	1.415(2)	1.410(2)	0.005	0.00283	1.767	3.122
C_5—C_6	1.516(3)	1.528(3)	0.011	0.00424	2.594	6.731
C_6—C_7	1.510(3)	1.512(3)	0.002	0.00424	0.472	0.222
C_7—O_3	1.213(2)	1.216(2)	0.003	0.00283	1.060	1.124
C_7—O_4	1.392(2)	1.395(2)	0.003	0.00283	1.060	1.124
C_8—O_4	1.427(2)	1.429(2)	0.002	0.00283	0.707	0.499

Table 7.6. Statistical chi-square distribution table.

Probability	0.95	0.90	0.80	0.70	0.50	0.30	0.20	0.10	0.05	0.01	0.001
1	0.004	0.02	0.06	0.15	0.46	1.07	1.64	2.71	3.84	6.64	10.83
2	0.10	0.21	0.45	0.71	1.39	2.41	3.22	4.60	5.99	9.21	13.82
3	0.35	0.58	1.01	1.42	2.37	3.66	4.64	6.25	7.82	11.34	16.27
4	0.71	1.06	1.65	2.20	3.36	4.88	5.99	7.78	9.49	13.28	18.47
5	1.14	1.61	2.34	3.00	4.35	6.06	7.29	9.24	11.07	15.09	20.52
6	1.63	2.20	3.07	3.83	5.35	7.23	8.56	10.64	12.59	16.81	22.46
7	2.17	2.83	3.82	4.67	6 35	8.38	9.80	12.02	14.07	18.48	24.32
8	2.73	3.49	4.59	5.53	7.34	9.52	11.03	13.36	15.51	20.09	26.12
9	3.32	4.17	5.38	6.39	8.34	10.66	12.24	14.68	16.92	21.67	27.88
10	3.94	4.86	6.18	7.27	9.34	11.78	13.44	15.99	18.31	23.21	29.59

Degrees of freedom n (left axis); Chi-square χ^2 (right axis). Non-significant | Significant

the probability lies between 0.01 and 0.05 so that the differences are significant. While the results are satisfactory for the data presented in the tables of data, the discrepancy in the C_5–C_6 bond in the side-chain invites further experimental investigation.

7.5. Summary of Typical Parameters from a Crystal Structure Analysis

A study of the literature on crystal structure analyses shows that scientific journals vary in the totality of reported parameters. The following list contains the parameters that would probably appear in the details of a good crystal structure analysis.

Crystal

Chemical name, formula and relative molecular mass; crystallization solvent, crystal size and habit. In the crystallization process, two liquids are used frequently, one in which the material is very soluble and another in which it is far less soluble.

Data collection

Electronic image device; photographs, if used; diffractometer type and X-ray wavelength λ; method of collection and temperature; θ-range for unit-cell determination; unit-cell parameters $a, b, c, \alpha, \beta, \gamma$ and volume V_c, density D_x (experimental), all with esds; D_c (calculated) and number Z_c of molecules in the unit cell; space group; F(000).

Data analysis

Range of θ for data collection; total data collected; cut-off range, if applicable; total unique data (after merging) R, weighted R and R_{int}; percentage completeness and h, k, l ranges; absorption coefficient.

Structure solution

Structure-solving method and refinement procedure, $|\mathbf{F}|^2$ or $|\mathbf{F}|$ — full matrix expected; type of temperature factors; reflections/parameters ratio; constraints/restraints, if applicable; hydrogen atoms and their treatment (riding model); weighting scheme.

Correctness evaluation

R and R_w, and whether on all data or to an $n\sigma$ limit $(n = 2 - 4)$; goodness-of-fit parameter (on $|\mathbf{F}|^2$) Goof; average shift/error ratio; maximum shift/error ratio; maximum and minimum on difference Fourier map.

Structural features

Bond lengths, bond angles, torsion angles all with esds; molecular (full or partial) planarity and/or symmetry; hydrogen bonding, packing of molecules and steric interference.

7.6.   Neutron Diffraction Studies

In Section 2.2, the variation of symmetry with the nature of the examining probe was discussed and exemplified by a comparison of the scattering of X-rays and of neutrons from metallic chromium. Neutrons are electrically neutral particles of greater penetration power than X-rays. Whereas X-rays are scattered by electrons with a power that increases with increasing atomic number, neutrons are scattered by atomic nuclei in a manner that depends upon various factors, such as isotopic composition and spin. The relative scattering power of selected atomic species is illustrated below, where 'black' indicates a negative scattering factor:

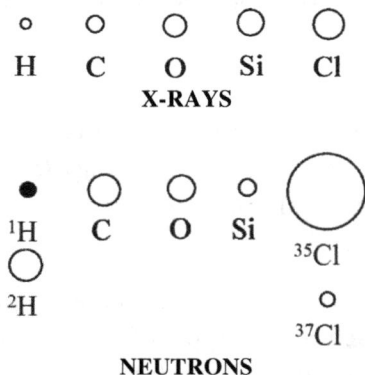

From this comparison, it may be seen that neutron diffraction studies can distinguish between isotopes of an element. However, the very important property of neutron diffraction in crystal structure analysis is its ability to resolve very light atoms in the presence of heavy species. The crystal structure of sodium hydride, shown in Fig. 7.3, was one of the first to be determined by neutron diffraction. It has the

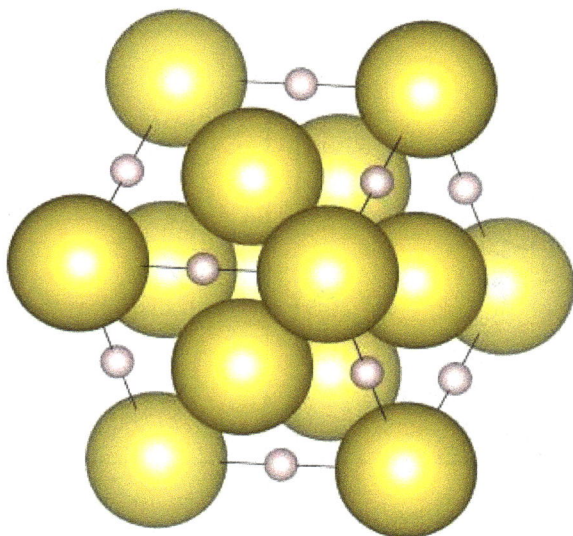

Fig. 7.3.   The crystal structure of sodium hydride NaH. The very small degree of X-ray scattering from the hydrogen atoms is swamped by that from the sodium atoms, so that hydrogen-atom contours are not observed on an electron density map.

sodium chloride structure type (Fig. 4.1c) and the unit-cell dimension a is 4.98 Å; it is not difficult to appreciate that the small scattering from the hydrogen atoms is swamped by that from sodium.

Of particular significance is the location of the positions of hydrogen atoms in molecules, particularly organic molecules and proteins, or in molecules of water of crystallization; another early example of a neutron analysis is that of potassium dihydrogen phosphate. In practice, deuterium may be substituted for hydrogen as the scattering cross-section for hydrogen is negative and these atoms are revealed by negative contours on a density map. In Fig. 7.4, illustration (a) shows the neutron scattering density map for potassium dihydrogen phosphate KH_2PO_4; the hydrogen atoms are revealed clearly and their density contours are negative. In (b), which is the corresponding difference Fourier map, but with hydrogen replaced by deuterium, deuterium atoms are well resolved, with positive contours, against an almost flat background.

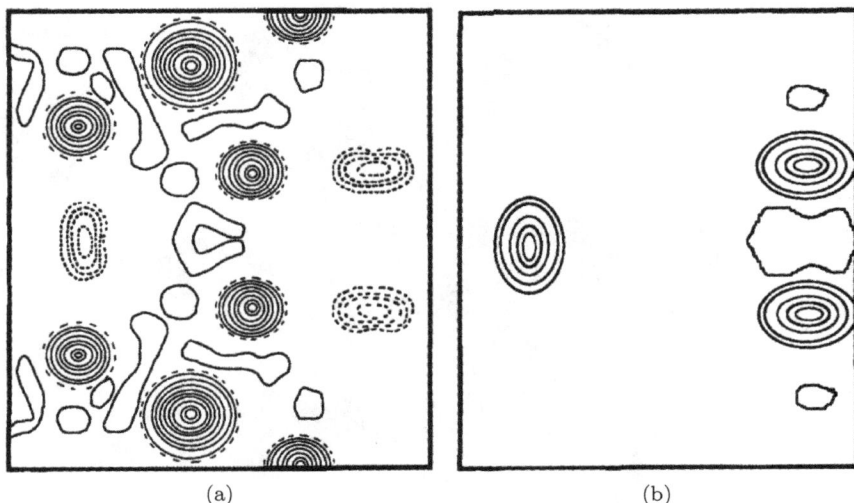

(a)   (b)

Fig. 7.4. Neutron diffraction study of potassium dihydrogen phosphate. (a) The (001) density map projection for KH_2PO_4. The hydrogen atom contours are clearly resolved, but are negative because the scattering cross-section for neutrons by hydrogen is a negative quantity. (b) The (001) *difference* density map of the same projection for KD_2PO_4. The deuterium atom (positive) density contours are clearly resolved against an almost featureless background; the neutron cross-section is positive for deuterium. [Reproduced with permission from Bacon GE, *Neutron Diffraction*, 3rd edn. Clarendon Press (1975).]

In general, neutron diffraction data collection and structure solution follow the same paths as does X-ray diffraction, although experimental arrangements differ in certain aspects. Frequently, X-ray studies are a precursor to the use of neutron diffraction, and in conjunction with X-ray analysis, neutron diffraction presents an important additional analytical tool. Space limitation permits only this brief mention of neutron diffraction; further detailed discussions of this subject are available in the literature [5–7].

7.7. Powder Diffraction

X-ray powder diffraction was mentioned briefly in Section 5.5.2 with the discussion on X-ray diffractometry, and an example of a modern X-ray powder diffractometer was shown in Fig. 5.26. The basic

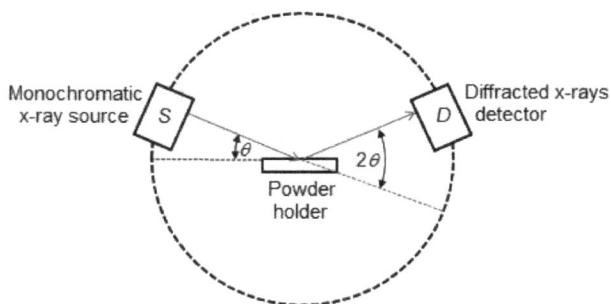

Fig. 7.5. The basic arrangement for X-ray powder diffraction using Debye–Scherrer geometry. An alternative arrangement is the Bragg–Brentano setting which is used in several powder diffractometers [8]. In the illustration, monochromatic X-rays from the source S strike the powder sample, and the diffracted beams are collected and measured by the detector D. The Bragg angle is θ and the scattering angle is 2θ.

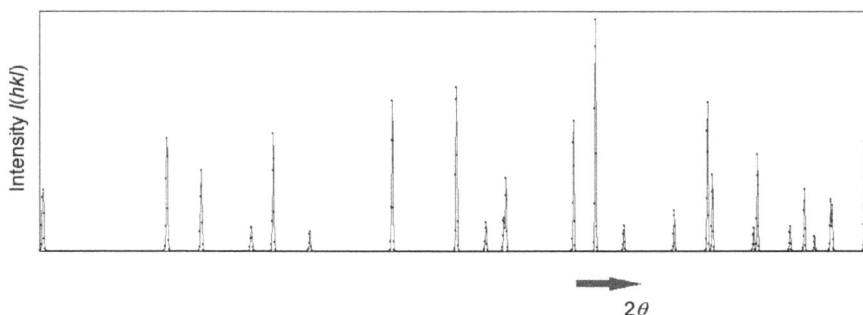

Fig. 7.6. X-ray diffractometer trace for finely powdered aluminium oxide Al_2O_3. The very good resolution of the peaks leads to precise values for the measured θ-angles.

arrangement for powder diffraction is straightforward, as illustrated in Fig. 7.5.

X-ray powder diffractometry (XRPD) is a sufficiently well-developed technique for crystal structure analysis to be carried out with materials that cannot be crystallized to a satisfactory single-crystal size. To date, a large number of structure analyses from powder data, including proteins, have been published. A typical trace from an X-ray powder diffractometer is shown in Fig. 7.6. The high resolution shown by the diagram permits a very good estimate of the value of θ, the only parameter given by this diffraction method.

Once the data have been collected, the first stage in structure analysis involves indexing the pattern. Several methods have been published for solving the indexing procedure. One such technique, based on Ito's method, is the Visser zone-indexing program [9, 10]. An executable copy of the Visser program (ITO12) is in the Program Suite (by kind permission of Dr J W Visser) together with a data set for potassium nitrate, which can be used to follow through the procedure for this particular method.

Data are input usually as values of either $10^4 Q(hkl)$ or $2\theta(hkl)$. Q-values are calculated in terms of the general equation for d^*:

$$d^{*2}(hkl) = h^2 a^{*2} + k^2 b^{*2} + l^2 c^{*2} + 2klb^* c^* \cos \alpha^*$$

$$+ 2lhc^* a^* \cos \beta^* + 2hka^* b^* \cos \gamma^* \qquad (7.26)$$

This equation is generally cast in terms of Q values, such that $Q(hkl) = d^{*2}(hkl), Q_A = a^{*2}, \ldots, Q_D = 2b^* c^* \cos \alpha^*, \ldots$, hence,

$$Q(hkl) = h^2 Q_A + k^2 Q_B + l^2 Q_C + kl Q_D + lh Q_E + hk Q_F \qquad (7.27)$$

and applied as $10^4 Q$ (hkl), for convenience. For example, if $c = 5.417$ Å, then $d^*_{001} = 0.18460$ Å^{-1} and $10^4 Q_{001} = 10^4 d^{*2}_{001} = 340.8$ Å^{-2}. Also,

$$d^{*2}(hkl) = \frac{1}{d^2(hkl)} = \frac{4 \sin^2 \theta}{\lambda^2} \qquad (7.28)$$

The program QVALS in the Program Suite calculates Q-values from unit-cell data according to the hkl values supplied.

Self-assessment 7.5. Potassium nitrate is orthorhombic, with $a = 6.425$ Å, $b = 9.171$ Å and $c = 5.417$ Å. Report the values of $10^4 Q(hkl)$, $d(hkl)$ and $2\theta(hkl)$ for the hkl data 010, 110 and $1\bar{2}\bar{3}$; $\lambda = 1.5406$ Å.

Readers who wish to pursue this topic are recommended to the useful and comprehensive literature now available [11–13].

7.8. Databases

A vast resource of crystal structure data is available through several different monitored and updated listings. The following are organizations' archived or published crystal structure data for the use and benefit of those whose work involves knowledge of accurate information on crystal structures.

American Mineralogist Crystal Structure Database
⟨http://rruff.geo.arizona.edu/AMS/amcsd.php⟩

Bilbao Crystallographic Server (BCC)
⟨http://www.cryst.ehu.es/⟩

Biological Macromolecule Crystallization Database (BMCD)
⟨https://www.hsls.pitt.edu/obrc/index.php?page=URL113319518
9⟩

Cambridge Crystallographic Database
⟨https://www.ccdc.cam.ac.uk/1176x176-Banner-0217(002).jpg-875
K.jpg⟩; 875,000 entries

Cambridge Structural Database
⟨https://www.ccdc.cam.ac.uk/⟩

Crystallography Open Database (COD)
⟨http://www.crystallography.net/cod/⟩;
386,734 entries

Inorganic Chemistry Structural Database (ICDB)
⟨https://icsd.fiz-karlsruhe.de/search/index.xhtml;jsessionid=B19A
71FB904933081BEDF74E0CA38C8D⟩;
187,000 entries (since 1913)

International Centre for Diffraction Data (ICCD)

⟨http://www.icdd.com/⟩
The Powder Diffraction File^{TM}& Related Products

Marseille Protein Crystallization Database (MPCD)

⟨http://www.cinam.univ-mrs.fr/mpcd/⟩

Pearson's Crystal Structure Database

⟨https://www.asminternational.org/materials-resources/online-dat
abases/-/journal_content/56/10192/6382084/DATABASE⟩;
304,000 entries

Predicted Powder Diffraction Database (P2D2)

⟨http://www.crystallography.net/pcod/P2D2/⟩;
1,000,000 entries

- ZEFSA II; 898,707 entries non-zeolites
- GRINSP Inorganic crystal structures generator

Protein Databank (PDB)

⟨https://www.rcsb.org/pdb/home/home.do⟩

- Protein Sequences; 42,645 entries
- Human Sequences; 37,730 entries
- Structures containing nucleic acids; 9,645 entries

Wyckoff RWG. *Crystal Structures*, Vols. 1–4, Supplement 5 (1961).
  The websites contain contact information of the various organizations for further details on the use of or comments on the databases.

Problems

7.1. (a) Find the scale and temperature factors with Wilson's
     method from the following five ranges of data:

$\sum_j f_j^2(hkl)/\langle|\mathbf{F}_{\text{obs},j}(hkl)|^2\rangle$	$\langle[\sin^2\theta_j(hkl)]/\lambda^2\rangle$
54.27	0.100
271.8	0.200
663.1	0.300
2841	0.400
11956	0.500

(b) Calculate the mean square displacement of the atoms corresponding to the value found for B.

7.2. The orthorhombic unit cell of 1-methyl-2-chlorobenzene has the dimensions $a = 7.210(4)$ Å, $b = 10.43(1)$ Å, $c = 15.22(2)$ Å, with $Z_c = 4$. The coordinates of the chlorine atoms are $\pm(1/4, y, z; 1/4, 1/2 + y, 1/2 + z)$. Given that $y(\text{Cl}) = 0.140(2)$ and $z(\text{Cl}) = 0.000(2)$, calculate the shortest Cl- - -Cl distance and its esd. (A sketch of the atomic positions may be helpful.)

7.3. Determine a set of Cartesian coordinates for the carbon atoms in benzene given that C–C $= 1.400$ Å and $\angle$C–C–C $= 120°$.

7.4. An atom J in an orthorhombic crystal has the following anisotropic thermal tensor:

$$\mathbf{U}_J = \begin{pmatrix} 0.0200 & 0 & 0 \\ 0 & 0.0300 & 0 \\ 0 & 0 & 0.0400 \end{pmatrix}$$

(a) What is the equivalent isotropic temperature factor for this atom? (b) What isotropic temperature correction to the atomic scattering factor for atom J arises for a Bragg angle of $30°$ and $\text{Cu}K\alpha$ radiation ($\lambda = 1.5418$ Å)?

7.5. The following coordinates were obtained from the crystal structure analysis for a tetracyclic compound of molecular formula $C_{16}H_9OBr$; the unit-cell dimensions of the crystal are

$a = 7.508(4)$ Å, $b = 5.959(5)$ Å, $c = 26.172(6)$ and $\beta = 92.55(2)°$. (a) Calculate the bond lengths and bond angles with the program MOLGOM.

(b) Sketch the molecular formula of $C_{16}H_9OBr$; in the list of coordinates below, atom 1 is bromine and atom 7 is oxygen.

Atom	Atom	Atom coordinates		
number	name	x	y	z
1	Br	0.7602	0.3152	0.4848
2	C1	0.7820	0.4187	0.3789
3	C2	0.7310	0.4951	0.4252
4	C3	0.6524	0.7075	0.4297
5	C4	0.6214	0.8413	0.3871
6	C4a	0.6794	0.7619	0.3406
7	O5	0.6520	0.9051	0.2995
8	C5a	0.6973	0.8329	0.2526
9	C6	0.6714	0.9397	0.2077
10	C6a	0.7384	0.7990	0.1678
11	C7	0.7401	0.8230	0.1150
12	C8	0.8078	0.6526	0.0858
13	C9	0.8766	0.4574	0.1079
14	C10	0.8731	0.4268	0.1605
15	C10a	0.8035	0.5954	0.1908
16	C10b	0.7767	0.6076	0.2454
17	C11	0.8064	0.4734	0.2850
18	C11a	0.7593	0.5475	0.3359

7.6. Figure P7.1 is the sharpened, origin-removed Patterson $hk0$ projection of the unit cell for a structure containing one nickel atom and two sulphur atoms in the asymmetric unit. (a) The space group of the crystal is $P2_12_12_1$; what is the plane group symmetry? (b) Determine the x and y coordinates for the nickel and sulphur atoms in the unit cell.

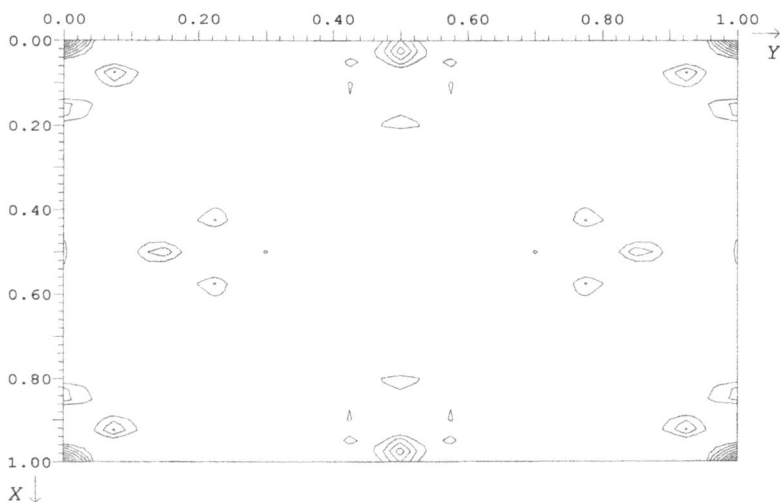

Fig. P7.1. Sharpened Patterson projection for the nickel *o*-phenanthroline compound.

7.7. A crystal, believed to be C_4HO_4K, was examined by X-rays and found to be monoclinic holosymmetric (highest symmetry of its system), with the unit-cell dimensions $a = 8.641$ Å, $b = 10.909$ Å, $c = 6.563$ Å and $\beta = 99.81°$. The density found by experiment ρ_x was 1.855 g cm^{-3}. What additional information can be derived from this data?

7.8. Continue the crystal structure analysis of the nickel *o*-phenanthroline derivative $C_{12}H_{14}N_6S_2Ni$ that was the subject of Problem 7.6, from which the x and y coordinates of the nickel and sulphur atoms were deduced. Use the method of successive Fourier synthesis, and least-squares refinement. *Note.* The electron density contours from the program XRAY are in intervals of 10 units. In order to find some of the atoms, it will be necessary to print the electron density field figures and draw in the level 5 contours.

7.9. The crystal structure analysis of the organic compound (A) gave the tabulated coordinates for the atoms (excluding hydrogen) of the following molecule:

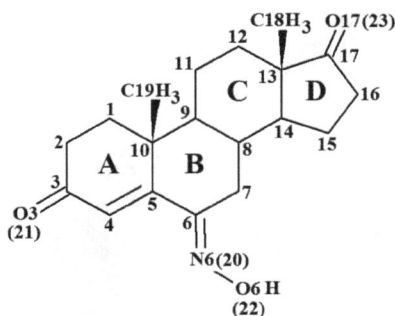

Molecular formula of compound A
(The numbers in parentheses refer to the atom (other than carbon) numbers for a MOLDAT.TXT file)

The crystals are orthorhombic, with the unit-cell dimensions $a = 6.2642$ Å, $b = 12.5737$ Å, $c = 20.0368$ Å, $\alpha = \beta = \gamma = 90°$.

(a) Determine the best-fit plane for the ring system of carbon atoms C_1 to C_{17} of the molecule, using the program PLANE. *Note that the inputs to the programs PLANE and MOLGOM do not require the atom names shown in column 2 of the data below.*

(b) Using the program MOLGOM, evaluate the bond lengths and bond angles for the molecule.

(c) Consider any significant departure from planarity, and determine any appropriate torsion angles with MOLGOM that could reveal the extent of non-planarity.

Atomic coordinates for compound A

Atom number	Atom	x	y	z
1	C1	0.295011	0.137395	0.401787
2	C2	0.337743	0.028395	0.371497
3	C3	0.274926	0.019936	0.299579
4	C4	0.224121	0.119153	0.264316
5	C5	0.271931	0.216405	0.288246

(*Continued*)

(*Continued*)

Atom number	Atom	x	y	z
6	C6	0.241911	0.312091	0.245609
7	C7	0.215768	0.419365	0.277361
8	C8	0.335198	0.432552	0.343117
9	C9	0.288289	0.336712	0.388941
10	C10	0.368846	0.230925	0.357541
11	C11	0.369240	0.352673	0.461051
12	C12	0.308975	0.460423	0.491836
13	C13	0.378119	0.549939	0.445548
14	C14	0.267293	0.534950	0.377526
15	C15	0.298253	0.642679	0.342358
16	C16	0.268350	0.723000	0.399470
17	C17	0.301849	0.660740	0.463476
18	C18	0.623167	0.558279	0.440136
19	C19	0.613876	0.229323	0.349011
20	N6	0.235947	0.294089	0.182705
21	O3	0.273384	−0.065560	0.270664
22	O6	0.196193	0.386977	0.145684
23	O17	0.268088	0.694083	0.519337

Further structure analyses

Additional practice with the structure-solving procedures discussed herein can be gained by working through any of the other data sets presented with the XSYST program system. In particular, the three-dimensional crystal structure of 2-S-methylthiouracil can be determined from the two data sets, SMTX and SMTY, provided for this compound. The molecular geometry of this compound is shown at the end of the Tutorial Solutions chapter.

The data set for nitroguanidine (NO2G) is very complete, and is a good example for the practice of direct methods. Note that the true a and b unit-cell dimensions are halved in the (001) projection of *Fdd*2. Also, with *ca.* 3.6 Å (cp. the van der Waals non-bonded distance), good resolution of the molecule can be expected in the (001) projection.

The correctness of the solution of any of the structures should be judged according to the criteria discussed in Section 7.3.

Answers to Self-Assessments

7.1. Since $B = 8\pi^2 \langle U^2 \rangle$, the mean displacement is $\sqrt{\langle U^2 \rangle} = \sqrt{\frac{3.5\ \text{Å}^2}{8\pi^2}} = 0.21$ Å.

7.2. Using Eqs. (7.19) and (7.20), simplified by symmetry, gives the following results:

1–2   1.365 Å;   2–3   1.404 Å;   3–4 1.382 Å.
1–2–3   120.7°;   2–3–4   120.6°. (The results could also be obtained with the program METTENS.)

7.3. From the program MOLGOM, $\chi_{1234} = 1.49°$, which implies that the four-atom fragment is almost planar; this would be expected for an aromatic system. (The results for Self-assessment 7.2 are also listed.)

7.4. From Section 7.4.1, $d_{AB} = (z_B - z_A)c = (0.5418 - 0.3712) \times 10.060$ Å $= 1.7162$ Å. Also,

$$\sigma^2_{d_{AB}} = [(0.5418 - 0.3712) \times (0.02\ \text{Å})]^2$$
$$+ [10.060\ \text{Å} \times 0.0002\ \text{Å}]^2 + [10.060\ \text{Å} \times 0.0003\ \text{Å}]^2$$
$$= 1.6067 \times 10^{-5}\ \text{Å}^2, \text{ so that the esd } \sigma = 0.0040.$$

The result would be written as $d_{AB} = 1.716(4)$ Å.

7.5. $Q = 10^4 d(hkl)^{*2} = 10^4 \left[\left(\frac{h^2}{a^2}\right) + \left(\frac{k^2}{b^2}\right) + \left(\frac{l^2}{c^2}\right) \right]$, hence

h	k	l	Q(hkl)	d(hkd)	2θ(hkl)
0	1	0	118.9	9.171	4.818
0	1	0	361.1	5.262	8.418
1	$\bar{2}$	$\bar{3}$	378.5	1.625	28.29

References

[1] Collaborative Computation Project (CCP4), *Software for Macromolecular X-ray Crystallography*. Online at ⟨http://www.ccp4.ac.uk/⟩.

[2] Wilson AJC, *Acta Crystallogr. A* **32**, 994 (1976).

[3] Fokin AA *et al.*, *J. Amer. Chem. Soc.* **134**, 13641 (2012).

[4] University of Baltimore. ⟨https://home.ubalt.edu/ntsbarsh/Business-stat/S tatistialTables.pdf⟩.

[5] Bacon GE, *Neutron Diffraction*, 3rd edn. Clarendon Press (1975).

[6] Kisi EH and Howard CJ, *Applications of Neutron Powder Diffraction*. Oxford University Press (2008).

[7] Wilson CC, *Single Crystal Neutron Diffraction from Molecular Materials*. World Scientific (2000).

[8] MIT. Online at ⟨http://prism.mit.edu/xray/oldsite/Basics%20of%20X-Ray %20Powder%20Diffraction.pdf⟩.

[9] Collaborative Computational Project. Online at ⟨http://www.ccp14.ac.uk/t utorial/crys/program/ito12.htm⟩.

[10] Cockcroft JK. Online at ⟨http://pd.chem.ucl.ac.uk/pdnn/unit2/zones.htm⟩.

[11] Dinnebier RE and Billinge SJL (eds), *Powder Diffraction: Theory and Practice*. ESC Publishing (2008).

[12] Will G, *Powder Diffraction*. Springer (2006).

[13] Pecharsky V and Zavalij P, *Fundamentals of Powder Diffraction and Structural Characterization of Materials*. Springer (2003).

Chapter 8

Computer-Assisted Studies in X-ray Crystallography

If we really understand the problem,
the answer will come out of it, because
the answer is not separate from the problem.
Jiddu Krishnamurti

Key Topics

- Program Suite outlined
- Plotting system
- Commands with the Python interpreter
- X-ray program system, with examples
- Other program systems, with examples
- Crystallographic tables

8.1. Introduction

The Program Suite has been devised and assembled for use with this book, since computational methods are an essential feature of any modern scientific endeavour. Computers do not teach; rather, they assist in both the study of scientific material and the solving of associated problems. The collection of programs in the Suite has been designed around the text, and can be accessed from the publisher's website: https://www.worldscientific.com/worldscibooks/10.1142/Q0188#t=suppl.

This chapter has been provided in order to assist in the best use of the program materials given, for both the study of the text and the solving of the associated problems and other exercises. The programs available in the Suite enable a simulation of various stages and calculations involved in crystal-structure analysis. Of particular interest is the interactive XSYST system that solves actual crystal structure data, albeit in two-dimensional projections.

The Program Suite comprises five sections:

- The plotting system: DSYST for X, Y graphing
- The general system: GSYST for a number of crystallographic calculations
- The powder system: PSYST for indexing powder diffraction data
- The tables system: TSYST for space group data
- The X-ray system: XSYST for solving structures from single-crystal data

The plotting program uses the Python interpreter [1, 2], whereas the other programs in the Suite are written in Fortran 90 [3, 4] and supplied as IBM-compatible .exe files. It is suggested that the complete Suite be downloaded to a personal folder and a back-up copy made, although the Suite can be recovered from the website at any time. The programs are executed in a command window and their operation is mostly self-explanatory. Nevertheless, the following sections may assist in their best use. Each program is called into execution by entering its program name; thus, the input 'LSLI' will invoke the least-squares program LSLI.

8.2. The DSYST System

The DSYST plotting system uses Python, a line-by-line interpreter, and the plotting program is GRFN. Its coding format is case sensitive, and it is necessary to work with Python commands, except for certain items such as labelling or other strings, in *lower case*; data-file names are not case sensitive, but must have a .txt suffix.

8.2.1. Python interpreter set-up

The *Enthought* distribution may be used for obtaining a copy of the Python interpreter:

(a) Go to https://www.enthought.com/products/canopy. As a student or academic, *Register* for an Academic licence and follow the screen instructions; use your e-mail address when registering for the first time. Enter a username and password for subsequent use as required.

(b) Click on *Get Canopy* and choose the *Express* version; be patient, the download is lengthy.

(c) Install Python GUI (graphical user interface) by following the screen instructions; for simplicity, use the default location. In Windows, the Windows Installer is needed in order to download *.msi* files.

(d) Open the Canopy window and sign in with your username and password. Once you have logged in, you will be remembered on subsequent occasions when opening Canopy.

(e) At the bottom right of the *Welcome* screen, the Canopy version is listed; install any update that is available.

There will be now a sufficient Python environment on your computer for the work associated with this book. It is suggested that a dedicated folder named WORK FILE, or similar, be set up and the files *grfn.py* and *parb.txt* copied to it.

A correct installation should lead to the Canopy icon on the taskbar; if not, generate one there for subsequent easy access. Click on the Python logo: a 'Hi, "your name" welcome to Canopy' message will appear. Click on the *Editor* and minimize the *Welcome* window.

If the Editor shows only a single widow, click on *View* and tick *Python* in the drop-down menu; repeat this procedure any time the Python *shell* (lower pane) is not present. The Editor (upper pane) is used for organizing programs; the process of typing correct Python commands is assisted by the appearance of a red line below an erroneous entry.

Note that this discussion is not concerned with writing programs, nor teaching Python, but rather with using correctly the plotting program already in the Suite. After use, the *Navigation* (left-hand) pane will carry a name, e.g. 'John' and a list of recent files used. If the *Navigation* pane is not present, click *View* and tick *File, Browser*.

8.2.2. Plotting and labelling in Python (GRFN)

The program **GRFN** plots a graph from an x, y data set contained in a user-named .txt data file. The file may be given any name; it must carry the .txt suffix, but the suffix is not entered.

If an error is made in typing a command, it will be indicated. Use the *up-arrow* key and then the *back-arrow* key to return to the fault and correct it. The axes of the plot are labelled X and Y by default, but can be changed.

To change a label, the following commands are example instructions that can be used:

xlabel(r'T') writes T as an abscissa label.

xlabel(r'$(1/T)/\rm J\,mol^{-1}\,K^{-1}$') writes $(1/T)/$ K^{-1} as an abscissa label.

ylabel(r'$C_p/\rm J\,mol^{-1}\,K^{-1}$') writes $C_p/$J mol^{-1} K^{-1} as an ordinate label.

To insert points on a drawn graph, the following procedure may be followed. Suppose three x, y pairs are [4, 21], [6, 43], [8, 73], then the following commands will plot each point as a • on the graph:

x = [4, 6, 8]
y = [21, 43, 73]
plot(x, y, 'o')

To draw a line (axis) from a point p, r to a point q, s, the following command may be used:

plt.plot([p, q], [r, s])

To draw an x-axis at the level $y = q$, the following command may be used ($q = 20$):

plt.plot([0, p], [20, 20]) where p marks the end of the plotted x-axis.

To plot two functions, f1 and f2, on one graph, the following command may be used:
grfn.graf(['f1'], ['f2'])
whereas
grfn.graf(['f1', 'f2'], True) plots the two functions and their sum on one graph; **True** is the Python indicator to form the sum, whereas **False**, meaning no sum, is the default value. *Note that the two files (f1 and f2) to be summed must be exactly equal in length.*

8.2.3. Python test example

For the present work, click on *Select files from your computer* (they will be in the WORK FILE); click on *Create a new file* only if writing a new program. Then DC (double click) on the required program name. The cursor should appear in the shell. Now, press the *Enter* key; the flashing cursor line should appear in the shell as 'ln [2]' or an other number depending on how many times Enter is pressed.

Note

(a) Pressing the Enter key in the Editor will simply add an unwanted blank line to the program.
(b) Ensure that commands are entered in the shell. Normally, any keystroke will go first to the Editor pane, which may introduce an error into the program.

If the cursor is not in the shell, then:

(c) Click the ∇ on the lower right-hand side of the shell and ensure that *Keep directory synced to editor* is ticked; or
(d) Click *Run* → *Restart kernel*; or
(e) Click *File* → *Close* and recall the program; or
(f) Click *File* → *Exit* and start again.

Assuming that the program has been now called and that the cursor is flashing in the shell, carry out the following commands (lower case), noting the format carefully. The Enter key is required in order to complete each command: **bold** font is used here to highlight the commands (lower case).

import grfn

Enter

grfn.graf(['parb'])

Enter

A plot will be indicated by $/\backslash/\backslash/$ in the taskbar, and will be that of the parabola $y = x^2 + x + 1$; a file GRFN.PYC will have been generated in the Python folder, but is of no concern and may be deleted.

More details and examples occur in this chapter and elsewhere in the book.

8.2.4.　Further notes on Python

1. The default font for a label is *italic*.
2. \rm followed by a *spacebar space* converts the font to Roman font; \it followed by a *spacebar space* reverts the font to italic.
3. \, is a *space within text* as in $\mathbf{J\backslash,K\{-1\}}$ = J K^{-1}. Spacebar spaces *alone* can be added for clarity; they cause no action.
4. '**\$.........\$**' writes everything between the two **\$** characters.
5. $\char94$ is the superscript indicator and _ indicates a subscript. If a superscript/subscript contains more than one character it must be enclosed by { } as in **{-1}**.
6. Ensure that *Keep directory synced to editor* (see above) is always ticked.
7. If a problem arises, it may be necessary to restart the shell: Click *View → Restart kernel*.
8. Further notes on Python may be obtained from the *Help* facility in the Python shell, namely, **help(plot)**.

8.3.　The GSYST System

This section comprises a number of general crystal structure and other programs that should be useful in various sections of the text and in problem solving. A double click on the .EXE file initiates execution.

8.3.1.   Derivation of point groups (EULR)

This program describes a derivation of point groups based upon the law of combination of rotations, which was discussed in Section 2.3. The program is non-interactive; it proceeds in several stages; the procedure used is described in the literature [8].

Input

Data as specified on the monitor screen.

Output

1. MONITOR (to check progress of the program).
2. ANGOUT.TXT file, containing results of all procedures carried out.

   Also, the reader is invited to work through some interesting exercises based on the program.

8.3.2.   One-dimensional Fourier summation (FOUR)

This program calculates a one-dimensional Fourier summation $\rho(X)$ over the period 0 to a.

Input

1. KEYBOARD Data as specified below:
Number NX of subdivisions X along the repeat period a; maximum 100.
Number NH of data lines; maximum 100.
Number $HMAX$ of the maximum value of h.
Scale factor NS; normally 1 is adequate.
2. USER-NAMED .TXT FILE, comprising the NH lines of h, $A(h)$ and $B(h)$.
NB. For centrosymmetric crystals with the centre of symmetry (2 in two dimensions) at the origin, $B(h)$ is entered as zero for all h.

Output

1. RHOX.TXT file, containing X and $\rho(X)$ in a form for plotting with the program GRFN.

8.3.3. Internal and Cartesian coordinates (INTXYZ)

This program converts a given molecular geometry to Cartesian coordinates. The example data set is based on the following molecular fragment: carbon atoms 1-2-3-4-5 form a planar configuration, and the torsion angle 2-3-4-6 is $-30°$. The geometry of the first atom (atom code 0) is always set as 0, 0.0, 0.0, 0.0. In subsequent lines, the code of the current atom is the atom number of a previous atom to which the current atom is linked.

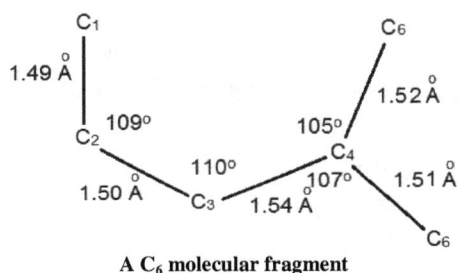

A C_6 molecular fragment

Input

1. <u>CART.TXT</u> file, comprising

Title.

Number N of atoms followed by the molecular geometry as follows (only the numeric data).

Atom name	Atom code	Bond angle/°	Torsion angle/°	Bond length/Å
C1	0	0.0	0.0	0.0
C2	1	0.0	0.0	1.49
C3	2	109.0	0.0	1.50
C4	3	110.0	0.0	1.54
C5	4	107.0	180.0	1.51
C6	4	107.0	−30.0	1.52

Output

1. METRIC.TXT file, containing the Cartesian coordinates.

8.3.4. Linear least squares (LSLI)

This program determines the best-fit straight line to a set of data of at least three in number and with a maximum of 100; unit weights are assumed. The output is self-explanatory; Appendix A11 outlines the least-squares method.

Input

1. USER-NAMED.TXT file, comprising:
Title.
Number N of data lines.
The N data points x, y.

Output

1. MONITOR
2. LSLN.TXT file, containing the least-squares parameters and deviations.

8.3.5. Matrix operations (MATOPS)

This program accepts an input of two 3×3 matrices $\mathbf{A}$ and $\mathbf{B}$ and form $\mathbf{A} + \mathbf{B}$, $\mathbf{A} - \mathbf{B}$, $\mathbf{A} \times \mathbf{B}$, $\mathbf{A}^{\mathrm{T}}$, $\mathbf{B}^{\mathrm{T}}$, Trace($\mathbf{A}$), Trace($\mathbf{B}$), Det($\mathbf{A}$), Det($\mathbf{B}$), Cof($\mathbf{A}$), Cof($\mathbf{B}$), $\mathbf{A}^{-1}$ and $\mathbf{B}^{-1}$. If results are required for only one matrix $\mathbf{A}$, then matrix $\mathbf{B}$ is set to unity by the program.

Input

1. KEYBOARD Data as specified on the monitor screen.

Output

1. MATOUT.TXT file, containing the results of the matrix operations.

8.3.6. Metric tensor operations (METTENS)

This program calculates the metric tensor G, discussed in Section 3.8.1, for a crystal unit cell and it can be used to determine distances and angles. One form of calculation with the metric tensor may be written generally in the form

$$[u\,v\,w]_{\text{row}}\ G\,[u'v'w']_{\text{column}}$$

where the multiplications follow the rules for matrices. For a calculation of the magnitude of a vector, $[u\,v\,w] = [u'v'w']$. For a calculation of an angle θ made by the two vectors, two such distances are evaluated followed by a calculation of the angle using the two different vectors:

$$\cos\theta = \frac{1}{p\,q}[u\,v\,w]_{\text{row}}\ G\,[u'v'w']_{\text{column}}$$

where p and q are the magnitudes of the two vectors obtained through the first equation.

Input
1. <u>KEYBOARD</u> Data as specified on the monitor screen.

Output
1. <u>MATS.TXT</u> file, containing *inter alia* the magnitude of the vector $[u\,v\,w]$, the metric tensor G and the unit-cell volume.

8.3.7. Molecular geometry (MOLGOM)

The program calculates bond lengths, bond angles and torsion angles for a molecular structure.

Input
1. <u>KEYBOARD</u> User-named data file name (=TITLE; max 30 characters) comprising:

Unit-cell parameters a, b, c, α, β, γ.
Number N of atoms.
Atom name (Max. 4 characters) and x, y, z coordinates for the N atoms.
2. <u>LIMIT FACTOR</u> Enter when requested from the monitor screen.

Output
1. <u>GEOM.TXT</u> file, containing the calculated molecular geometry.

8.3.8.   Best-fit plane (PLANE)

The program PLANE obtains the best fit to a set of X, Y, Z data. The form of the equation is

$$aX + bY + cZ = 1$$

which is solved by least squares so as to obtain the unknowns, a, b and c.

Input
1. <u>KEYBOARD</u>
Compound name.
Unit-cell parameters a, b and c on Cartesian (orthogonal) axes.
Number $NMAX$ of equations forming the input data matrix.
The prepared user-named file containing the $NMAX$ data lines.
2. The <u>USER-NAMED.TXT</u> file, containing the $NMAX$ lines of x, y and z values.
Then, enter 0 or 1 as requested from the monitor screen.

Output
1. <u>PLNDEV.TXT</u> file lists either the basic results of the equation and deviations, or these results and the intermediate stages in the calculation, according the keyboard input parameter of 0 or 1.

8.3.9.   Reciprocal unit cell (RECIP)

This program determines the parameters of the reciprocal unit cell from those of a real unit cell or *vice versa*, and also the volumes of both cells; the appropriate wavelength is also an input datum.

Input

1. KEYBOARD

Unit cell or reciprocal unit-cell data as specified on the monitor screen.

Output

1. MONITOR

The direct and reciprocal unit-cell dimensions and their volumes are given on the monitor screen.

8.3.10.   Point-group recognition (SYMM)

This program uses an interactive procedure to assist in the recognition of the point group of a structural model, be it a solid or a molecular model. It is based on a method detailed in the literature [8], particularly Sections 2.8 and 2.10. Section A3.3 of this book contains further notes on model-making for use with this program.

Input

1. KEYBOARD

Data as specified on the monitor screen.

Output

1. MONITOR

If an error is made, the user is directed to the point of error for further consideration.

8.3.11.   Zone symbols/Miller indices (ZONE)

This program calculates the Miller indices of a plane from the input of two zone symbols, or a zone symbol from the input of two Miller indices.

Input

1. KEYBOARD

Data, hkl or UVW, as required by the monitor screen.

Output

1. MONITOR

The output is UVW or hkl, according to the input.

8.3.12. *Q*-values (QVALS)

This program calculates Q-values for a crystal structure.

Input
1. KEYBOARD
Title; Unit-cell dimensions a, b, c, α, β, γ and wavelength λ; Miller indices h, k, l values, as per the monitor screen. There are two modes of output, with and without intermediate results, according to the LIMIT parameter.
2. LIMIT PARAMETER; 0 or 1, as per the monitor screen.

Output
1. QDAT file, containing $10^4 Q(hkl)$ and $d(hkl)$ values.

8.4. Powder System (PSYST)

This section contains a powder indexing program that uses the Ito zone indexing procedure; it was made available by the kindness of Dr J W Visser, University of Delft.

8.4.1. Powder indexing (ITO12)

Input
1. ITOINP.DAT input data file for potassium nitrate KNO_3 (i.e. a six-letter name: ***INP.DAT).
2. MWAINP.DAT problem input data file for magnesium tungstate $MgWO_4$ (20 powder lines).
3. MWBINP.DAT problem input data file for magnesium tungstate $MgWO_4$ (40 powder lines).
NB. *The blank lines in the input data are essential.*

Output
1. ITOINP.DOC A left-over from the days of punched card input (ignore).
2. ITOINP.LAT The 4 most probable solutions.
3. ITOINP.OUT Instruction manual.
4. ITOINP.SMY Indexing and final cell choice.

8.5. Data Source Tables (TSYST)

This section contains references to the International Tables.

8.5.1. International Tables for Crystallography, Vol. A

Input

1. DC on the ITCA link in TSYST.

8.5.2. International Tables for X-ray Crystallography, Vol. I

Input

1. DC on (or PASTE) the ITX1 link in TSYST.

8.6. The XSYST System

The XSYST system comprises the program XRAY and its accompanying graphical routines together with a number of example data sets that can be solved to give electron density maps each of which is interpretable in terms of a chemical structure, albeit in projection on to a principal plane. The program is interactive, and the basic structure-solving techniques can be practised. Although the messages on the screen indicate the procedures to be carried out, it may be helpful here to discuss the program features briefly.

Each data set provides information about the unit cell, crystal symmetry in the given projection, wavelength of X-radiation, space group, $|\mathbf{F}_{\mathrm{obs}}(hk0)|$ data and other relevant information appertaining to each data set.

In two dimensions, the symbols used are a, b, γ, h, k, x, y and so on. For a projection other than (001), adjustments may be made to the data set; thus, true y and z are interpreted as x and y in the system.

The following calculations are available:

1. Patterson function;
2. Superposition (Buerger minimum) function;

3. Structure factor calculation;
4. Least-squares refinement;
5. Electron density calculation;
6. Direct phasing methods;
7. Calculation of E-maps;
8. Distance and angle calculation;
9. Scale and temperature factors;
10. **E**-values calculation.

Each X-ray data set carries a four-letter name with a .TXT suffix, such as NIOP.TXT, but only the four-letter name is input to the program. For convenience, the user-named *output file* name can be entered with a .TXT suffix; alternatively, output with WORD.

8.6.1. Patterson functions and maps

A Patterson function $P(u\,v)$ may be calculated using $|\mathbf{F}_{\text{obs}}(hk0)|^2$, or in a sharpened mode using sharpened coefficients as described in Section 6.6.3, giving a sharp, origin-removed Patterson vector map. Sharpening may introduce a small number of spurious maxima because it elevates the high-order intensities. Patterson (and electron density) maps may be printed by the system and contoured by hand as necessary. Atomic coordinates can be read from the screen, thus enhancing the interactive qualities of the system.

8.6.2. Superposition function

This routine calculates a minimum function $M(x,\,y)$ at each grid point:

$$M(x,\,y) = \min[P(x + \Delta u_1,\, y + \Delta v_1),$$
$$P(x + \Delta u_2,\, y + \Delta v_2), \ldots, P(x + \Delta u_n,\, y + \Delta v_n)]$$

$$(8.1)$$

where the Δu and Δv terms are n displacement vectors. This technique has not been discussed in the text, but the procedure can be followed through with the aid of the instructions on the monitor screen. A starting point is the recognition of a $H - H$ vector, particularly if H is a relatively heavy atom [5, 7].

8.6.3. Structure factor calculations

Each atom in the structure factor calculation requires the following data:

1. Atom type identity number;
2. Fractional x and y coordinates;
3. Population parameter: 1, unless the atom is in a special position;
4. Temperature factor, initially an overall B-value, which can be altered from routine 9.

The coordinates may be entered from a file or at the keyboard. If from a file, it must be named XYS.TXT and the first line must be the number of atoms to follow. After an $|\mathbf{F}_{\text{calc}}(hk0)|$ calculation and least-squares refinement, the current coordinates are retained in the file COORDS.TXT. This file may be used in a subsequent calculation, or edited if desired. The file XYS.TXT is unaltered by the program.

8.6.4. Least-squares refinement

This routine uses the *diagonal* least-squares approximation. The x, y coordinates are refined and a B-factor is applied to each individual atom and refined. The changes δx, δy and δB are applied to each atom, and the R-factor listed. The cycles can be repeated until no further improvement occurs. In the diagonal approximation, 60% of the calculated shifts are applied; this number can be varied at will. The scale factor is determined as $\sum |\mathbf{F}_{\text{calc}}(hk0)| / \sum |\mathbf{F}_{\text{obs}}(hk0)|$ and may be applied at the end of any refinement cycle. Towards the end of the refinement, it is sometimes advantageous to reduce the shifts to a smaller percentage than 60. At each cycle, the new coordinates are stored in COORDS.TXT.

8.6.5. Electron density maps

As phases become available, they are combined with the $|\mathbf{F}_{\text{obs}}(hk0)|$ data to calculate an electron density map. This map may reveal atoms not yet included. Then another structure factor calculation will reveal the degree of improvement and the cycles repeated. This procedure is known as successive Fourier syntheses refinement. When R is less than 0.3, a difference Fourier map can be calculated, which may show new atom positions or indicate other desirable changes.

8.6.6. Direct phasing methods

The normalized structure factor $|\mathbf{E}|$ has been given by the equation

$$|\mathbf{E}(hk0)|^2 = \frac{K^2 |\mathbf{F}_{\text{obs}}(hk0)|^2}{\varepsilon \sum_j f_j^2 \exp[-2B \sin^2 \theta) \lambda^2]} \tag{8.2}$$

where ε is the point-group dependent epsilon factor. However, an overall B-factor may not represent the structure well, and an alternative procedure for calculating $|\mathbf{E}|$-values is given by

$$|\mathbf{E}(hk0)|^2 = \frac{K(s) |\mathbf{F}_{\text{obs}}(hk0)|^2}{\varepsilon \sum_j f_j^2} \tag{8.3}$$

where $K(s)$ now includes the scaling of $|\mathbf{F}_{\text{obs}}(hk0)|$ and temperature factor adjustments. A number of n ranges are set up of equal increments of s^2, where $s = (\sin \theta)/\lambda$. In each range, $K(s)$ is calculated as $\sum \varepsilon \sigma^2 / \sum |\mathbf{F}_{\text{obs}}(hk0)|^2$, with each value of $|\mathbf{F}_{\text{obs}}(hk0)|$ given its multiplicity; $\sigma^2 = \sum_j f_j^2$. The $K(s)$ is interpolated [6] so as to derive $|\mathbf{F}_{\text{obs, corr}}(hk0)|$ for each reflection. Then,

$$|\mathbf{E}(hk0)|^2 = \frac{K(s) |\mathbf{F}_{\text{obs, corr}}(hk0)|^2}{\varepsilon \sigma^2} \tag{8.4}$$

It is best to enter only the s and $K(s)$ values corresponding to the end of the data range, especially at the low values of θ, since extrapolation can be uncertain. In the program, the results are written to a file named EVALS together with some statistics of the $|\mathbf{E}|$-values' distribution. $|\mathbf{E}|$-values' data greater than a chosen limit ELIM are written to output, and a $\sum_2$ listing set up in the file SIG2. The file EDATA contains the $|\mathbf{E}|$-values $\geq$ ELIM. The program halts at this stage so that the $\sum_2$ listing can be printed and some signs developed; only centrosymmetric structures are provided for the direct methods' practice in the XSYST system. In practice, it may be necessary to lower the value of ELIM if too few data are obtained at this stage. After a set of signs (± 1) has been developed from the Σ_2 listing, re-open the XRAY program, go to routine 7 and follow the directions on the monitor screen.

8.6.7. Calculation of $|\mathbf{E}|$-maps

An $|\mathbf{E}|$-map is an electron density map calculated with $|\mathbf{E}|$-values and their phases, or $\pm |\mathbf{E}|$ with centrosymmetric structures. As $|\mathbf{E}|$-values are sharpened coefficients, some spurious peaks can arise. Once a sensible chemical entity has been identified, electron density calculations should be carried out with phased $|\mathbf{F}_{\text{obs}}(hk0)|$ values. In the direct methods routine, signs can be modified as required. A zero sign for the sign s indicates an unsigned reflection which does not contribute to the $|\mathbf{E}|$-map.

8.6.8. Bond lengths and bond angles

This routine calculates bond lengths according to the limit of an input value (in Å). Usually, there is no need to set this limit greater than the van der Waals radius of the largest species present. In the projection, the distances may be slightly distorted from their expected values on account of their orientation to the plane of projection.

8.6.9. Scale and temperature factors

These results are obtained initially through Wilson's method. They are refined with the atomic parameters at a later stage in the structure analysis. Accidental absences should be included at $0.55|\mathbf{F}_{\text{obs, local min}}(hk0)|$ for a centric distribution, $0.66|\mathbf{F}_{\text{obs, local min}}(hk0)|$ for an acentric distribution and

$0.59|\mathbf{F}_{\text{obs, local min}}(hk0)|$ if the centricity is not known. Some of the sets do not have their accidental absences included; however, the data set NO2G is complete in this respect.

8.6.10. Calculation of $|\mathbf{E}|$-values

An $|\mathbf{E}|$-value may be calculated by the equation

$$|\mathbf{E}_{\text{calc}}| = \sqrt{\frac{(A_Z^2 + B_Z^2)}{\varepsilon\sigma}} \tag{8.5}$$

where ε is the epsilon factor for the $(hk0)$ reflection, and σ is given by

$$\sigma = \sqrt{\sum_j Z_j^2} \tag{8.6}$$

where Z_j is the atomic number of the jth atom, and A_Z and B_Z are given by

$$A_Z = \sum_j Z_j \cos 2\pi(hx_j + ky_j)$$

$$B_Z = \sum_j Z_j \sin 2\pi(hx_j + ky_j) \tag{8.7}$$

where the sums are taken over all N atoms in the unit cell. Note that a temperature factor is not involved since $f_j = Z_j$ for a point atom. It follows that $|\mathbf{E}(000)|$ is given by

$$|\mathbf{E}(000)| = \frac{\sum_j Z_j}{\sqrt{\sum_j Z_j^2}} \tag{8.8}$$

which is equal to $\sqrt{N}$ for identical atoms. The values of $|\mathbf{E}_{\text{calc}}|$ are listed with the $|\mathbf{E}|$-values in the results files ECALC and EVALS, respectively.

8.7. Crystal Structure Analysis Problems

Not all structure-solving routines operate equally well with all example data sets, and some guidance is given here for the sets available.

- Calculated bond lengths and bond angles may deviate from standard values on account of the data being a projection.
- Fourier maps will not necessarily be true to scale and will not present a true β angle in oblique projections. When the axis of projection is not normal to the plane of projection, the true axes of the projection should be modified. For example, for a monoclinic unit cell projected on to (100), the axes are b and $c\sin\beta$. The $\sin\beta$ term will be significant if the β angle is very different from $90°$. For the projection of a triclinic unit cell, the axes are $a\sin\beta$ and $b\sin\alpha$. However, since coordinates are always refined by least squares, these corrections are not very significant in practice.
- It is rare to locate hydrogen atoms by X-ray diffraction, but they may be positioned from a difference electron-density map, or by geometrical considerations.
- Notes about each example structure follow.

8.7.1.   Nickel *o*-phenanthroline complex (NIOP)

This structure occurs as a Problem to Chapter 7, and has been discussed sufficiently there and in Tutorial Solution 7.8.

8.7.2.   2-Amino-4,6-dichloropyrimidine(CL2P)

The asymmetric unit contains two chlorine atoms that are not related by symmetry. List the coordinates of the Patterson vectors for the given plane group. Then, find the chlorine atoms by Patterson syntheses and proceed with the structure determination. This structure and that for CL1P produce satisfactory Wilson plots. It may help to print more than one copy of the Patterson map and join them such that the origin is at the centre of the composite. This structure $C_4H_3N_3Cl_2$ refines to $R = 12.1\%$.

8.7.3.   2-Amino-4-methyl-chloropyrimidine (CL1P)

Consider the following data:

	2-Amino-4-methyl-6-chloropyrimidine	2-Amino-4,6-di-chloropyrimidine
$a/\text{Å}$	16.426	16.447
$b/\text{Å}$	4.000	3.845
$c/\text{Å}$	10.313	10.283
β	109.13°	107.97°
Z	4	4
Space group	$P2_1/a$	$P2_1/a$

The two structures are isomorphous: one chlorine atom of 2-amino-4, 6-dichloropyrimidine has been replaced by a methyl group in this structure. One approach, therefore, is to set up trial coordinates by comparison with the CL1P structure, and calculate structure factors and a first electron density map.

Alternatively, the isomorphous character can be used. Let the structure factors of the two forms be $F1$ for CL1P and $F2$ for CL2P; the difference between them is $f_{Cl} - f_{CH_3} = \Delta f$, say 8–10 electrons; remember that f is a function of θ. For a given $hk0$, if $F2$ is greater than $F1$, then $F2$ takes a negative sign because Δf is positive. In this way, signs can be allocated. Uncertainty can arise for small magnitudes of $|F|$. The following diagram illustrates the procedure:

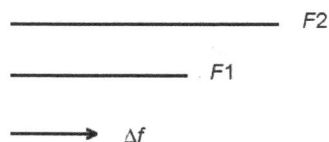

$$\begin{array}{ll} \text{————————————} & F2 \\ \text{——————} \cdot & F1 \\ \longrightarrow \quad \Delta f & \end{array}$$

Isomorphous replacement: the centrosymmetric case

where $(F_2 - F_1) = \Delta f$. This structure solves to $R = 12.3\%$.

8.7.4. *m*-Tolidine dihydrochloride (MTOL)

This substance has been reported in space group $I2$, a non-standard setting of $C2$, with $Z_c = 2$. The plane group of the projection is $p2$, and solves to R of *ca.* 22% with this data set.

8.7.5.   Nitroguanidine (NO2G)

Nitroguanidine $C(NH_2)_2 NNO_2$ crystallizes in space group $Fdd2$ with $a = 17.639$ Å, $b = 24.873$ Å, $c = 3.5903$ Å and $Z = 16$. The small c dimension means that good resolution will arise in the (001) projection. This structure can be solved by direct methods (review the notes on the $|\mathbf{E}|$-map calculation given in Section 8.6.7). Two reflections fix the origin in two dimensions, one from each of two different parity groups: h even, k odd or h odd, k even or h odd, k odd. The plane group is $p2gg$ and the sign relationships may be summarized as $s_{hk} = s_{\bar{h}\bar{k}} = (-1)^{h+k} s_{\bar{h}k}$. In some cases, letter symbols may be used. They will evolve during the procedure or by calculating E-maps. This structure solves to *ca.* 5%.

8.7.6.   Bis(6-sulphanyloxy)-1,3,5-triazin-2(1H)-one (BSTO)

This compound crystallizes in space group $P2_1/m$ with $Z_c = 2$, so that the molecules occupy special equivalent positions in the space group. The Patterson (010) projection indicates more than one peak of similar height in the asymmetric unit, so that it may be necessary to investigate both in order to obtain a good trial structure. This example can be used with the minimum function (superposition technique), and refines to $R = 15.5\%$.

8.7.7.   2-*S*-methylthiouracil (SMTX) and (SMTY)

2-*S*-methylthiouracil is triclinic, space group $P\bar{1}$ and $Z_c = 2$. Data sets SMTX and SMTY are for the (100) and (010) projections, respectively. This structure may present a little more difficulty than previous examples because several peaks of similar height occur in the Patterson maps. The correct structure refines to *ca.* 13%. The carbon atom attached to the sulphur atom is not clearly resolved in projection. (How might this have arisen?) For the (100) projection, the axes marked x and y are strictly $y \sin \gamma$ and $z \sin \beta$. By solving also the (010) projection, a three-dimensional representation of the molecule can be obtained.

Overall, the given selection of data sets provides practice in the important methods of structure solving. Other data sets may be devised, and suitable data may be found in the early volumes of *Acta Crystallographica*. It is essential to adopt the exact format shown by the examples; the program MAKDAT.TXT can be used for preparing new data sets.

Problems

Problems relating to this chapter arise throughout the text itself. Solutions have been provided for the data set NIOP (Section 8.7.1) and SMTX/SMTY (Section 8.7.7) in relation to problems for Chapter 7. The solutions obtained for the problems of other structure analyses may be judged by application of the tests for correctness given in Section 7.3.

References

[1] Swoop CH, *A Byte of Python*. Free E-book. Online at ⟨http://www.swaroopch.com/notes/python/⟩ (2012).
[2] Lutz M, *Learning Python*, 5th edn. O'Reilly Media (2013).
[3] Chapman SJ, *Fortran 90/95 for Scientists and Engineers*. WCB/McGraw-Hill (1998).
[4] Dowling B, *Interfacing Python with Fortran*. Online at ⟨http://ucs.cam.ac.uk/docs/course-notes/unix-courses/pythonfortran/files/f2py.pdf⟩.
[5] Buerger MJ, *Vector Space*. Wiley, NY (1959).
[6] Ladd MFC, *Z. Kristallogr.* **147**, 279 (1978).
[7] Ladd M and Palmer R, *Structure Determination by X-ray Crystallography*, 5th edn. Springer (2013).
[8] Ladd M, *Symmetry of Crystals and Molecules*. Oxford University Press (2014).

Appendix A1

Vectors and Matrices

A1.1. Introduction

This appendix is a brief introduction to vector algebra for the purpose of defining the use of vectors and matrices in the study of the geometry and symmetry of crystals and molecules. The reader who wishes for additional mathematical details on vector algebra should refer to a standard text on this subject [1–3].

A1.2. Vectors

Ordinary numbers and their representations are *scalar* quantities; they have the magnitude expressed by the quantity itself. A vector has both magnitude and direction. Thus, the number 30 and the variable x (written in *italic* font) are examples of scalars, whereas the bearing $30°\,\mathrm{N}\,40°\,\mathrm{W}$ and the variable $\mathbf{X}$ (written in **bold** font) represent vectors. *Unit vectors* may be defined as $\mathbf{i}, \mathbf{j}$ and $\mathbf{k}$, where $|\mathbf{i}| = |\mathbf{j}| = |\mathbf{k}| = 1 = \frac{\mathbf{X}}{|\mathbf{X}|}$; normally, $\mathbf{i}, \mathbf{j}$ and $\mathbf{k}$ are parallel, respectively, to x, y and z reference axes. In practice, the unit vectors are usually, but not necessarily, mutually perpendicular.

A1.2.1. Sum, difference and scalar product of two vectors

In Fig. A1.1, **a** and **b** are two vectors drawn from the origin O; QS is perpendicular to OP. In Pythagorean notation,

$$(PQ)^2 = (OQ)^2 + (OP)^2 - 2(OP)(OS)$$

or

$$(PQ)^2 = (OQ)^2 + (OP)^2 - 2(OP)(OQ)\cos\widehat{QOP} \qquad (A1.1)$$

which can be written as

$$(PQ)^2 = r^2 = a^2 + b^2 - 2ab\cos\widehat{\mathbf{a}\,\mathbf{b}} \qquad (A1.2)$$

where $\widehat{\mathbf{a}\mathbf{b}} = \widehat{QOP} = \theta$. In vector notation,

$$\mathbf{PQ} = \mathbf{b} - \mathbf{a} \qquad (A1.3)$$

Since the algebraic manipulations of addition and subtraction of vectors must take place along a line, with the vectors resolved as necessary, a scalar product term is modified by the cosine of the angle between the forward directions of the vectors; thus, the scalar product of **PQ** with itself, written as $\mathbf{PQ}\cdot\mathbf{PQ}$, or as $\mathbf{r}\cdot\mathbf{r}$, is given by

$$\mathbf{PQ}\cdot\mathbf{PQ} = \mathbf{r}\cdot\mathbf{r} = (\mathbf{b}-\mathbf{a})\cdot(\mathbf{b}-\mathbf{a})$$

or

$$(PQ)^2 = r^2 = a^2 + b^2 - 2(\mathbf{a}\cdot\mathbf{b}) \qquad (A1.4)$$

Comparing Eq. (A1.4) and Eq. (A1.2),

$$\mathbf{a}\cdot\mathbf{b} = ab\cos\widehat{\mathbf{a}\mathbf{b}} \qquad (A1.5)$$

which equation defines the *scalar product* of two vectors, and may be written for Fig. A1.1 as

$$\mathbf{a}\cdot\mathbf{b} = ab\cos\theta \qquad (A1.6)$$

where θ is the angle between the *forward directions* of **a** and **b**; a and b are the magnitudes of the vectors, and may be written also as $|\mathbf{a}|$ and $|\mathbf{b}|$. It follows that the scalar product, also called the *dot*

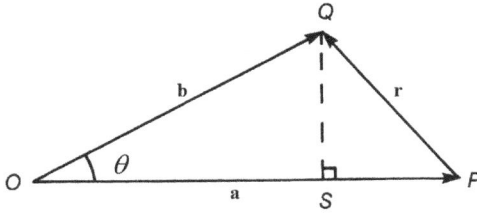

Fig. A1.1.   The $\triangle OPQ$ in which $\mathbf{r} = \mathbf{b} - \mathbf{a}$.

product, of a vector with itself is the square of the magnitude of that vector:

$$\mathbf{a} \cdot \mathbf{a} = |\mathbf{a}|^2 = a^2 \qquad (A1.7)$$

since the angle $\widehat{\mathbf{a}\,\mathbf{a}}$ is zero. Further, easily-verified results with dot products are as follows:

$$\left.\begin{array}{l} \mathbf{a} \cdot \mathbf{b} = \mathbf{b} \cdot \mathbf{a} \\ \mathbf{c} \cdot (\mathbf{a} \cdot \mathbf{b}) = \mathbf{c} \cdot \mathbf{a} + \mathbf{c} \cdot \mathbf{b} \\ p(\mathbf{a} \cdot \mathbf{b}) = (p\mathbf{a}) \cdot \mathbf{b} = \mathbf{a} \cdot (p\mathbf{b}) \end{array}\right\} \qquad (A1.8)$$

where p is a scalar constant.

A1.2.2.   Vector product of two vectors

Consider first the two vectors $\mathbf{a}$ and $\mathbf{b}$ from the origin O in Fig. A1.2. The *vector product*, also called *cross product*, is defined as

$$\mathbf{a} \times \mathbf{b} = \mathbf{k}\, a\, b \sin \theta \qquad (A1.9)$$

where $a\, b \sin \theta$ is the area of the parallelogram formed by the vectors $\mathbf{a}$ and $\mathbf{b}$, and $\mathbf{k}$ is a unit vector normal to the plane containing $\mathbf{a}$ and $\mathbf{b}$. The sequence $\mathbf{a}$, $\mathbf{b}$, $\mathbf{k}$ is the same as that for the crystallographic reference axes x, y, z. Thus, the direction of $\mathbf{k}$ in Fig. 1.2 is that of z; the three vectors $\mathbf{a}$, $\mathbf{b}$ and $\mathbf{k}$ form a right-handed set of directions. It follows from the definition of the cross product that

$$\mathbf{b} \times \mathbf{a} = -\mathbf{a} \times \mathbf{b} \qquad (A1.10)$$

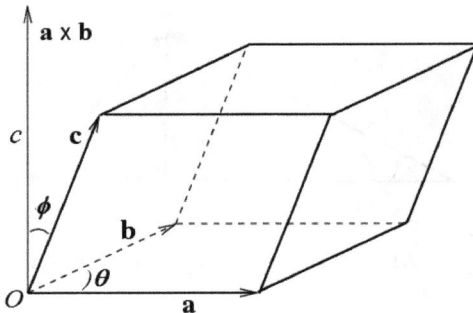

Fig. A1.2. A parallelepiped **a**, **b**, **c**; the cross product $\mathbf{a} \times \mathbf{b} = ab \sin\theta$ and lies in the direction $\mathbf{c}\cos\phi$, perpendicular to the **a**_**b** plane.

A1.2.3.   Vector product in coordinate notation

Let **c** be a vector with components c_1, c_2, c_3 with respect to orthogonal[a] directions **a**, **b** and c. Then,

$$c_1 = (a_2b_3 - a_3b_2), \quad c_2 = (a_3b_1 - a_1b_3), \quad c_3 = (a_1b_2 - a_2b_1)$$

Also,

$$
\begin{aligned}
\mathbf{a} \times \mathbf{b} = {} & (\mathbf{i}a_1 + \mathbf{j}a_2 + \mathbf{k}a_3) \times (\mathbf{i}b_1 + \mathbf{j}b_2 + \mathbf{k}b_3) \\
& + (a_1b_1\mathbf{i} \times \mathbf{i} + a_1b_2\mathbf{i} \times \mathbf{j} + a_1b_3\mathbf{i} \times \mathbf{k}) \\
& + (a_2b_1\mathbf{j} \times \mathbf{i} + a_2b_2\mathbf{j} \times \mathbf{j} + a_2b_3\mathbf{j} \times \mathbf{k}) \\
& + (a_3b_1\mathbf{k} \times \mathbf{i} + a_3b_2\mathbf{k} \times \mathbf{j} + a_3b_3\mathbf{k} \times \mathbf{k}) \\
= {} & (a_2b_3 - a_3b_2)\mathbf{i} + (a_3b_1 - a_1b_3)\mathbf{j} \\
& + (a_1b_2 - a_2b_1)\mathbf{k}
\end{aligned}
\tag{A1.11}
$$

where **i**, **j** and **k** are unit vectors along a, b and c, and $\mathbf{i} \times \mathbf{j} = \mathbf{k}$, $\mathbf{j} \times \mathbf{i} = -\mathbf{k}$ and $\mathbf{i} \times \mathbf{i} = 0$, and cyclic permutations thereof. Thus, if $\mathbf{a} \times \mathbf{b} = \mathbf{c}$, the scalar components of the vector **c** will be $c_1 = (a_2b_3 - a_3b_2)$, $c_2 = (a_3b_1 - a_1b_3)$, and $c_3 = (a_1b_2 - a_2b_1)$.

---

[a] Mutually perpendicular.

A1.3. Volume of a Parallelepiped

An important application of the vector product arises in determining the volume of a general parallelepiped formed by three sides a, b, c, parallel to the x-, y- and z-axes, respectively, and the three interaxial angles α, β and γ, defined as in Section 1.2.

Referring again to Fig. A1.2, the volume of a parallelepiped is the area of the base multiplied by the perpendicular height, that is, the magnitude $|\mathbf{a} \times \mathbf{b}|$ multiplied by $|\mathbf{c}| \, |\cos \phi|$. Note that if $\mathbf{a}$ and $\mathbf{b}$ are interchanged, then $\mathbf{a} \times \mathbf{b}$ would point downward: the angle ϕ would be larger than $\pi/2$, and $\cos \phi$ would be negative. Thus, the volume V is written as

$$V = |\mathbf{c} \cdot (\mathbf{a} \times \mathbf{b})| \tag{A1.12}$$

or $|\mathbf{b} \cdot (\mathbf{c} \times \mathbf{a})|$ or $|\mathbf{a} \cdot (\mathbf{b} \times \mathbf{c})|$. In order to evaluate V, $\mathbf{a}$, $\mathbf{b}$ and $\mathbf{c}$ are expressed in terms of a set of orthogonal unit vectors $\mathbf{i}$, $\mathbf{j}$ and $\mathbf{k}$:

$$\mathbf{a} = a_1\mathbf{i} + a_2\mathbf{j} + a_3\mathbf{k}$$
$$\mathbf{b} = b_1\mathbf{i} + b_2\mathbf{j} + b_3\mathbf{k} \tag{A1.13}$$
$$\mathbf{c} = c_1\mathbf{i} + c_2\mathbf{j} + c_3\mathbf{k}$$

Then, using Eq. (A1.12) with expansion of the vectors in terms of unit vectors $\mathbf{i}$, $\mathbf{j}$ and $\mathbf{k}$ in the directions of $\mathbf{a}$, $\mathbf{b}$ and $\mathbf{c}$, respectively, and recalling that products such as $\mathbf{i} \times \mathbf{i} = 0$, $\mathbf{i} \times \mathbf{j} = \mathbf{k}$ and $\mathbf{j} \times \mathbf{i} = -\mathbf{k}$,

$$V = (c_1\mathbf{i} + c_2\mathbf{j} + c_3\mathbf{k}) \cdot (a_1 b_2\mathbf{k} - a_1 b_3\mathbf{j} - a_2 b_1\mathbf{k} + a_2 b_3\mathbf{i}$$
$$+ a_3 b_1\mathbf{j} - a_3 b_2\mathbf{i}) \tag{A1.14}$$

which, after simplification, may be expressed as the *determinant*

$$\begin{vmatrix} a_1 & a_2 & a_3 \\ b_1 & b_2 & b_3 \\ c_1 & c_2 & c_3 \end{vmatrix} \tag{A1.15}$$

Since rows and columns of a determinant can be interchanged without altering its value,

$$V^2 = \begin{vmatrix} a_1 & a_2 & a_3 \\ b_1 & b_2 & b_3 \\ c_1 & c_2 & c_3 \end{vmatrix} \begin{vmatrix} a_1 & b_1 & c_1 \\ a_2 & b_2 & c_2 \\ a_3 & b_3 & c_3 \end{vmatrix} \tag{A1.16}$$

Multiplying the determinants, according to the rules for matrices, Sections A1.4.7 and A1.4.9, leads to

$$V^2 = \begin{vmatrix} a_1a_1 + a_2a_2 + a_3a_3 & a_1b_1 + a_2b_2 + a_3b_3 & a_1c_1 + a_2c_2 + a_3c_3 \\ b_1a_1 + b_2a_2 + b_3a_3 & b_1b_1 + b_2b_2 + b_3b_3 & b_1c_1 + b_2c_2 + b_3c_3 \\ c_1a_1 + c_2a_2 + c_3a_3 & c_1b_1 + c_2b_2 + c_3b_3 & c_1c_1 + c_2c_2 + c_3c_3 \end{vmatrix} \tag{A1.17}$$

which may be expressed in vector notation as the determinant

$$V^2 = \begin{vmatrix} \mathbf{a \cdot a} & \mathbf{a \cdot b} & \mathbf{a \cdot c} \\ \mathbf{b \cdot a} & \mathbf{b \cdot b} & \mathbf{b \cdot c} \\ \mathbf{c \cdot a} & \mathbf{c \cdot b} & \mathbf{c \cdot c} \end{vmatrix} \tag{A1.18}$$

Evaluating

$$V^2 = a^2b^2c^2 + ab\cos\gamma\, bc\cos\alpha\, ca\cos\beta + ac\cos\beta\, ba\cos\gamma\, bc\cos\alpha$$
$$- ca\cos\beta\, b^2ca\cos\beta - bc\cos\alpha\, a^2bc\cos\alpha - ab\cos\gamma c^2ab\cos\gamma$$

simplifies to

$$V = abc(1 - \cos^2\alpha - \cos^2\beta - \cos^2\gamma + 2\cos\alpha\cos\beta\cos\gamma)^{1/2} \tag{A1.19}$$

It may be seen that V may be given also as $\sqrt{\det(G)}$, where G is the metric tensor (see Section 3.8.1).

A1.4. Matrices

A matrix is an ordered rectangular array of numbers or variables. The matrix may be symbolized by $\mathbf{A}$ and its size as $m \times n$. Its elements

a_{ij} run from $i = 1$ to m down the columns and from $j = 1$ to n along the rows.

A1.4.1.   General matrix

Thus, in the 4×3 matrix $\mathbf{A}$,

$$
\begin{array}{c}
\quad\quad\quad \rightarrow j \\
\mathbf{A} = \begin{pmatrix} a_{11} & a_{12} & a_{13} \\ a_{21} & a_{22} & a_{23} \\ a_{31} & a_{32} & a_{33} \\ a_{41} & a_{42} & a_{43} \end{pmatrix} \downarrow i
\end{array}
\tag{A1.20}
$$

where i runs from 1 to 4 and j runs from 1 to 3. In most cases, the matrices of interest herein will be 3×3 size matrices; a matrix in which $m = n$ is a *square* matrix.

A1.4.2.   Row matrix

A row matrix consists of a single line array, dimensions $1 \times n$. The array

$$
\mathbf{R} = (a_{11}\ a_{12}\ a_{13})
\tag{A1.21}
$$

is a row matrix, or *row vector*, of dimensions 1×3.

A1.4.3.   Column matrix

The matrix

$$
\mathbf{C} = \begin{pmatrix} a_{11} \\ a_{21} \\ a_{31} \\ a_{41} \end{pmatrix}
\tag{A1.22}
$$

is a column matrix, or *column vector*, of dimensions 4×1.

A1.4.4. Symmetric, skew-symmetric, equal and identity matrices

A *symmetric* matrix $\mathbf{A}$ has elements $a_{ij} = a_{ji}$ for all i and j; otherwise, it is *skew-symmetric*; the matrices $\mathbf{A}$ and $\mathbf{B}$ below are both skew-symmetric. Note that a symmetric matrix is equal to its transpose (Section A1.4.6). *Equal* matrices have both the same dimensions and equal corresponding elements; such matrices need not be square. The *identity* matrix $\mathbf{I}$ has diagonal elements of unity and zero elements otherwise.

$$\mathbf{A} = \begin{pmatrix} 1 & 0 & 1 \\ 0 & 1 & 2 \\ \bar{2} & 1 & 1 \end{pmatrix}, \quad \mathbf{B} = \begin{pmatrix} 1 & 1 & 2 \\ 0 & \bar{1} & 0 \\ 1 & 0 & 0 \end{pmatrix}, \quad \mathbf{I} = \begin{pmatrix} 1 & 0 & 0 \\ 0 & 1 & 0 \\ 0 & 0 & 1 \end{pmatrix}$$

$$(A1.23)$$

For neatness, a negative sign is placed above the digit to which it refers, as with Miller indices.

A1.4.5. Addition and subtraction of matrices

Two matrices $\mathbf{A}$ and $\mathbf{B}$ may be added or subtracted if $\mathbf{A}$ has both the same number of columns and same number of rows as $\mathbf{B}$. Thus, in $\mathbf{C} = \mathbf{A} + \mathbf{B}$, $c_{ij} = a_{ij} + b_{ij}$:

$$\begin{pmatrix} 1 & 0 & 1 \\ 0 & 1 & 2 \\ \bar{2} & 1 & 1 \end{pmatrix} + \begin{pmatrix} 1 & 1 & 2 \\ 0 & \bar{1} & 0 \\ 1 & 0 & 0 \end{pmatrix} = \begin{pmatrix} 2 & 1 & 3 \\ 0 & 0 & 2 \\ \bar{1} & 1 & 1 \end{pmatrix}$$
$$\mathbf{A} \qquad\qquad \mathbf{B} \qquad\qquad \mathbf{C}$$

or $\mathbf{C} = \mathbf{A} - \mathbf{B}$, $c_{ij} = a_{ij} - b_{ij}$

$$\begin{pmatrix} 1 & 0 & 1 \\ 0 & 1 & 2 \\ \bar{2} & 1 & 1 \end{pmatrix} - \begin{pmatrix} 1 & 1 & 2 \\ 0 & \bar{1} & 0 \\ 1 & 0 & 0 \end{pmatrix} = \begin{pmatrix} 0 & \bar{1} & \bar{1} \\ 0 & 2 & 2 \\ \bar{3} & 1 & 1 \end{pmatrix}$$
$$\mathbf{A} \qquad\qquad \mathbf{B} \qquad\qquad \mathbf{C}$$

A1.4.6.   Transposition of a matrix

A transposed matrix has a_{ij} replaced by a_{ji} for all i and j. Thus, if a matrix $\mathbf{A}$ is

$$\mathbf{A} = \begin{pmatrix} 1 & 0 & 1 \\ 0 & 1 & 2 \\ \bar{2} & 1 & 1 \end{pmatrix} \tag{A1.24}$$

its transpose $\mathbf{A}^{\mathrm{T}}$ is given by

$$\mathbf{A}^{\mathrm{T}} = \begin{pmatrix} 1 & 0 & \bar{2} \\ 0 & 1 & 1 \\ 1 & 2 & 1 \end{pmatrix} \tag{A1.25}$$

A1.4.7.   Multiplication of matrices

If a matrix is multiplied by a scalar quantity s, then all elements a_{ij} of the matrix are transformed to $s\,a_{ij}$. Two matrices $\mathbf{A}$ and $\mathbf{B}$ can be multiplied if and only if $\mathbf{A}$ has the dimensions $m \times p$ and $\mathbf{B}$ has the dimensions $p \times n$, so that their product $\mathbf{C}$ has the dimensions $m \times n$. The element c_{ij} of $\mathbf{C}$ is given by multiplying the ith row of $\mathbf{A}$ and the jth column of $\mathbf{B}$ element by element. In general,

$$c_{ij} = \sum_{k=1}^{p} \mathbf{A}_{ik}\mathbf{B}_{kj} \tag{A1.26}$$

For a 3×3 matrix, $c_{12} = a_{11}b_{12} + a_{12}b_{22} + a_{13}b_{32}$. Thus, if c_{12} is formed from matrices $\mathbf{A}$ and $\mathbf{B}$ in Eq. (A1.23), $c_{12} = (1 \times 1) + (0 \times \bar{1}) + (1 \times 0) = 1$. Note that the position of c_{ij} in $\mathbf{C}$ is at the junction of the ith row and jth column.

The multiplication of $\mathbf{A}$ and $\mathbf{B}$ from Eq. (A1.23) to give $\mathbf{C}$ is written concisely as

$$\mathbf{C} = \mathbf{A}\,\mathbf{B} \tag{A1.27}$$

In general, $\mathbf{BA} \neq \mathbf{AB}$. Thus, using the matrices from Eq. (A1.23),

$$\mathbf{A\,B} = \begin{pmatrix} 2 & 1 & 2 \\ 2 & \bar{1} & 0 \\ \bar{1} & 3 & \bar{4} \end{pmatrix} \quad \text{and} \quad \mathbf{B\,A} = \begin{pmatrix} \bar{3} & 3 & 5 \\ 0 & \bar{1} & \bar{2} \\ 1 & 0 & 1 \end{pmatrix} \qquad (A1.28)$$

It may be noted that modern compilers usually include matrix operations as intrinsic functions.

A1.4.8.   Some multiplicative properties of matrices

The following properties of matrices are useful from time to time; the list is not exhaustive.

$$\left. \begin{aligned} &\mathbf{A(BC)} = \mathbf{(AB)C} \\ &\mathbf{(A + B)C} = \mathbf{AC} + \mathbf{BC} \\ &s\mathbf{(A + B)} = s\mathbf{A} + s\mathbf{B} \\ &\text{where } s \text{ is a scalar constant} \\ &\mathbf{(A + B)}^{\mathrm{T}} = \mathbf{A}^{\mathrm{T}} + \mathbf{B}^{\mathrm{T}} \\ &\mathbf{(AB)}^{\mathrm{T}} = \mathbf{B}^{\mathrm{T}} + \mathbf{A}^{\mathrm{T}} \end{aligned} \right\} \qquad (A1.29)$$

A1.4.9.   Determinant value of a matrix

The determinant of a matrix is a mathematical tool of importance in many applications; here, the concern is mainly with its use in evaluating the inverse of matrix. Since an inverse exists only for a square matrix, the same condition applies to the evaluation of a determinant.

A1.4.9.1. 2×2 *and* 3×3 *matrices*

Consider the two-dimensional matrix

$$\mathbf{a} = \begin{pmatrix} a_{11} & a_{12} \\ a_{21} & a_{22} \end{pmatrix}$$

The determinant, written variously as $\det(\mathbf{a})$, $|\mathbf{a}|$ or Δ, is given by

$$\det(\mathbf{a}) = \begin{vmatrix} a_{11} & a_{21} \\ a_{12} & a_{22} \end{vmatrix} = a_{11}a_{22} - a_{12}a_{21} \qquad (A1.30)$$

Its evaluation is straightforward; a 3×3 matrix is a little more involved.

Let the general 3×3 matrix be

$$\mathbf{A} = \begin{pmatrix} a_{11} & a_{12} & a_{13} \\ a_{21} & a_{22} & a_{23} \\ a_{31} & a_{32} & a_{33} \end{pmatrix} \qquad (A1.31)$$

One could write two such matrices side-to-side, delete the final column and cross-multiply in a manner somewhat similar to that used in evaluating zone symbols, as in Example 1.1 of Section 1.5. A more general procedure is given by

$$\det(\mathbf{A}) = \begin{vmatrix} a_{11} & a_{12} & a_{13} \\ a_{21} & a_{22} & a_{23} \\ a_{31} & a_{32} & a_{33} \end{vmatrix} = a_{11} \begin{vmatrix} a_{22} & a_{23} \\ a_{32} & a_{33} \end{vmatrix}$$

$$-a_{12} \begin{vmatrix} a_{21} & a_{23} \\ a_{31} & a_{33} \end{vmatrix} + a_{13} \begin{vmatrix} a_{21} & a_{22} \\ a_{31} & a_{32} \end{vmatrix} \qquad (A1.32)$$

If either procedure is applied to the matrix $\mathbf{A}$ from Eq. (A1.23), then $\det(\mathbf{A})$ is formed as

$$\det(\mathbf{A}) = 1 \begin{vmatrix} 1 & 2 \\ 1 & 1 \end{vmatrix} - 0 \begin{vmatrix} 0 & 2 \\ 2 & 1 \end{vmatrix} + 1 \begin{vmatrix} 0 & 1 \\ 2 & 1 \end{vmatrix} = -1 + 2 = 1 \qquad (A1.33)$$

Some useful properties of determinants are as follows:

- A matrix and its transpose have the same determinant value, *vide* Eq. (A1.16).
- Interchanging two columns of a determinant multiplies its value by -1.
- If any row or column is zero, the determinant value is zero.

A1.4.10. Inverse of a matrix

The inverse of a *square* matrix may be obtained through its *cofactor* matrix. Using matrix $\mathbf{A}$ from Eq. (A1.23) as an example, the *cofactor* matrix $\mathbf{D}$ is obtained with elements d_{ij} given by

$$d_{ij} = (-1)^{i+j} M_{ij} \qquad (A1.34)$$

where M_{ij} is the *minor* determinant of $\mathbf{A}$, obtained by striking out the row and column containing the ij element; thus,

$$d_{12} = - \begin{vmatrix} 0 & 2 \\ \bar{2} & 1 \end{vmatrix} = -4$$

Proceeding in this manner,

$$\mathbf{D} = \begin{pmatrix} \bar{1} & \bar{4} & 2 \\ 1 & 3 & \bar{1} \\ \bar{1} & \bar{2} & 1 \end{pmatrix} \qquad (A1.35)$$

The transpose of the cofactor matrix is the *adjoint* matrix $\mathbf{A}^{\dagger}$, where $a_{ij}^{\dagger} = d_{ji}$, but there is no need to use it explicitly here. Finally, the inverse of $\mathbf{A}$, written as $\mathbf{A}^{-1}$, is given by

$$\mathbf{A}^{-1} = \frac{1}{\det(\mathbf{A})} \mathbf{D}^{\mathrm{T}} = \frac{1}{1} \begin{pmatrix} \bar{1} & 1 & \bar{1} \\ \bar{4} & 3 & \bar{2} \\ 2 & \bar{1} & 1 \end{pmatrix} = \begin{pmatrix} \bar{1} & 1 & \bar{1} \\ \bar{4} & 3 & \bar{2} \\ 2 & \bar{1} & 1 \end{pmatrix} \qquad (A1.36)$$

A simpler, equivalent procedure is to form the elements of $\mathbf{A}^{-1}$ directly as

$$a_{ij}^{-1} = (-1)^{i+j} M_{ji} \qquad (A1.37)$$

where M_{ji} is the minor determinant formed by deleting the row and column containing the ji element of the matrix $\mathbf{A}$. Thus, the a_{23}^{-1} element of the inverse matrix is given by Eq. (A1.34), but with M_{ji} obtained by striking out the row and column of its 3,2 element:

$$a_{23}^{-1} = \frac{1}{\det(\mathbf{A})} (-1)^5 \begin{vmatrix} 1 & 1 \\ 0 & 2 \end{vmatrix} = -2$$

A1.4.11. Orthogonal and unitary matrices

A matrix that fulfils the condition

$$\mathbf{A}^{-1} = \mathbf{A}^{\mathrm{T}} \tag{A1.38}$$

is said to be an *orthogonal* matrix. All orthogonal matrices are square but not necessarily symmetrical, as in the following example:

$$\mathbf{A} = \begin{pmatrix} \cos\theta & \overline{\sin\theta} & 0 \\ \sin\theta & \cos\theta & 0 \\ 0 & 0 & 1 \end{pmatrix}, \quad \mathbf{A}^{-1} = \begin{pmatrix} \cos\theta & \sin\theta & 0 \\ \overline{\sin\theta} & \cos\theta & 0 \\ 0 & 0 & 1 \end{pmatrix},$$

$$\mathbf{A}^{\mathrm{T}} = \begin{pmatrix} \cos\theta & \sin\theta & 0 \\ \overline{\sin\theta} & \cos\theta & 0 \\ 0 & 0 & 1 \end{pmatrix}$$

A matrix $\mathbf{A}$ is *unitary* if its adjoint is equal to its inverse

$$\mathbf{A}^{\dagger} = \mathbf{A}^{-1}$$

that is, $a_{ij}^{\dagger} = a_{ij}^{-1}$ for all i and j; matrix $\mathbf{A}$ below is unitary:

$$\mathbf{A} = \begin{pmatrix} 1 & 0 & 0 \\ 0 & 0 & e^{i} \\ 0 & e^{-2i} & 0 \end{pmatrix}, \quad \mathbf{A}^{\dagger} = \begin{pmatrix} 1 & 0 & 0 \\ 0 & 0 & e^{2i} \\ 0 & e^{-i} & 0 \end{pmatrix},$$

$$\mathbf{A}^{-1} = \begin{pmatrix} 1 & 0 & 0 \\ 0 & 0 & e^{2i} \\ 0 & e^{-i} & 0 \end{pmatrix}$$

If a matrix $\mathbf{A}$ is complex, then the adjoint matrix $\mathbf{A}^{\dagger}$ is the complex conjugate of the transpose, $(\mathbf{A}^{T})^{*}$, as can be seen in the matrices above.

A1.4.12. Matrices, rows and columns

In certain transformations (see Section 3.8), row and column matrices are involved. Thus, operating on a row $\mathbf{x}$ by a matrix $\mathbf{A}$ gives the result for $\mathbf{x}'$ as

$$\mathbf{xA} = \mathbf{x}'$$

For example,

$$(1 \quad 2 \quad 3) \begin{pmatrix} 1 & 0 & 1 \\ 0 & 2 & 1 \\ 1 & 3 & \bar{1} \end{pmatrix} = (4 \quad 13 \quad 0)$$

$$\quad\quad \mathbf{x} \quad\quad\quad\quad \mathbf{A} \quad\quad\quad\quad\quad \mathbf{x}'$$

whereas if $\mathbf{x}$ is treated as a column,

$$\begin{pmatrix} 1 & 0 & 1 \\ 0 & 2 & 1 \\ 1 & 3 & \bar{1} \end{pmatrix} \begin{pmatrix} 1 \\ 2 \\ 3 \end{pmatrix} = \begin{pmatrix} 4 \\ 7 \\ 4 \end{pmatrix}$$

$$\quad \mathbf{A} \quad\quad\quad \mathbf{x} \quad\quad \mathbf{x}''$$

However,

$$\begin{pmatrix} 1 & 0 & 1 \\ 0 & 2 & 3 \\ 1 & 1 & \bar{1} \end{pmatrix} \begin{pmatrix} 1 \\ 2 \\ 3 \end{pmatrix} = \begin{pmatrix} 4 \\ 13 \\ 0 \end{pmatrix}$$

$$\quad \mathbf{A}^{\mathrm{T}} \quad\quad\quad \mathbf{x} \quad\quad \mathbf{x}'$$

or

$$\mathbf{A}^{\mathrm{T}}\mathbf{x} = \mathbf{x}'$$

Note also that $(\mathbf{A}^{\mathrm{T}})^{-1}\mathbf{x} = \mathbf{x}(\mathbf{A}^{-1})^{\mathrm{T}}$ and $(\mathbf{A}^{-1})\mathbf{x} = \mathbf{x}(\mathbf{A}^{\mathrm{T}})^{-1}$.

A1.5. Normal to a Crystal Plane

In Fig. A1.3, (hkl) is any plane in a lattice, making intercepts a/h, b/k and c/l with the x-, y- and z- axes, respectively. A normal $\mathbf{n}$ to the plane is, from Section A1.2.2, given by any cross product such

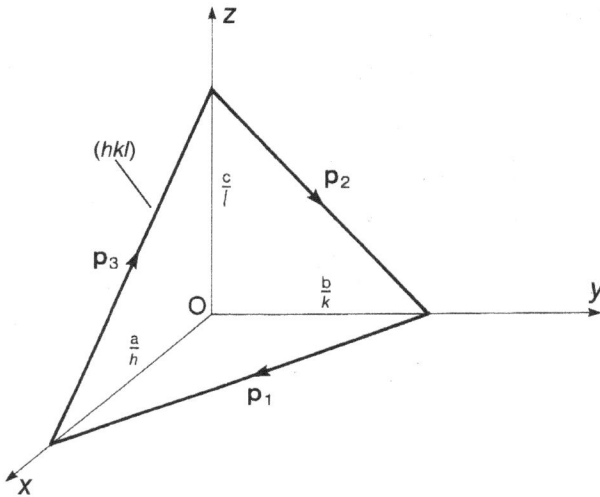

Fig. A1.3. A plane (hkl) making intercepts a/h, b/k and c/l with the x-, y- and z-axes, respectively.

as $\mathbf{p}_1 \times \mathbf{p}_2$. Thus, remembering that $\mathbf{a} \times \mathbf{b} = -\mathbf{b} \times \mathbf{a}$ and $\mathbf{b} \times \mathbf{b} = 0$, $\mathbf{n}$ may be written as

$$
\begin{aligned}
\mathbf{n} &= \left(\frac{\mathbf{a}}{h} - \frac{\mathbf{b}}{k}\right) \times \left(\frac{\mathbf{b}}{k} - \frac{\mathbf{c}}{l}\right) \\
&= \frac{1}{hk}\mathbf{a} \times \mathbf{b} + \frac{1}{kl}\mathbf{b} \times \mathbf{c} + \frac{1}{lh}\mathbf{c} \times \mathbf{a}
\end{aligned}
\tag{A1.39}
$$

Multiplying by hkl yields

$$
\mathbf{n} = h(\mathbf{b} \times \mathbf{c}) + k(\mathbf{c} \times \mathbf{a}) + l(\mathbf{a} \times \mathbf{b}) \tag{A1.40}
$$

A1.6. Solution of Linear Simultaneous Equations

Matrices are used conveniently in Cramer's method for solving systems of simultaneous equations; the following example illustrates the procedure:

$$\left.\begin{array}{r} x_1 + x_2 - x_3 = 6 \\ x_1 - x_2 + x_3 = 2 \\ x_1 - 2x_3 = 4 \end{array}\right\} \qquad (A1.41)$$

Writing Eq. (A1.41) in the following matrix form yields

$$\underset{\mathbf{S}}{\begin{pmatrix} 1 & 1 & \bar{1} \\ 1 & \bar{1} & 1 \\ 1 & 0 & \bar{2} \end{pmatrix}} \underset{\mathbf{x}}{\begin{pmatrix} x_1 \\ x_2 \\ x_3 \end{pmatrix}} = \underset{\mathbf{x}'}{\begin{pmatrix} 6 \\ 2 \\ 4 \end{pmatrix}} \qquad (A1.42)$$

In general, $x_i = \frac{\det(\mathbf{S}_i)}{\det(\mathbf{S})}$, where $\mathbf{S}_i$ is the appropriate determinant in which the column $\mathbf{x}'$ replaces the ith column of the matrix $\mathbf{S}$. Thus, in the given example,

$$\det(\mathbf{S}) = \begin{vmatrix} 1 & 1 & \bar{1} \\ 1 & \bar{1} & 1 \\ 1 & 0 & \bar{2} \end{vmatrix} = 4, \quad \det(\mathbf{S}_1) = \begin{vmatrix} 6 & 1 & \bar{1} \\ 2 & \bar{1} & 1 \\ 4 & 0 & \bar{2} \end{vmatrix} = 16$$

$$\det(\mathbf{S}_2) = \begin{vmatrix} 1 & 6 & \bar{1} \\ 1 & 2 & 1 \\ 1 & 4 & \bar{2} \end{vmatrix} = 8, \quad \det(\mathbf{S}_3) = \begin{vmatrix} 1 & 1 & 6 \\ 1 & \bar{1} & 2 \\ 1 & 0 & 4 \end{vmatrix} = 0$$

Hence, $x_1 = 4$, $x_2 = 2$, $x_3 = 0$.

A1.7.   Useful Matrices

It may be noted first that from Appendix A4, the *trace* of the rotation matrix, Eq. (A4.6), that is, the sum of the diagonal elements, is $(1 + 2\cos\phi)$, for either orthogonal or non-orthogonal axes; ϕ is the angle of rotation for a symmetry axis in a crystal. Since $|\cos\phi| \le 1$, an *integral* value of the trace can have only the values $-1, 0, 1, 2$ or 3. Hence, the following results are obtained:

$(1 + 2\cos\phi)$	ϕ	R-fold axis
-1	$180°$	2
0	$120°$	3
1	$90°$	4
2	$60°$	6
3	$0°(360°)$	1

This result may be compared with that from Section 3.6.

The set of matrices that follows is useful in working with point groups and space groups. Some of them will be self-evident from the work that has already been studied. Others may be prepared by suitable matrix combinations. For example, consider the stereogram in Fig. A1.4. Rotation of a point x, y, z, about [111] has the following actions:

$$x,\, y,\, z \xrightarrow{\ ^{3}[111]\ } z,\, x,\, y \xrightarrow{\ ^{3}[111]\ } y,\, z,\, x$$
$$(1) \qquad\qquad (2) \qquad\qquad (3)$$

But what of a three-fold rotation of x, y, z about $[11\bar{1}]$? From the stereogram, it may be seen that the sequence of symmetry

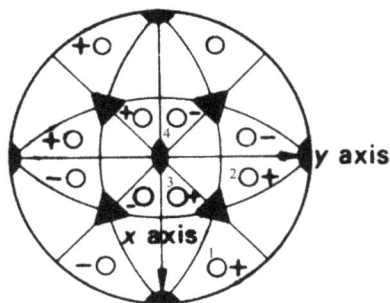

Fig. A1.4. Stereogram of the cubic point group 23 showing the general form $\{hkl\}$. The points marked 1–4 may be ignored for now; their significance will arise in Section A1.7.

operations

$$x,\, y,\, z \xrightarrow{\;3^2_{[111]}\;} y,\, z,\, x \xrightarrow{\;2_{[0y0]}\;} \bar{y},\, z,\, \bar{x}$$

$$(1) \qquad\qquad (3) \qquad\qquad (4)$$

describes a rotation of the point x, y, z about $[11\bar{1}]$. In matrix notation, the complete sequence of operations may be written as

$$\begin{pmatrix} \bar{1} & 0 & 0 \\ 0 & 1 & 0 \\ 0 & 0 & \bar{1} \end{pmatrix} \begin{pmatrix} 0 & 0 & 1 \\ 1 & 0 & 0 \\ 0 & 1 & 0 \end{pmatrix} \begin{pmatrix} 0 & 0 & 1 \\ 1 & 0 & 0 \\ 0 & 1 & 0 \end{pmatrix} = \begin{pmatrix} 0 & \bar{1} & 0 \\ 0 & 0 & 1 \\ \bar{1} & 0 & 0 \end{pmatrix}$$

$$\quad\; 2_{[0y0]} \qquad\quad 3_{[111]} \qquad\quad 3_{[111]} \qquad\qquad 3_{[11\bar{1}]}$$

The trace of each matrix for a three-fold rotation is zero, as expected from the above analysis.

   The matrices given below are designated mostly by a $[uvw]$ symbol, which is the direction of the corresponding symmetry operator. Where $m_{[uvw]}$ is listed, the plane normal to the direction $[uvw]$ is implied. The symbolism $r'(x',\, y',\, z') = R\,r(x,\, y,\, z)$ means that the vector $r(x',\, y',\, z')$ results from multiplying the vector $r(x,\, y,\, z)$ by the matrix R, in the order indicated.

A1.7.1.   Rotation and reflection matrices on orthogonal axes

A1.7.1.1.   *Two-fold rotation and mirror ($\bar{2}$) reflection*

$$2_{[100]} = \begin{pmatrix} 1 & 0 & 0 \\ 0 & \bar{1} & 0 \\ 0 & 0 & \bar{1} \end{pmatrix}, \quad 2_{[010]} = \begin{pmatrix} \bar{1} & 0 & 0 \\ 0 & 1 & 0 \\ 0 & 0 & \bar{1} \end{pmatrix}, \quad 2_{[001]} = \begin{pmatrix} \bar{1} & 0 & 0 \\ 0 & \bar{1} & 0 \\ 0 & 0 & 1 \end{pmatrix}$$

$$2_{[011]} = \begin{pmatrix} \bar{1} & 0 & 0 \\ 0 & 0 & 1 \\ 0 & 1 & 0 \end{pmatrix}, \quad 2_{[101]} = \begin{pmatrix} 0 & 0 & 1 \\ 0 & \bar{1} & 0 \\ 1 & 0 & 0 \end{pmatrix}, \quad 2_{[110]} = \begin{pmatrix} 0 & 1 & 0 \\ 1 & 0 & 0 \\ 0 & 0 & \bar{1} \end{pmatrix}$$

$$\mathbf{2}_{[01\bar{1}]} = \begin{pmatrix} \bar{1} & 0 & 0 \\ 0 & 0 & \bar{1} \\ 0 & \bar{1} & 0 \end{pmatrix}, \quad \mathbf{2}_{[\bar{1}01]} = \begin{pmatrix} 0 & 0 & \bar{1} \\ 0 & \bar{1} & 0 \\ \bar{1} & 0 & 0 \end{pmatrix}, \quad \mathbf{2}_{[1\bar{1}0]} = \begin{pmatrix} 0 & \bar{1} & 0 \\ \bar{1} & 0 & 0 \\ 0 & 0 & \bar{1} \end{pmatrix}$$

From the foregoing, the matrices for mirror reflection $\mathbf{m}(\equiv \mathbf{2}_{\perp m})$ across the planes normal to [100], [010], [001], [011], [101], [110], (01$\bar{1}$), ($\bar{1}$01) and (1$\bar{1}$0) may be obtained by negating *all* non-zero elements of the above six matrices. Thus, from the first example, the matrix $\mathbf{2}_{[100]}$, which is equivalent to $\mathbf{m}_{[100]}$, refers to the plane (0yz) and is obtained by negating the non-zero elements of the matrix $\mathbf{2}_{[100]}$; thus,

$$\mathbf{m}_{(0yz)} = \begin{pmatrix} \bar{1} & 0 & 0 \\ 0 & 1 & 0 \\ 0 & 0 & 1 \end{pmatrix}$$

A1.7.1.2. *Three-fold rotation inclined to x, y, z at* $\cos^{-1}(1/\sqrt{3})$, *approximately 54.74°*

$$\mathbf{3}_{[111]} = \begin{pmatrix} 0 & 0 & 1 \\ 1 & 0 & 0 \\ 0 & 1 & 0 \end{pmatrix}, \quad \mathbf{3}_{[\bar{1}11]} = \begin{pmatrix} 0 & \bar{1} & 0 \\ 0 & 0 & 1 \\ \bar{1} & 0 & 0 \end{pmatrix},$$

$$\mathbf{3}_{[1\bar{1}1]} = \begin{pmatrix} 0 & \bar{1} & 0 \\ 0 & 0 & \bar{1} \\ 1 & 0 & 0 \end{pmatrix}, \quad \mathbf{3}_{[11\bar{1}]} = \begin{pmatrix} 0 & 1 & 0 \\ 0 & 0 & \bar{1} \\ \bar{1} & 0 & 0 \end{pmatrix}$$

A1.7.1.3. *Four-fold rotation*

$$\mathbf{4}_{[100]} = \begin{pmatrix} 1 & 0 & 0 \\ 0 & 0 & \bar{1} \\ 0 & 1 & 0 \end{pmatrix}, \quad \mathbf{4}_{[010]} = \begin{pmatrix} 0 & 0 & 1 \\ 0 & 1 & 0 \\ \bar{1} & 0 & 0 \end{pmatrix},$$

$$4_{[001]} = \begin{pmatrix} 0 & \bar{1} & 0 \\ 1 & 0 & 0 \\ 0 & 0 & 1 \end{pmatrix}, \quad \bar{4}_{[001]} = (-1)\,4_{[001]} = \begin{pmatrix} 0 & 1 & 0 \\ \bar{1} & 0 & 0 \\ 0 & 0 & \bar{1} \end{pmatrix}$$

A1.7.1.4. *Six-fold rotation*

$$6_{[001]} = 2_{[001]}\,3^2_{[001[} = \begin{pmatrix} \bar{1} & 0 & 0 \\ 0 & \bar{1} & 0 \\ 0 & 0 & 1 \end{pmatrix} \begin{pmatrix} -1/2 & -\sqrt{3}/2 & 0 \\ \sqrt{3}/2 & -1/2 & 0 \\ 0 & 0 & 1 \end{pmatrix}$$

$$\times \begin{pmatrix} -1/2 & -\sqrt{3}/2 & 0 \\ \sqrt{3}/2 & -1/2 & 0 \\ 0 & 0 & 1 \end{pmatrix} = \begin{pmatrix} 1/2 & -\sqrt{3}/2 & 0 \\ \sqrt{3}/2 & 1/2 & 0 \\ 0 & 0 & 1 \end{pmatrix}$$

A1.7.2. Rotation and reflection matrices on rhombohedral axes

$$3_{[111]} = \begin{pmatrix} 0 & 0 & 1 \\ 1 & 0 & 0 \\ 0 & 1 & 0 \end{pmatrix}, \quad m_{(0yz)} = \begin{pmatrix} 1 & 0 & 0 \\ 0 & 0 & 1 \\ 0 & 1 & 0 \end{pmatrix}$$

(i) The matrix $3_{[111]}$ here is identical with that for $3_{[111]}$ on orthogonal axes (Appendix A1.7.1.2).

(ii) Other points related by m symmetry may be obtained by the matrix combinations $3_{[111]}\,m_{(0y\,z)}$.

A1.7.3. Rotation and reflection matrices on hexagonal axes

$$2_{[120]} = \begin{pmatrix} \bar{1} & 1 & 0 \\ 0 & 1 & 0 \\ 0 & 0 & \bar{1} \end{pmatrix}, \quad m_{[120]} = \begin{pmatrix} 1 & \bar{1} & 0 \\ 0 & \bar{1} & 0 \\ 0 & 0 & 1 \end{pmatrix} \quad [\equiv m||(x,0,z)]$$

$$2_{[210]} = \begin{pmatrix} 1 & 0 & 0 \\ 1 & \bar{1} & 0 \\ 0 & 0 & \bar{1} \end{pmatrix}, \quad m_{[210]} = \begin{pmatrix} \bar{1} & 0 & 0 \\ \bar{1} & 1 & 0 \\ 0 & 0 & 1 \end{pmatrix} \quad [\equiv m||(0, y, z)]$$

$$3_{[0001]} = \begin{pmatrix} 0 & \bar{1} & 0 \\ 1 & \bar{1} & 0 \\ 0 & 0 & 1 \end{pmatrix}, \quad 6_{[001]} = \begin{pmatrix} 1 & \bar{1} & 0 \\ 1 & 0 & 0 \\ 0 & 0 & 1 \end{pmatrix}$$

References

[1] Fleisch DA, *A Student's Guide to Vectors and Tensors*. Cambridge University Press (2012).
[2] Joag PS, *Vectors, Vector Operations and Vector Analysis*. Cambridge University Press (2016).
[3] Sands DE, *Vectors and Tensors in Crystallography*. Dover Publications Inc (1982).

Appendix A2

Analytical Geometry of Direction Cosines

A2.1.   Direction Cosines of a Line

In Fig. A2.1, P_1 is the point x_1, y_1, z_1 referred to as x, y and z orthogonal axes. Lines from P_1 perpendicular to the x-, y- and z-axes cut them at A, B and C, respectively. Thus, $OA = x_1$, $OB = y_1$ and $OC = z_1$. The direction cosines of OP_1 are given by $\cos \chi_1 = x_1/OP_1$, $\cos \psi_1 = y_1/OP_1$ and $\cos \omega_1 = z_1/OP_1$. Hence,

$$\cos^2 \chi_1 + \cos^2 \psi_1 + \cos^2 \omega_1 = (x_1^2 + y_1^2 + z_1^2)/OP_1^2 \qquad \text{(A2.1)}$$

Since x_1, y_1 and z_1 are the projections of OP_1 on to the x-, y- and z-axes, it follows that

$$x_1^2 + y_1^2 + z_1^2 = OP_1^2 \qquad \text{(A2.2)}$$

Hence,

$$\cos^2 \chi_1 + \cos^2 \psi_1 + \cos^2 \omega_1 = 1 \qquad \text{(A2.3)}$$

A2.1.1.   Alternative, vector derivation

Properties of vectors are introduced in Appendix A1. In Fig. A2.1, let $OP_1 = \mathbf{d}_1$, $OA = x$, $OB = y$ and $OC = z$. Then, $\cos \chi_1 = \frac{\mathbf{d}_1 \cdot \mathbf{i}}{|\mathbf{d}_1|} = \frac{(\mathbf{i}x + \mathbf{j}y + \mathbf{k}z) \cdot \mathbf{i}}{|\mathbf{d}_1|} = \frac{x}{d}$, and similarly for ψ and ω *mutatis mutandis*. Thus,

$$\cos^2 \chi_1 + \cos^2 \psi_1 + \cos^2 \omega_1 = \frac{x^2}{d^2} + \frac{y^2}{d^2} + \frac{z^2}{d^2} = 1 \qquad \text{(A2.4)}$$

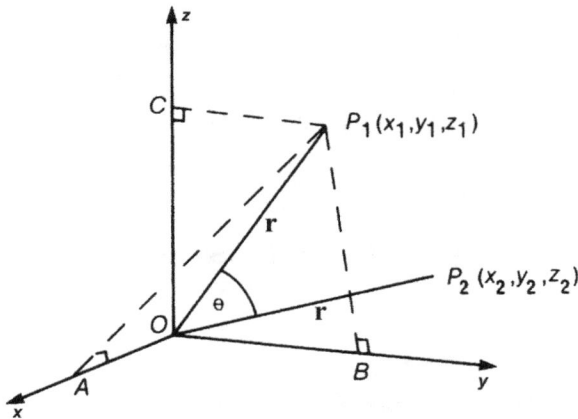

Fig. A2.1.   The direction cosines of OP_1 and OP_2 refer to the rectangular axes x, y and z, where $\chi_n = \widehat{AOP_n}$, $\psi_n = \widehat{BOP_n}$, $\omega_n = \widehat{COP_n}(n = 1, 2)$.

A2.2.   Angle Between Two Lines

The point $P_2(x_2, y_2, z_2)$ is constructed on Fig. A2.1 such that $OP_2 = OP_1 = r$. The length d of the vector $P_1 \rightarrow P_2$ is given by $\mathbf{d} = \mathbf{OP_2} - \mathbf{OP_1}$. Then,

$$
\begin{aligned}
d^2 &= (x_2 - x_1)^2 + (y_2 - y_1)^2 + (z_2 + z_1)^2 \\
&= r^2(\cos^2\chi_1 + \cos^2\chi_2 - 2\cos\chi_1\cos\chi_2 \\
&\quad + \cos^2\psi_1 + \cos^2\psi_2 - 2\cos\psi_1\cos\psi_2 \\
&\quad + \cos^2\omega_1 + \cos^2\omega_2 - 2\cos\omega_1\cos\omega_2)
\end{aligned} \tag{A2.5}
$$

Using Eq. (A2.3) with Eq. (A2.4),

$$d^2 = 2r^2[1 - (\cos\chi_1\cos\chi_2 + \cos\psi_1\cos\psi_2 + \cos\omega_1\cos\omega_2)] \tag{A2.6}$$

In $\triangle OP_1P_2$

$$d/2 = r\sin(\theta/2) \tag{A2.7}$$

Therefore,

$$d^2/4r^2 = \sin^2(\theta/2) = (1 - \cos\theta)/2$$

or

$$d^2/2r^2 = 1 - \cos\theta \qquad\qquad (A2.8)$$

Comparing Eq. (A2.6) and Eq. (A2.8), it follows that the angle θ between OP_1 and OP_2 is given by

$$\cos\theta = \cos\chi_1 \cos\chi_2 + \cos\psi_1 \cos\psi_2 + \cos\omega_1 \cos\omega_2 \qquad (A2.9)$$

Example A2.1. Let two points on the diagram below be $A(4\mathbf{i}, 0\mathbf{j}, 7\mathbf{k})$ and $B(-2\mathbf{i}, \mathbf{j}, 3\mathbf{k})$. (a) The angle θ, (b) the length d of the line AB and (c) the direction cosines of the line AB will be determined.

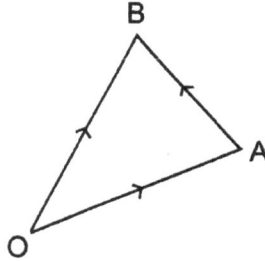

(a) $\cos\theta = \dfrac{\mathbf{OA} \cdot \mathbf{OB}}{OA\ OB}$

$= \dfrac{(4\mathbf{i} + 0\mathbf{j} + 7\mathbf{k}) \cdot (-2\mathbf{i} + \mathbf{j} + 3\mathbf{k})}{\sqrt{(4\mathbf{i} + 0\mathbf{j} + 7\mathbf{k}) \cdot (4\mathbf{i} + 0\mathbf{j} + 7\mathbf{k})}\sqrt{(-2\mathbf{i} + \mathbf{j} + 3\mathbf{k}) \times (-2\mathbf{i} + \mathbf{j} + 3\mathbf{k})}}$

$= \dfrac{13}{\sqrt{65}\sqrt{4}} = 0.430946$, so that $\theta = 64.47°$

(b) $\mathbf{d} - \mathbf{OB} - \mathbf{OA} = -6\mathbf{i} + \mathbf{j} - 4\mathbf{k}$, so that $d^2 = \mathbf{d} \cdot \mathbf{d} = (-6\mathbf{i} + \mathbf{j} - 4\mathbf{k}) \cdot (-6\mathbf{i} + \mathbf{j} - 4\mathbf{k}) = 53$, and $d = 7.280$. (Check d with the cosine rule.)

(c) $\cos\chi = \dfrac{x_2 - x_1}{|\mathbf{d}|} = \dfrac{-6}{OAOB} = \dfrac{-6}{30.166}$; similarly $\cos\psi\dfrac{1}{|\mathbf{d}|} = \dfrac{1}{30.166}$ and $\cos\omega = \dfrac{-4}{30.166}$.

Appendix A3

Stereoviews, Crystal Models and Point-Group Recognition

A3.1. Stereoviews and Stereoviewing

Stereoviews have been used to illustrate the three-dimensional character of crystal structures since 1926, and computer programs are available for preparing the two views needed for producing the desired three-dimensional image [1, 2].

Two diagrams of a given object are needed in order to form a three-dimensional image. In viewing stereoscopic diagrams, each eye should see only the appropriate half of the complete illustration. The simplest procedure is direct viewing with a *stereoviewer*, whereupon the three-dimensional image appears centrally between the stereoscopic pair of diagrams. A relatively inexpensive stereoviewer is available from *3Dstereo.com. Inc. 1930 Village Center Circle, #3-333, Las Vegas, NV 89134, USA*, or via the 'Amazon' company online.

Alternatively, one may be able to train the unaided eyes to see a stereo image. Focus the eyes on infinity, and allow each eye to see only the appropriate diagram. The eyes must be relaxed and look straight ahead; it may help to close the eyes for a moment, then open them wide and allow them to relax without consciously focusing on the diagram. Varying the distance from the eyes to the diagram may help in obtaining the stereoview. Normally, three images appear; the central image is the stereoview. The viewing process may be aided by holding a white card edgeways between the two drawings. If difficulty is experienced with this direct viewing method, a stereoviewer will be needed (see Section A3.1.1).

A3.1.1.   Construction of a Stereoviewer

The essence of this construction rests in two convex lenses separated by the interocular distance of *ca.* 63 mm.

A pair of plano-convex or biconvex lenses, each of focal length *ca.* 100 mm and diameter *ca.* 30 mm, is mounted between two opaque cards, with the centres of the lenses *ca.* 63 mm apart. The card frame must be so shaped that the lenses may be brought close to the eyes. Figure A3.1 illustrates the construction of the stereoviewer. It may be helpful to obscure a segment S of each lens, closest to the nose region N, of size approximately 25% of the lens diameter.

A3.2.   Crystal Models

In this section, instructions are given for making crystal models that may be helpful in studying crystal symmetry. Other sets of instructions can be found in the literature [4–7].

A3.2.1.   Cube

On a thin card, draw a square of side 40 mm. On each side of this square, draw another identical square. Lightly score the edges of the first square and fold the other four to form five faces of a cube, and fasten the edges with Sellotape. There is an advantage in leaving the sixth face of the cube open, as indicated in the next section, and the presence of this sixth face can be imagined.

A3.2.2.   Tetrahedron

On another similar card, draw an equilateral triangle of side $39.5\sqrt{2}$ mm; this measure is just less than the length of the face diagonal of the cube described above. On each side of the triangle, draw another identical triangle. Lightly score the edges of the first triangle, fold the other three triangles in the same sense to meet at an apex, and seal the apex with glue or with Sellotape. Note that on

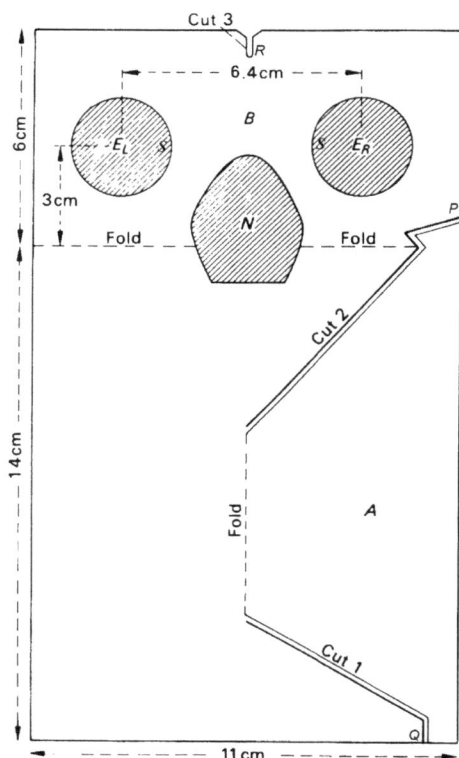

Fig. A3.1. To construct a stereoviewer, prepare two pieces of the thin card as shown in the figure; cut out and discard the shaded portions. Make cuts 1 and 2 along the double line PQ and cut 3 at R. Glue the two cards together with the lenses E_L and E_R in position. To use, fold the portions A and B backwards, and set the projection next to P into the cut at R; strengthen the fold with sellotape, if necessary. View the stereo diagrams from the side marked B.

placing the tetrahedron inside the cube such that an edge of the tetrahedron represents a face diagonal of the cube, the assembly aligns the symmetry elements common to both models.

A3.2.3. Model with $\bar{4}$-symmetry

A four-fold inversion axis is one of the more difficult symmetry elements to appreciate from a drawing. To construct a model that exhibits *inter alia* a $\bar{4}$-axis, mark out a thin card in accordance with

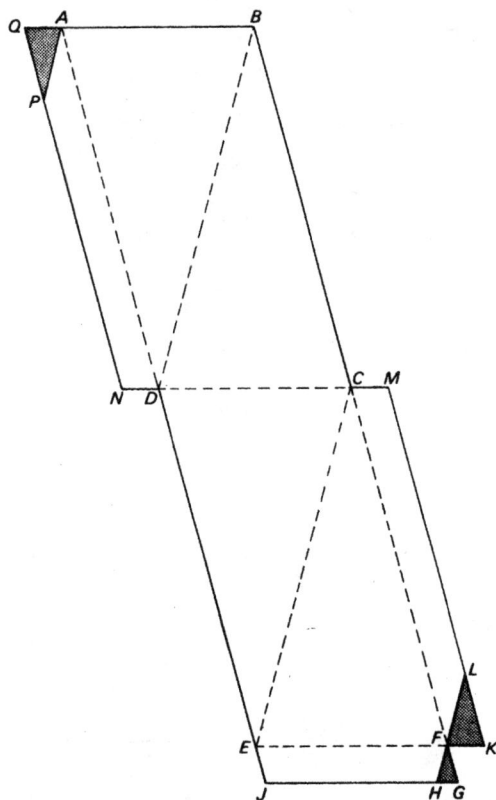

Fig. A3.2. Data for constructing a crystal with a four-fold inversion axis, $\bar{4}$. $NQ = AD = BD = BC = DE = CE = CF = KM = 10$ cm; $AB = CD = EF = GJ = 5$ cm; $AP = PQ = FL = KL = 2$ cm; $AQ = DN = CM = FK = FG = FH = EJ = 1$ cm; the length HG is defined by the line LFH.

the dimensions listed in the legend to Fig. A3.2; then cut along the solid lines, discarding the shaded portions. Make folds in the same direction along the dotted lines; the sections $ADNP$ and $CFLM$ are glued internally, and $EFHJ$ is glued externally. Figure A3.3 is a stereogram that shows a $\bar{4}$-symmetry operation.

Other directions for making models with other point-group symmetries are given in the literature [5, 6]. With these models, the diagrams given are best glued on to thin card for best results before cutting out the shapes.

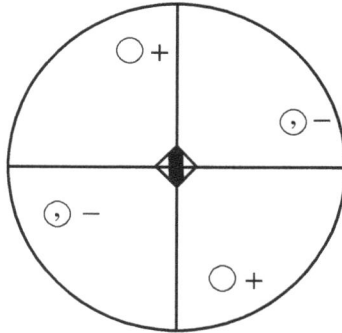

Fig. A3.3.   Stereogram illustrating a $\bar{4}$-symmetry operation.

A3.3.   Point-Group Recognition

If the models are used with the program SYMM in the Program Suits, allocate model number **7** for the cube, **19** for the tetrahedron and **86** for the model with the $\bar{4}$-axis. The identification of the point group of each model then proceeds along the lines indicated by the block diagram of the program SYMM. The following table shows a model number (bold font) appropriate to each point group:

Crystal system	Point group with corresponding model number in bold type
Triclinic	1, **91**; $\bar{1}$, **78**
Monoclinic	2, **77**; $m\ (\equiv \bar{2})$, **83**; $\frac{2}{m}$, **68**
Orthorhombic	222, **67**; $mm2$, **16**; mmm, **59**
Trigonal	3, **93**; $\bar{3}$, **48**; 32, **43**; $3m$, **42**; $\bar{3}m$, **38**
Tetragonal	4, **85**; $\bar{4}$, **95**; $\frac{4}{m}$, **56**; 422, **55**; $4mm$, **96**; $\bar{4}2m$, **57**; $\frac{4}{m}mm$,**50**
Hexagonal	6, **88**; $\bar{6}$, **98**; $\frac{6}{m}$, **37**; 622, **36**; $6mm$, **81**; $\bar{6}m2$, **44**; $\frac{6}{m}mm$, **29**.
Cubic	23, **27**; $m\bar{3}$, **22**; 432, **26**; $\bar{4}3m$, **17**; $m\bar{3}m$, **2**
Infinity groups	∞, **101**; $\infty/m (\equiv \bar{\infty})$, **102**

References

[1] *ORTEP*. Online at ⟨http://www.chem.gla.ac.uk/~louis/software/ortep3/⟩.

[2] *MERCURY*. Online at ⟨https://www.ccdc.cam.ac.uk/solutions/csd-system/components/mercury/⟩.

[3] Barthelmy D, *Crystal Forms*. Online at ⟨http://webmineral.com/help/Forms.shtml⟩.

[4] *Models of Crystal Shapes*. Online at ⟨http://learn-science.20m.com/student_crystals.html⟩.

[5] IUCr, *Introduction to the Crystal Class Models*. Online at ⟨https://www.iucr.org/___data/assets/pdf_file/0009/3123/Class.pdf⟩.

[6] McHenry E, *Cut-And-Assemble Mineral Crystal Shapes*. Online at ⟨http://www.ellenjmchenry.com/homeschool-freedownloads/earthscience-games/documents/Crystalshapes.pdf⟩.

[7] Ladd M, *Symmetry of Crystals and Molecules*. Oxford University Press (2014).

Appendix A4

Derivation of a General, Normal to a Plane Rotation Matrix

In this appendix, a matrix $\boldsymbol{\Psi}$ is derived for the anticlockwise rotation of a point X, Y, Z by an angle ϕ about an axis normal to the plane of x and y, with an angle γ between the x- and y-axes. Since the rotation axis is normal to the plane, the Z-coordinate of the point remains unchanged; the axes are a right-handed set. PQ is drawn parallel to the y-axis.

In Fig. A4.1, $\widehat{PQN} = \pi - \gamma$ and $\widehat{QPN} = \gamma - \pi/2$. Then,

$$X = r \cos\theta - Y \cos\gamma$$
$$Y = r \sin\theta / \sin\gamma \tag{A4.1}$$

It follows that

$$X' = r \cos(\theta + \phi) - Y' \cos\gamma$$
$$Y' = r \sin^{-1}\gamma \sin(\theta + \phi) \tag{A4.2}$$

Expanding Eq. (A4.2), substituting for $r \sin\theta$ and $r \cos\theta$ from Eq. (A4.1), and rearranging leads to

$$X' = X(\cos\phi - \cos\gamma \sin\phi / \sin\gamma) - Y(\sin\gamma \sin\phi - \cos^2\gamma \sin\phi / \sin\gamma)$$
$$\tag{A4.3}$$

$$Y' = X(\sin\phi / \sin\gamma) + Y(\cos\phi + \cos\gamma \sin\phi / \sin\gamma) \tag{A4.4}$$

Written concisely,

$$\mathbf{X'} = \boldsymbol{\Psi}\mathbf{X} \tag{A4.5}$$

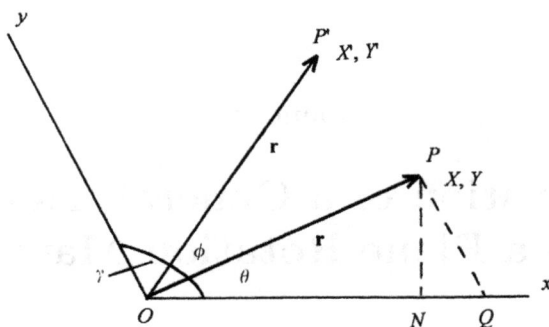

Fig. A4.1.   Vector $\boldsymbol{OP}$ of length $|\mathbf{r}|$ at a general angle θ to the x-axis; $\boldsymbol{OP'}$ is that vector after being rotated anticlockwise (right-handed rotation) by an angle ϕ from the direction of $\boldsymbol{OP}$; the rotation axis z is normal to the x-y plane. The general angle between the x- and y-axes is γ. For three-fold symmetry (on hexagonal axes) and six-fold symmetry, γ has the value of $120°$, and for four-fold symmetry it is $90°$.

where the matrix $\boldsymbol{\Psi}$ is given by

$$\boldsymbol{\Psi} = \begin{pmatrix} (\cos\phi - \cos\gamma \sin\phi/\sin\gamma) & (-\sin\gamma \sin\phi - \cos^2\gamma \sin\phi/\sin\gamma) & 0 \\ (\sin\phi/\sin\gamma) & (\cos\phi + \cos\gamma \sin\phi/\sin\gamma) & 0 \\ 0 & 0 & 1 \end{pmatrix}$$

(A4.6)

The matrix (A4.6) will suffice for all rotational operations that are encountered in studying point groups where the rotation axis, z in this discussion, is *normal to the x-y plane*. For three-fold rotation in the cubic system, the stereogram for point group 432 in Fig. 2.11 shows that a four-fold anticlockwise rotation about the x-axis followed by a four-fold anticlockwise rotation about the z-axis is equivalent to a three-fold anticlockwise rotation about the direction [111].

A general equation for the rotation of $\mathbf{X}(X, Y, Z)$ is $\mathbf{X}' = \boldsymbol{\Psi}\mathbf{X}$. Thus, from Eq. (A4.6) for a four-fold rotation $\boldsymbol{\Psi}_X$ about the x-axis in the cubic system ($\gamma = 90°$), by interchange of axes,

$$\underset{\boldsymbol{\Psi}_X}{\begin{pmatrix} 1 & 0 & 0 \\ 0 & 0 & \bar{1} \\ 0 & 1 & 0 \end{pmatrix}} \underset{\mathbf{X}}{\begin{pmatrix} X \\ Y \\ Z \end{pmatrix}} = \underset{\mathbf{X}'}{\begin{pmatrix} X \\ \bar{Z} \\ Y \end{pmatrix}}$$

(A4.7)

and from a successive rotation $\boldsymbol{\Psi}_Z$ about the z-axis,

$$
\begin{pmatrix} 0 & \bar{1} & 0 \\ 1 & 0 & 0 \\ 0 & 0 & 1 \end{pmatrix} \begin{pmatrix} X \\ \bar{Z} \\ Y \end{pmatrix} = \begin{pmatrix} Z \\ X \\ Y \end{pmatrix} \tag{A4.8}
$$

$$\qquad \boldsymbol{\Psi}_Z \qquad \mathbf{X'} \qquad \mathbf{X''}$$

Hence, the matrix for the rotation $\mathbf{3}_{[111]}$ in the cubic system is obtained by the product $\boldsymbol{\Psi}_Z\,\boldsymbol{\Psi}_X$:

$$
\begin{pmatrix} 0 & \bar{1} & 0 \\ 1 & 0 & 0 \\ 0 & 0 & 1 \end{pmatrix} \begin{pmatrix} 1 & 0 & 0 \\ 0 & 0 & \bar{1} \\ 0 & 1 & 0 \end{pmatrix} = \begin{pmatrix} 0 & 0 & 1 \\ 1 & 0 & 0 \\ 0 & 0 & 1 \end{pmatrix} \tag{A4.9}
$$

$$\qquad \boldsymbol{\Psi}_Z \qquad\qquad \boldsymbol{\Psi}_X \qquad\qquad \boldsymbol{\Psi}_{[111]}$$

$$
\begin{pmatrix} 0 & 0 & 1 \\ 1 & 0 & 0 \\ 0 & 0 & 1 \end{pmatrix} \begin{pmatrix} X \\ Y \\ Z \end{pmatrix} = \begin{pmatrix} Z \\ X \\ Y \end{pmatrix} \tag{A4.10}
$$

$$\qquad \boldsymbol{\Psi}_{[111]} \qquad \mathbf{X} \qquad \mathbf{X''}$$

The same results may be obtained graphically. For example, with points Y and Z in the plane of the paper and with orthogonal right-handed axes, the $+x$-axis projects towards the observer.

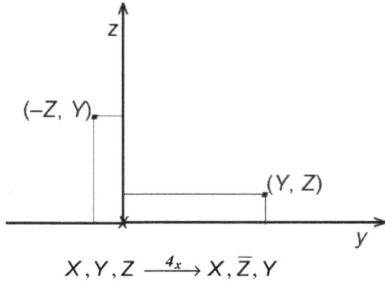

$$X,Y,Z \xrightarrow{\;4_x\;} X,\bar{Z},Y$$

Appendix A5

Spherical Trigonometry

A5.1. Spherical Triangle

Figure A5.1 shows a spherical triangle ABC on the partial surface of a sphere described by the intersections of great circles of which AB, AC and BC are parts (great circles pass through the centre O of the sphere). The arcs a, b, c are the *sides* of the triangle, and A, B, C are its *angles*; the length a is measured by $\widehat{COB}$. The angle A, formed by the angle between the tangents at A to the arcs AB and AC, is the plane angle $\widehat{QAP}$. The other four elements of triangle OQP are defined similarly, *mutatis mutandis*. The following equations for spherical triangles and are proved readily by trigonometry:

$$\cos a = \cos b \cos c + \sin b \sin c \cos A \qquad (A5.1)$$

$$\sin A/\sin a = \sin B/\sin b = \sin C/\sin c \qquad (A5.2)$$

Other examples of Eq. (A5.1) are obtained by cyclic permutation.

A5.2. Polar Triangle

The polar triangle and equations derived from it are important for the solution of triangles arising when applying Euler's theorem on the combination of rotations to the derivation of point groups [1].

In Fig. A5.2, ABC is a spherical triangle. The arc $B'C'$ is drawn such that all points on it are 90° away from A; thus, A is the *pole* of the great circle of which $B'C'$ is an arc. Similarly, B and C are the

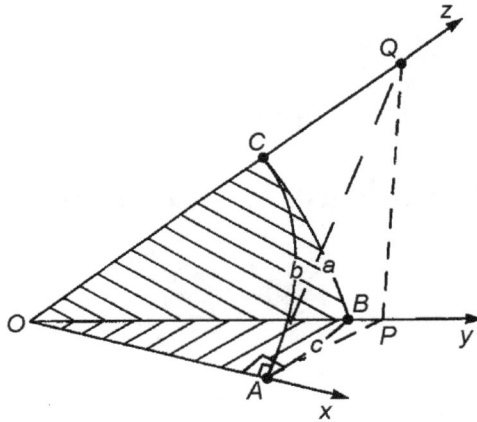

Fig. A5.1. Spherical triangle ABC on the partial surface of a sphere of centre O.

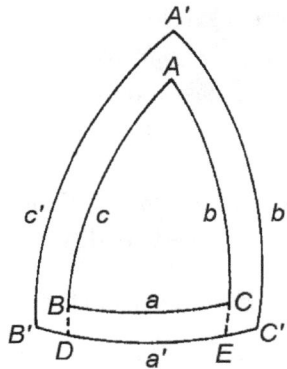

Fig. A5.2. Polar triangle $A'B'C'$ of the triangle ABC; equally, triangle ABC is the polar triangle of triangle $A'B'C'$. All points on the arc $B'C'$ are 90° away from A, the pole of the great circle through $B'C'$; similar relationships apply to the poles B and C.

poles of the arcs $A'C'$ and $A'B'$, respectively: $A'B'C'$ is defined as the *polar* triangle of triangle ABC. By reciprocity, ABC is the polar triangle of triangle $A'B'C'$.

   The *great circle* arcs AB and AC are produced to cut $B'C'$ in D and E, respectively. The spherical angle at A is measured by arc DE. Also, from the construction given above, D is on the great circle arc AB, so that $C'D = \pi/2$. Similarly, $B'E = \pi/2$. Since $a' = B'C' =$

$B'E + C'D - DE = \pi - DE = \pi - A$. Thus,

$$a' = \pi - A \qquad \text{(A5.3)}$$

and since the triangles ABC and $A'B'C$ are polar

$$a = \pi - A' \qquad \text{(A5.4)}$$

with similar relationships for b, c, B and C'.

Reference

[1] Todhunter I, *Spherical Trigonometry*. Macmillan (1886); 5th edn. co-authored Leatham J G, Macmillan (1911).

Appendix A6

Angles of the Reciprocal Unit Cell

The metric tensor in real space is given by Eq. (3.25). Its inverse G^{-1} (aka G^*) for reciprocal space is given, after some manipulation, by

$$G^{-1} = \frac{abc}{V^2} \begin{pmatrix} bc\sin^2\alpha/a & c(\cos\alpha\cos\beta - \cos\gamma) & b(\cos\alpha\cos\gamma - \cos\beta) \\ c(\cos\alpha\cos\beta - \cos\gamma) & ac\sin^2\beta/b & a(\cos\beta\cos\gamma - \cos\alpha) \\ b(\cos\alpha\cos\gamma - \cos\beta) & a(\cos\beta\cos\gamma - \cos\alpha) & ab\sin^2\gamma/c \end{pmatrix}$$

$$(A6.1)$$

but the inverse tensor may be written also as

$$G^{-1} = \begin{pmatrix} a^{*2} & a^*b^*\cos\gamma^* & c^*a^*\cos\beta^* \\ a^*b^*\cos\gamma^* & b^{*2} & b^*c^*\cos\alpha^* \\ c^*a^*\cos\beta^* & b^*c^*\cos\alpha^* & c^{*2} \end{pmatrix} \qquad (A6.2)$$

Thus, a one-to-one correspondence exists between the elements of the G^{-1} matrices shown by Eqs. (A6.1) and (A6.2).

From Eq. (A1.5),

$$\mathbf{a}^* \cdot \mathbf{b}^* = a^*b^*\cos\gamma^* \qquad (A6.3)$$

which, from Eq. (3.15), leads to

$$a^*b^*\cos\gamma^* = \frac{bc\sin\alpha}{V}\frac{ca\sin\beta}{V}\cos\gamma^* \qquad (A6.4)$$

But by comparing the $G_{2,1}^{-1}$ elements in Eqs. (A6.1) and (A6.2)

$$\frac{bc\sin\alpha}{V}\frac{ca\sin\beta}{V}\cos\gamma^* \ (\equiv a^*b^*\cos\gamma^*) = \frac{abc}{V^2}[c(\cos\alpha\cos\beta - \cos\gamma)]$$

$$(A6.5)$$

so that

$$\cos\gamma^* = \frac{abc}{V^2}[c(\cos\alpha\,\cos\beta - \cos\gamma)]\frac{V^2}{bc\sin\alpha\,ca\,\sin\beta}$$

or

$$\cos\gamma^* = \frac{(\cos\alpha\cos\beta - \cos\gamma)}{\sin\alpha\sin\beta} \qquad (A6.6)$$

Similar arguments may be used to deduce the equations for $\cos\alpha^*$ and $\cos\beta^*$, as shown in Eq. (3.16).

Appendix A7

Identities in Plane Trigonometry

The following formulae are often useful in manipulating a structure factor equation so as to obtain the geometrical structure factor equation and in other similar circumstances.

$$\cos A + \cos B = 2\cos \frac{A+B}{2} \cos \frac{A-B}{2} \qquad (\text{A7.1})$$

$$\cos A - \cos B = -2\sin \frac{A+B}{2} \sin \frac{A-B}{2} \qquad (\text{A7.2})$$

$$\sin A + \sin B = 2\sin \frac{A+B}{2} \cos \frac{A-B}{2} \qquad (\text{A7.3})$$

$$\sin A - \sin B = 2\cos \frac{A+B}{2} \sin \frac{A-B}{2} \qquad (\text{A7.4})$$

$$\cos(A+B) = \cos A \cos B - \sin A \sin B \qquad (\text{A7.5})$$

$$\cos(A-B) = \cos A \cos B + \sin A \sin B \qquad (\text{A7.6})$$

$$\sin(A+B) = \sin A \cos B + \cos A \sin B \qquad (\text{A7.7})$$

$$\sin(A-B) = \sin A \cos B - \cos A \sin B \qquad (\text{A7.8})$$

$$\sin 2A = 2\sin A \cos A \qquad (\text{A7.9})$$

$$\cos 2A = \cos^2 A - \sin^2 A = 2\cos^2 A - 1$$
$$= 1 - 2\sin^2 A \qquad (\text{A7.10})$$

$$\sin^2(2\pi n/4) = 0 \quad \text{for } n \text{ even or 1 for } n \text{ odd} \qquad (A7.11)$$

$$\cos^2(2\pi n/4) = 1 \quad \text{for } n \text{ even or 0 for } n \text{ odd} \qquad (A7.12)$$

$$\cos 2\pi(\theta + n/2) \text{ is crystallographically} \qquad (A7.13)$$

equivalent to $\cos 2\pi(\theta - n/2)$ where n

is an integer.[a]

---

[a] Any coordinate x can always be changed by $\pm 1/2$ to a crystallographically equivalent value x.

Appendix A8

Intensity Statistics

A8.1.   Introduction

The distributions of X-ray diffraction intensity data are useful often in determining the presence of a centre of symmetry in a crystal where this datum is not revealed by the limiting conditions (systematic absences) for the diffraction record. The distributions of both $|\mathbf{F}|$ and $|\mathbf{E}|$ have characteristic values, some of which will be described herein.

The mean value $\langle x \rangle$ for any distribution $f(x)$ is given by

$$\langle x \rangle = \frac{\int x \, f(x) \mathrm{d}x}{\int f(x) \mathrm{d}x} \tag{A8.1}$$

If the distribution is normalized, then the denominator of Eq. (A8.1) is unity.

A8.2.   Distributions of $|\mathbf{F}|$

In the *acentric* distribution, typified by space group $P1$, the A' and B' components are considered separately. Writing A' for a reflection $\mathbf{h}$, ignoring for convenience the θ subscript to g_j:

$$A' = \sum_{j=1}^{N} g_j \cos 2\pi(\mathbf{h} \cdot \mathbf{r}) \tag{A8.2}$$

Its average value $\langle A' \rangle$ over j atoms is zero, as discussed in Section 5.6.4. Following Eq. (5.70),

$$\langle A'^2 \rangle = \sum_{j=1}^{N} g_j^2 \langle \cos^2 2\pi(\mathbf{h} \cdot \mathbf{r}) \rangle = 1/2 \sum \qquad (A8.3)$$

In a similar manner,

$$\langle B'_\mathbf{h} \rangle = 0 \quad \text{and} \quad \langle B'^2_\mathbf{h} \rangle = 1/2 \sum \qquad (A8.4)$$

The probability that a given value of A' lies between A' and $A' +$ dA', assuming a normal probability distribution of the type $P(x) = (2\pi\sigma^2)^{-1/2} \exp[-(x - \langle x \rangle)/2\sigma^2]$, is

$$P_\mathrm{a}(A')\mathrm{d}A' = 1/\left(\pi \sum\right)^{1/2} \exp(-A'^2)/\sum \mathrm{d}A' \qquad (A8.5)$$

Similarly, for the B' component,

$$P_\mathrm{a}(B')\mathrm{d}B' = 1/\left(\pi \sum\right)^{1/2} \exp(-B'^2)/\sum \mathrm{d}B' \qquad (A8.6)$$

An area defined by dA'dB' is an infinitesimal portion of the annular ring, as shown in Fig. A8.1, distant $|\mathbf{F}|$ from the origin. Since A' and B' are not correlated, the joint probability that $|\mathbf{F}|$ lies between $|\mathbf{F}|$ and $|\mathbf{F}| + \mathrm{d}|\mathbf{F}|$ is the product

$$P_\mathrm{a}(|\mathbf{F}|)\mathrm{d}|\mathbf{F}| = P_\mathrm{a}(A')P_\mathrm{a}(B')\mathrm{d}A'\mathrm{d}B'$$

$$= \left[1/\left(\pi \sum\right)\right] \exp[(A'^2 + B'^2)]/\sum |\mathrm{d}\mathbf{S}| \quad (A8.7)$$

$$= \left[1/\left(\pi \sum\right)\right] \exp\left(-|\mathbf{F}|^2/\sum\right) |\mathrm{d}\mathbf{S}|$$

where d$|\mathbf{S}|$ is the area dA' dB' on the Argand diagram, and is equal to $2\pi|\mathbf{F_h}|\mathrm{d}|\mathbf{F_h}|$. Then, the joint probability is the area of the annular ring of radii $|\mathbf{F}|$ and $|\mathbf{F}| + \mathrm{d}|\mathbf{F}|$, namely,

$$P_a|\mathbf{F}|\mathrm{d}|\mathbf{F}| = \left(2|\mathbf{F}|/\sum\right)\exp\left(-|\mathbf{F}|^2/\sum\right)\mathrm{d}|\mathbf{F}| \qquad (A8.8)$$

or as

$$P_a|\mathbf{F}| = \left(2|\mathbf{F}|/\sum\right)\exp\left(-|\mathbf{F}|^2/\sum\right) \qquad (A8.9)$$

A similar equation can be derived for the *centric* distribution. In this case, with the centre of symmetry at the origin, the B' component is zero and

$$|\mathbf{F}| = A' = 2\sum_{j=1}^{N/2} g_j \cos 2\pi(\mathbf{h}\cdot\mathbf{r}) \qquad (A8.10)$$

Again, $\langle|\mathbf{F}|\rangle = 0$ and $\langle A'^2\rangle$ is the sum of $N/2$ terms of the form $4g_j^2\langle\cos^2 2\pi(\mathbf{h}\cdot\mathbf{r})\rangle$, which evaluates to the distribution parameter $\sum$. Hence, the centric probability would be

$$P_c|\mathbf{F}|\mathrm{d}|\mathbf{F}| = 1/\left(2\pi\sum\right)^{1/2}\exp\left[-|\mathbf{F}|^2/\left(2\sum\right)\right]\mathrm{d}|\mathbf{F}| \qquad (A8.11)$$

but since *amplitudes* are being considered, this equation must be multiplied by 2, as an amplitude $|\mathbf{F}|$ derives from both $+|\mathbf{F}|$ and $-|\mathbf{F}|$, thus transforming Eq. (8.11) to

$$P_c|\mathbf{F}|\mathrm{d}|\mathbf{F}| = 2/\left(2\pi\sum\right)^{1/2}\exp\left[-|\mathbf{F}|^2/\left(2\sum\right)\right]\mathrm{d}|\mathbf{F}| \qquad (A8.12)$$

or as

$$P_c|\mathbf{F}| = \left[2/\left(\pi\sum\right)\right]^{1/2}\exp\left[-|\mathbf{F}|^2/\left(2\sum\right)\right] \qquad (A8.13)$$

(see Fig. A8.1).

A8.3. Distributions of $|\mathbf{E}|$

The normalized structure factor $|\mathbf{E}|$ is defined by Eq. (5.74). Then, from Eq. (A8.8) and with $\varepsilon = 1$ for the typical space group $P1$, the

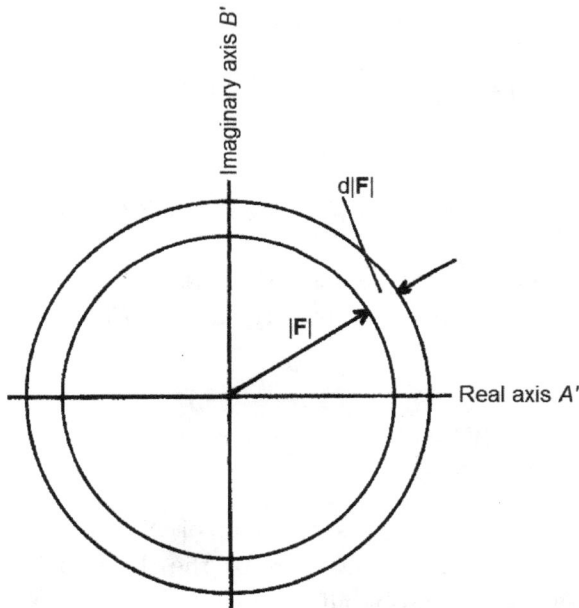

Fig. A8.1. Argand diagram showing the annular region for a structure amplitude $|\mathbf{F}|$ lying between the values $|\mathbf{F}|$ and $|\mathbf{F}| + \mathrm{d}|\mathbf{F}|$.

acentric distribution in terms of $|\mathbf{E}|$ follows as

$$P_\mathrm{a}|\mathbf{E_h}|\mathrm{d}|\mathbf{E_h}| = 2|\mathbf{E_h}|\exp(-|\mathbf{E_h}|^2)\mathrm{d}|\mathbf{E_h}| \qquad (A8.14)$$

or

$$P_\mathrm{a}|\mathbf{E_h}| = 2|\mathbf{E_h}|\exp(-|\mathbf{E_h}|^2) \qquad (A8.15)$$

A similar equation can be developed for the centric distribution of $|\mathbf{E}|$ values (see Problem 5.10).

Appendix A9

Gamma Function $\Gamma(n)$

The gamma function is useful in handling integrals of the type

$$\int_0^\infty x^n \exp(-ax^2)\,dx \qquad (A9.1)$$

where a is a constant; they occur in studying *inter alia* atomic scattering factors and intensity statistics. The gamma function $\Gamma(n)$ may be represented by the equation

$$\Gamma(n) = \int_0^\infty t^{n-1} \exp(-t)\,dt \qquad (A9.2)$$

The following results are useful:

for $n > 0$ and integral

$$\Gamma(n) = (n-1)! \qquad (A9.3)$$

for $n > 0$

$$\Gamma(n+1) = n\Gamma(n) \qquad (A9.4)$$

and if n is also integral

$$\Gamma(n+1) = n! \qquad (A9.5)$$

Also,

$$\Gamma\left(\frac{1}{2}\right) = \sqrt{\pi} \tag{A9.6}$$

$$\Gamma(1) = 0! = 1 \tag{A9.7}$$

Example A9.1. Consider the solution of the integral

$$\text{Int} = \int_0^\infty x^4 \exp(-x^2/2)\,dx$$

Let $x^2/2 = t$, so that $x = (2t)^{1/2}$ and $dx = (2t)^{-1/2}dt$. Then

$$\text{Int} = 2\sqrt{2}\int_0^\infty t^{3/2}\exp(-t)dt = 2\sqrt{2}\int_0^\infty t^{3/2}\exp(-t)dt$$

$$= 2\sqrt{2}\Gamma(5/2) = 2\sqrt{2}\Gamma(3/2+1) = 2\sqrt{2}(3/2)\Gamma(3/2)$$

$$= 3\sqrt{2}(1/2)\Gamma(1/2) = (3/\sqrt{2})\sqrt{\pi}$$

$$= 3\sqrt{\pi/2}$$

A reduction formula can be used also when working with these integrals:

$$\int x^m \exp(ax)dx = \frac{x^m \exp(ax)}{a} - \frac{m}{a}\int x^{m-1}\exp(ax)\,dx \tag{A9.8}$$

Appendix A10

Fourier Transforms in Crystal Structure Analysis

A10.1. Introduction

Referring to the discussion of Section 6.4, the structure factor equation

$$\mathbf{F}(h) = \frac{1}{a} \int_0^a \rho(X) \exp(2\pi h X/a) \mathrm{d}X \qquad (\text{A}10.1)$$

and the electron density equation

$$\rho(X) = \sum_{h=-\infty}^{\infty} |\mathbf{F}_h| \exp(-2\pi h X/a) \qquad (\text{A}10.2)$$

are known as Fourier transforms of each other. This topic is examined here insofar as it relates to the diffraction of X-rays from a crystal and the recombination of the scattered radiation to form an image of the electron density of the crystal.

A10.2. Generalized Fourier Transform

Consider the scattering diagram in Fig. 5.6. Let the three-dimensional electron density for the body at the point A distant $|\mathbf{r}|$ from the origin O be $\rho(\mathbf{r})$. The electron content of a small volume δV around the point A is then $\rho(\mathbf{r})\delta V$, and its phase with respect to the origin is $2\pi \mathbf{r} \cdot \mathbf{S}$. The contribution of the quantity $\rho(\mathbf{r})\delta V$ to scattering in the direction θ is $\rho(\mathbf{r}) \exp[\mathrm{i}2\pi(\mathbf{r} \cdot \mathbf{S})\delta V]$. Hence, the total scattering for the body of volume V is given by

$$\mathbf{G}(\mathbf{S}) = \int_V \rho(\mathbf{r}) \exp[\mathrm{i}2\pi(\mathbf{r} \cdot \mathbf{S})\mathrm{d}V \qquad (A10.3)$$

where $\mathbf{G}(\mathbf{S})$ is the *Fourier transform* of the electron density of the body, and $\mathbf{S}$ is defined in Section 5.4.1.

A10.3. Fourier Transform of a Number of Atoms

Consider a number n of atoms with coordinates $x_j, y_j, z_j (j = 1 - n)$ and scattering factors f_j. The jth atom is distant $\mathbf{r}_j$ from the origin, and from Eq. (A10.3) the wave scattered by the jth atom at a distance $\mathbf{r}$ from it is $\int_V \rho(\mathbf{r}) \exp[\mathrm{i}2\pi(\mathbf{r} + \mathbf{r}_j) \cdot \mathbf{S}]\mathrm{d}V$ or $\int_V \rho(\mathbf{r}) \exp[\mathrm{i}2\pi(\mathbf{r} \cdot \mathbf{S})]\mathrm{d}V \exp[\mathrm{i}2\pi(\mathbf{r}_j \cdot \mathbf{S})]$ which may be written as $f_j \exp[\mathrm{i}2\pi(\mathbf{r}_j \cdot \mathbf{S})]$. Then, the total wave from all n individual atoms is its Fourier transform, namely,

$$\mathbf{G}(\mathbf{S}) = \sum_{j=1}^{n} f_j \exp[\mathrm{i}2\pi(\mathbf{r}_j \cdot \mathbf{S})] \qquad (A10.4)$$

A10.4. Fourier Transforms and X-ray Reflections

Let the n atoms be now the contents of a unit cell of translation vectors $\mathbf{a}$, $\mathbf{b}$ and $\mathbf{c}$. Then, from Eq. (5.26), $\mathbf{r}_j \cdot \mathbf{S} = hx_j + ky_j + lz_j$ so that from Eq. (A10.4) the Fourier transform of the unit cell is

$$\mathbf{G}(\mathbf{S}) = \sum_{j=1}^{n} f_j \exp[\mathrm{i}2\pi(hx_j + ky_j + lz_j)] \qquad (A10.5)$$

Thus, it is established that the Fourier transform of the unit cell is equivalent to the structure factor equation (5.29). The essential difference between Eqs. (5.29) and (A10.5) is that the Fourier transform of a unit cell can be observed only at the (integral) reciprocal lattice points, whereas $\mathbf{G}(\mathbf{S})$ can be calculated for all values of the exponential term. The process of X-ray diffraction from a crystal may be considered pictorially as its reciprocal lattice mapped on to the

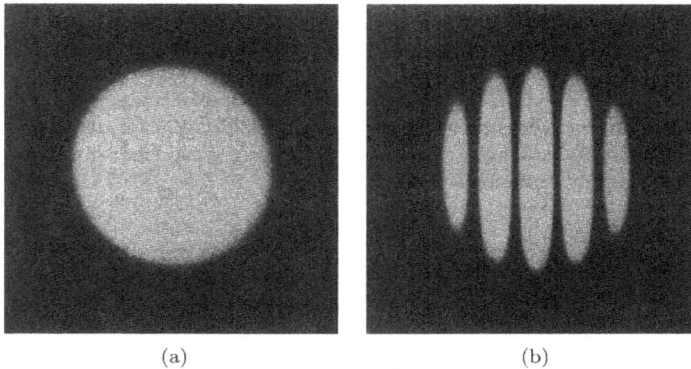

(a)   (b)

Fig. A10.1.   Diffraction patterns from holes. (a) Single hole of diameter 1 mm. (b) Two holes, each of diameter 1 mm, separated horizontally by a distance of *ca.* 3 mm. [Reproduced with permission from Taylor CA and Lipson H, *Optical Transforms*, Bell (1964).]

Fourier transform of the n atoms in the correct orientation. Then, the reciprocal lattice points that lie within the Ewald sphere and obey the limiting conditions of the space group give rise to the observed reflections.

A10.5.   Fourier Transforms and Diffraction Masks

Figure A10.1 illustrates the diffraction patterns of masks that are opaque screens with either a single hole as in (a), or two identical holes as in (b). The patterns are the Fourier transforms of the objects.

It is reasonable to suppose that if a mask is made of the diffraction pattern of a crystal, and used as an object, then its Fourier transform will be a reconstruction of the object. This principle is just a reiteration of the phase problem, which is still present with optical diffractometer, as shown in Fig. A10.2.

However, in the well-known example of the platinum derivative of phthalocyanine, the platinum atom is sufficiently heavy such that it controls the phases of the reflections. Since, the platinum atom lies at the origin in space group $P\frac{2_1}{c}$, the phases are zero. On account of a short b unit-cell dimension of 3.97 Å (the van der Waals radius for carbon is *ca.* 3.7 Å), the structure is well resolved in the $h0l$

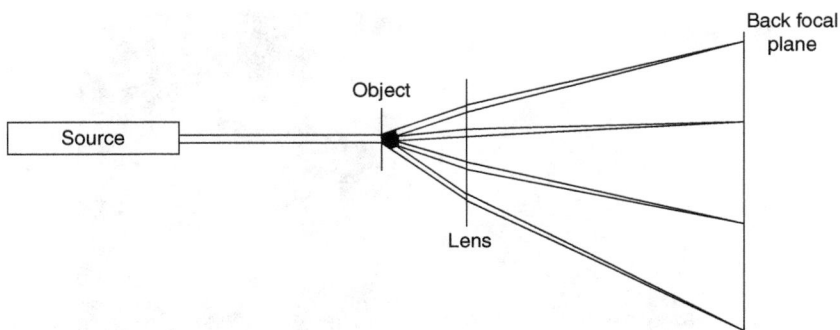

Fig. A10.2. Schematic arrangement of an optical diffractometer. Radiation from a He–Ne source impinges on the mask-object, and the Fraunhofer diffraction pattern is formed in the back focal plane of the lens. [Reproduced with permission from Taylor CA and Lipson H, *Optical Transforms*, Bell and Sons (1964).]

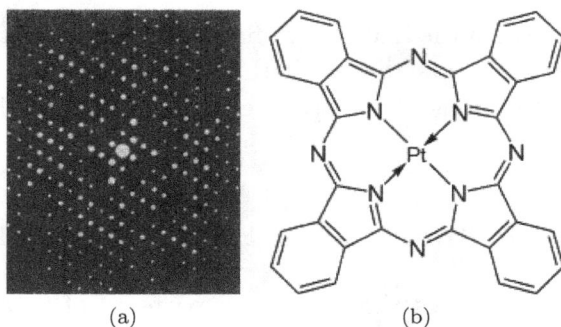

(a)   (b)

Fig. A10.3. Platinum phthalocyanine: (a) The $h0l$ weighted reciprocal lattice (diffraction pattern). (b) The structure of the molecule as determined by X-ray structure analysis.

projection; the diffraction pattern and the molecular formula are shown in Figs. A10.3 and A10.4.

Five masks showing increasing portions of the diffraction pattern of Fig. A10.3(a) are illustrated in Fig. A10.4 together with the Fourier transforms that were produced with the optical diffractometer of Fig. A10.2. Referring back to Fig. A10.1(b), it can be seen that introducing the second pinhole leads to a fringe system in the diffraction pattern and the fringes run in a direction that is normal to an imaginary line joining the centres of the two holes. As more holes are added in Fig. A10.4, representing more spots from the diffraction pattern of Fig. A10.3(a), the number of fringe systems increases. In the final transform pictured here, corresponding to mask (e), sufficient pattern has been taken for the transform to be recognizable as the platinum phthalocyanine molecule. The Fourier transform of the structure is thus a superposition of fringe systems, just like those of Fig. A10.3(b). The solution of a crystal structure from its diffraction pattern or transform is then the diffraction pattern of the diffraction pattern or the transform of the transform.

A10.6.   Change of Unit-Cell Origin

If the unit-cell origin is changed by an amount $\mathbf{u}$, then Eq. (A10.4) becomes

$$\mathbf{G}'(\mathbf{S}) = \sum_{j=1}^{n} f_j \exp[\mathrm{i}2\pi(\mathbf{r}_j + \mathbf{u}) \cdot \mathbf{S})] = \mathbf{G}(\mathbf{S}) \exp[\mathrm{i}2\pi(\mathbf{u} \cdot \mathbf{S})]$$

$$(A10.6)$$

which shows that $\mathbf{G}(\mathbf{S})$ is modified by the fringe function $\exp[\mathrm{i}2\pi(\mathbf{u} \cdot \mathbf{S})]$. From a practical point of view, $(\mathbf{u} \cdot \mathbf{S})$ is integral (n) at points where the transform can be observed, so that $\exp(\mathrm{i}2\pi n) = 1$. Hence, the amplitude and intensity are invariant under translation, as already noted, whereas phases are dependent upon position.

A10.7.   Limiting Conditions (Systematic Absences)

Limiting conditions and systematic absences have been discussed in Section 5.4.4, and may be explained here through Fourier transforms. The C-centred unit cell discussed in Section 5.4.4.2 involves the translational component $\frac{a}{2} + \frac{b}{2}$. If the vector $\mathbf{u}$ in Eq. (A10.6) is replaced by $\frac{\mathbf{a}+\mathbf{b}}{2}$, the equation becomes

$$\mathbf{G}'(\mathbf{S}) = \sum_{j=1}^{n} f_j \exp[\mathrm{i}\pi(\mathbf{a} + \mathbf{b}) \cdot \mathbf{S})] \qquad (A10.7)$$

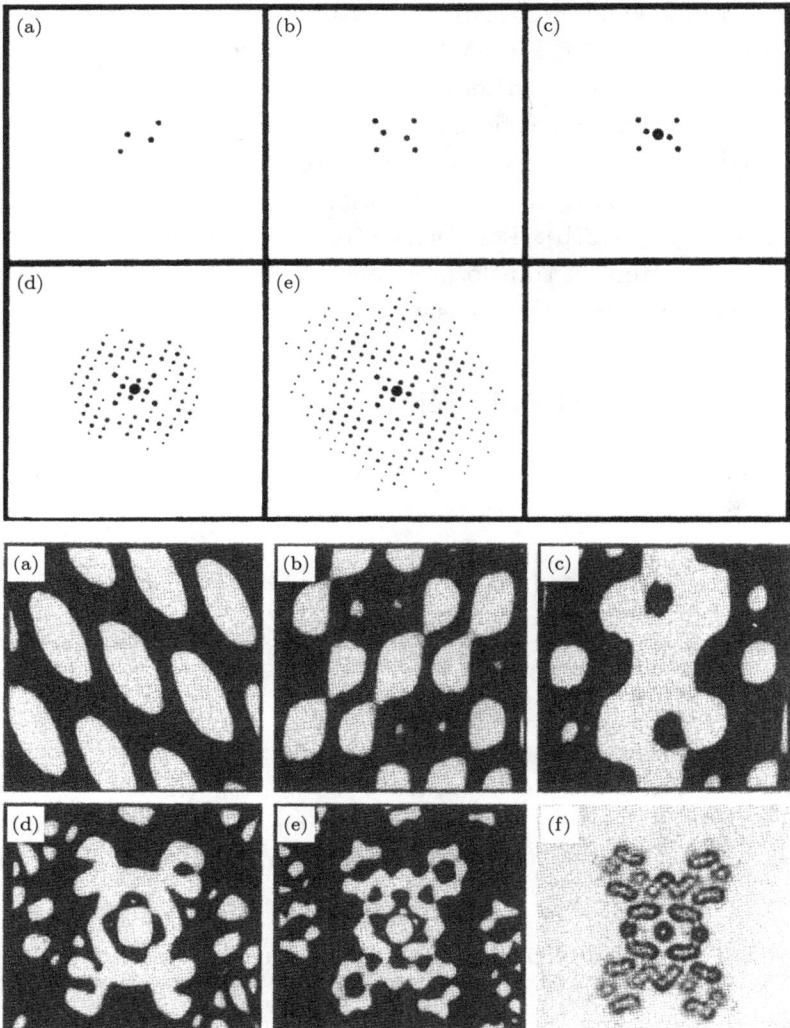

Fig. A10.4.   Masks and corresponding diffraction patterns for varying portions of the $h0l$ weighted reciprocal lattice of platinum phthalocyanine. The masks (a)–(e) show increasing portions of the diffraction pattern of Fig. A10.3(a) moving outwards from the centre. The transforms follow in sequence below. The sixth picture in this series is a two-dimensional electron density contour map $\rho(xz)$ of the phthalocyanine molecule. [Reproduced with permission from Taylor CA and Lipson H, *Optical Transforms*, Bell and Sons (1964).]

In Eq. (5.26), $\mathbf{S} = \mathbf{d}^* = (h\mathbf{a}^* + k\mathbf{b}^* + l\mathbf{c}^*)$, so that $(\mathbf{a} + \mathbf{b}) \cdot \mathbf{S}) = (\mathbf{a} + \mathbf{b}) \cdot (h\mathbf{a}^* + k\mathbf{b}^* + l\mathbf{c}^*) = h + k$, and Eq. (A10.7) becomes

$$\mathbf{G}'(\mathbf{S}) = \mathbf{G}(\mathbf{S})\{1 + \exp[i2\pi(h + k)]\} \qquad (A10.8)$$

which leads immediately to the limiting conditions hkl: $h + k = 2n$, for which the result is $2\mathbf{G}(\mathbf{S})$. In terms of systematic absences, the corresponding condition is hkl: $h + k = 2n + 1$.

A10.8. Diffraction and the Delta Function

A normal or Gaussian distribution $f(x)$ can be represented by the equation

$$f(x) = \frac{1}{k} \exp\left(-\frac{\pi x^2}{k^2}\right) \qquad (A10.9)$$

where k is termed the width of the function. Consider next a limiting case of Eq. (A10.9):

$$\delta(x) = \lim_{k \to 0} \left[\frac{1}{k} \exp\left(-\frac{\pi x^2}{k^2}\right)\right] \qquad (A10.10)$$

It has the properties

$$\delta(x) = 0 \text{ for } x \neq 0 \text{ and } \delta(x) = \infty \text{ for } x = 0$$

and

$$\int_{-\infty}^{\infty} \delta(x)\mathrm{d}x = 1 \qquad (A10.11)$$

and is known as the Dirac δ-function. It represents an infinitely sharp line of unit weight at the origin, similar to Fig. 6.3, but with a height of 1.0 instead of π. As $\pi \to 0$ so $f(x) \to \delta(x)$; at $\pi = 0 f(x) = \delta(x) = 1$. In other words, the Fourier transform of a δ-function is unity, as can be seen from the following general integral, given for the x dimension:

$$\int_{-\infty}^{\infty} f(x)\delta(x)\mathrm{d}x = f(0) \tag{A10.12}$$

Let $f(x) = \exp(i2\pi Sx)$, then

$$\int_{-\infty}^{\infty} \exp(i2\pi Sx)\delta(x)\mathrm{d}x = \exp(0) = 1 \tag{A10.13}$$

If the δ-function is located at $x = x_0$, then $f(x) = \delta(x - x_0)$. The Fourier transform of $f(x)$ is now

$$\mathbf{G}(\mathbf{S}) = \int_{-\infty}^{\infty} \delta(x - x_0)\exp(i2\pi Sx)\mathrm{d}x$$

$$= \int_{-\infty}^{\infty} \delta(x)\exp[i2\pi S(x + x_0)]\mathrm{d}x = \exp(i2\pi Sx_0)$$

$$\tag{A10.14}$$

Thus, the Fourier transform of a δ-function at a point x_0 is

$$\mathbf{G}(\mathbf{S}) = \exp(i2\pi S\,x_0) \tag{A10.15}$$

If the δ-function is at the origin, that is, $x_0 = 0$, then its Fourier transform is unity for all S. This one-dimensional argument can be extended to higher dimensions. Of particular note is the set of δ-functions that define a lattice. If the spacing of this lattice is $|\mathbf{a}|$, its Fourier transform is another set of δ-functions with the spacing $1/|\mathbf{a}|$, which is a one-dimensional reciprocal lattice (see Fig. A10.5).

A10.9. Convolution

A diffraction grating in the form of slits of a given width has a Fourier transform that can be made up of two parts, namely, the transform of a single slit together with that of an ideal grating; these two functions are shown in Fig. A10.5.

Consider next two functions $f(\mathbf{r})$ and $g(\mathbf{r})$ both spanning the real (Bravais) space of $\mathbf{r}$. Then the convolution of these functions

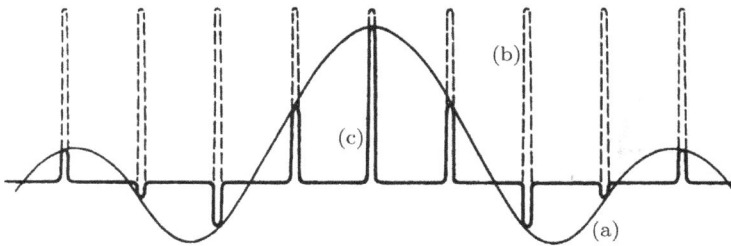

Fig. A10.5.   Product of transforms: (a) Transform of a single slit of given finite width (thin line). (b) Ideal slit grating (dashed line). (c) Grating of finite-width slits (bold line). The transform of the grating (c) is the product or *convolution* of the transforms (a) and (b). [Reproduced with permission from Taylor CA and Lipson H, *Optical Transforms*, Bell and Sons (1964).]

is defined as $c(\mathbf{r})$, given by

$$c(\mathbf{r}) = f(\mathbf{r}) \, ^* \, g(\mathbf{r}) = \int_{-\infty}^{\infty} f(\mathbf{r}')g(\mathbf{r} - \mathbf{r}')\mathrm{d}r' \qquad (A10.16)$$

The convolution theorem can be stated in two ways: either the Fourier transform of a product is the convolution of the Fourier transforms or the Fourier transform of a convolution is the product of the Fourier transforms.

A physical interpretation of convolution may be given by considering two functions, $f(\mathbf{r})$ and $g(\mathbf{r})$, separated in real space by the vector $\mathbf{r}$. The convolution $c(\mathbf{r})$ of these two functions is given by Eq. (A10.16). Let $f(\mathbf{r})$ represent a collection of identical molecules, while $g(\mathbf{r})$ represents a lattice that may be defined as

$$g(\mathbf{r}) = \sum_{\text{lattice}} \delta(\mathbf{r} - \mathbf{r}_{\text{lattice}}) \qquad (A10.17)$$

where δ is the delta function, normalized to unity and of infinitesimal width, $\mathbf{r}_{\text{lattice}}$ implies all points of the given lattice. In order to build a crystal structure, Eq. (A10.16) shows that the function $f(\mathbf{r})$ is multiplied by $g(\mathbf{r}_{\text{lattice}})$, that is, $f(\mathbf{r}') \, ^* \, g(\mathbf{r} - \mathbf{r}')$ at each lattice point before integration, and $\mathbf{r}'$ is a value in $\mathbf{r}$-space but different from $\mathbf{r}$ itself in general. This convolution process repeated at each lattice point leads

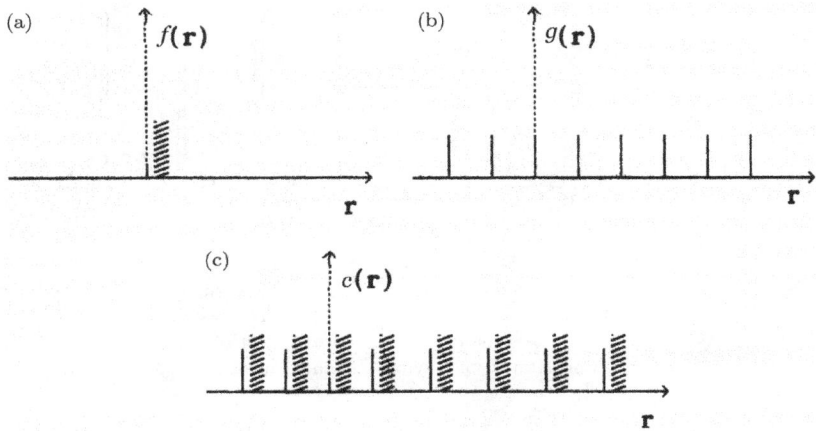

Fig. A10.6. The process of convolution: (a) Structural entity $f(\mathbf{r})$. (b) Infinite set of δ-functions $g(\mathbf{r})$. (c) The convolution $c(\mathbf{r}) = f(\mathbf{r})^*g(\mathbf{r})$. [*Structure Determination by X-ray Crystallography*, Mark Ladd and Rex Palmer, 5th edn. (2013). Reproduced by permission of Springer Science+Business Media, NY.]

to the crystal structure, as illustrated for a one-dimensional example in Fig. A10.6. The motif in illustration (a) represents a given molecule and the vertical lines in (b) represent the lattice. Then, picture (c) shows the convolution $c(\mathbf{r}) = f(\mathbf{r})^*g(\mathbf{r})$ or, in other words, the crystal structure. Thus, a crystal structure may be seen as a convolution of a unit cell including its contents, delineated by the translation vectors $\mathbf{a}$, $\mathbf{b}$ and $\mathbf{c}$, with a three-dimensional set of δ-functions of value unity at each Bravais lattice point and zero elsewhere.

The transform of the unit cell will involve generally more than one molecule, and is a continuous function in reciprocal space. However, in any diffraction experiment, the transform may be sampled only at the reciprocal lattice points because of the limitations imposed by the Bragg equation. The convolution of the unit cell and the δ-functions is the product of the transforms of the unit cell and the reciprocal lattice, and is revealed as the diffraction pattern. The diffraction pattern can be imagined as a map of the continuous transform of the unit-cell contents overlaid by the discrete reciprocal lattice, in the correct orientation. The weights of the transform at the reciprocal lattice points are proportional to the corresponding values of $|\mathbf{F}_{obs}|$, or $|\mathbf{F}_{obs}|^2$. The correct orientation is unknown initially, which is another way of stating the phase problem.

A10.10.   Fourier Transforms and Structure Solution

A Fourier transform of X-ray diffraction data from a crystal in terms of amplitude and phase converts the experimental data to the crystal structure. Several aspects of this procedure can be illuminated by transform theory.

A10.10.1.   Patterson function

The Patterson function is a phase-free transform of the diffraction intensities because it uses $|\mathbf{F}|^2$ as Fourier coefficients. A diffraction pattern is invariant under translation, but translation is coded into it since the translational property is revealed when the diffraction pattern is finally transformed; phase changes are not revealed by transformation because the Fourier coefficients here are intensities.

The transform of the intensities reproduces all possible pairs of scattering entities disposed symmetrically about the origin, and may be thought of as a superposition of many fringe systems all of which are positive at the origin of the unit cell. The transform of intensity may be represented by $\mathbf{G(S)G^*(S)}$, which is the transform of the electron density convoluted with the transform of the electron density inverted in the origin.

A four-atom array and its self-convolution are shown in Figs. A10.7(a) and A10.7(b). Four atoms give rise to 16 vectors, four of which, being of zero length, coincide at the origin; four vectors are single weight formed by symmetry 2 in this example, and eight vectors formed by symmetry m are of double weight. Figure A10.7(b) is a map of the vectors shown in Fig. A10.7(a), all taken to a common origin. In other words, Fig. A10.7(b) is Fig. A10.7(a) convoluted with itself inverted in the origin, the inversion being the same as the structure in a centrosymmetric crystal.

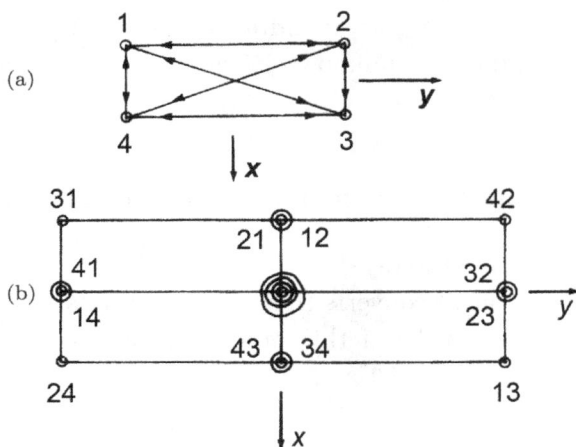

Fig. A10.7.   Convolution with four atoms: (a) A rectangular array of four atoms: 1, 2, 3, 4 and the interatomic vectors formed by them; '31' means atoms 3 and 1, and so on. (b) The convolution of the four-atom structure with itself inverted in the origin. Note that if a structure is centrosymmetric, the structure and its inversion are identical.

A10.10.2.   Heavy-atom technique

In the heavy-atom technique where the coordinates of the heavy atom H are known, the structure factors for the heavy atom contribution can be calculated in both amplitude and phase. Then, the Fourier transform in Eq. (A10.2) put in general terms becomes a series with coefficients of the form $|\mathbf{F}_{obs}| \exp(i\phi_{calc})$. Applying this result to an electron density calculation, the coefficients are of the form

$$|\mathbf{F}_{obs}| \exp(i\phi_{calc}) = \frac{|\mathbf{F}_{obs}|}{|\mathbf{F}_{calc}|}|\mathbf{F}_{calc}| \exp(i\phi_{calc}) \qquad (A10.18)$$

The right-hand side of this equation is the product of the transform of the heavy atom with that of the function $\frac{|\mathbf{F}_{obs}|}{|\mathbf{F}_{calc}|}$ which has a zero phase everywhere. The result is large peak at the origin, but on account of the inequality of $|\mathbf{F}_{obs}|$ and $|\mathbf{F}_{calc}|$, a bias towards the observed data is imposed on the heavy-atom transform. If the heavy-atom position is correct, or nearly so, the modified transform, or electron density map, will have peaks recognizable as probable atom positions. They can then be added to the heavy-atom contribution

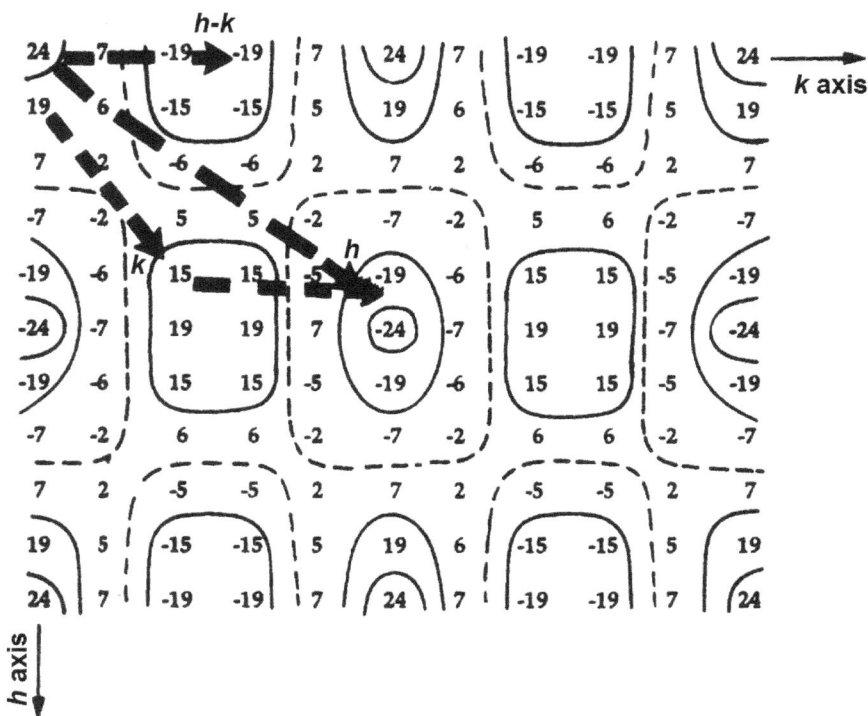

Fig. A10.8. Calculated transform of the four-atom group in Fig. A10.7(a), treating each atom as carbon (atomic number 6); contours are drawn at -10, 10 and 20. A particular TPR ($\mathbf{h}$, $\mathbf{k}$, $\mathbf{h-k}$) is indicated by arrows.

and the process repeated. This procedure is a structure solution by the Patterson/Fourier method discussed in Section 6.7.

A10.10.3. Sign relationships

Sign relationships were discussed in Section 6.10, and they form the basis of the direct methods of phase determination. It can be shown how the triple product sign relationship (TPR) of Eq. (6.34) may be deduced from a transform of a structure. Figure A10.8 is the calculated transform of the four-atom centrosymmetric structure of Fig. A10.7(a) for the ranges $h = 1$–10 and $k = 1$–10. Each atom is assumed to be carbon so that the maximum at the origin is 24 and the variation of f_C with θ has been neglected; the plot is contoured at the levels -10, 10 and 20.

Three vectors $\mathbf{h}$, $\mathbf{k}$ and $\mathbf{h}$–$\mathbf{k}$, where $\mathbf{h}$ is hkl and $\mathbf{k}$ is $h'k'l'$, form a triangle in reciprocal space with one vertex at the origin; these vectors correspond to three X-ray reflections. In the figure, the vectors terminate in regions of *large amplitude*, and all reflections are positive phase at the origin. By counting the number of times that a vector crosses zero boundaries, the sign of the transform at the reciprocal lattice points can be determined for the three vectors in Fig. A10.8 as follows:

Vector	hkl	Boundary crossings	Sign
$\mathbf{h}$	450	3	$-$
$\mathbf{k}$	420	2	$+$
$\mathbf{h}$–$\mathbf{k}$	030	1	$-$

In general,

$$s(\mathbf{h})\, s(\mathbf{k})\, s(\mathbf{h} - \mathbf{k}) = +1 \qquad (A10.19)$$

which is Sayre's equation. Reflections of large amplitude are essential: in regions of small magnitude in the transform, such as those around reflection 840, the TPR result is uncertain, because there exists in practice a difficulty in determining the number of zero boundaries that are crossed by the path of such a vector.

Appendix A11

Linear Least Squares

For a given function $y = \phi(x_n)n = 1, 2, 3, \ldots$, the principle of least squares defines the best fit of the n points to the function y by the minimization of the expression $(\sum_n \phi(x_n) - y)^2$.

A11.1.   Least-Squares Line

Let a line be of the form

$$y = ax + b \qquad \text{(A11.1)}$$

where a and b are constants to be determined. For any observation i

$$ax_i + b - y_i = e_i \qquad \text{(A11.2)}$$

where e_i is an error that will be assumed to be both random and to reside in the value of the dependent variable y_i, while the error in the independent variable x_i is relatively negligible. Applying the least-squares principle to N such equations, the best-fit values of a and b are those that minimize the sum of the squares of the e_i values, that is,

$$\min \left(\sum_{i=1}^{N} e_i^2 \right) = \min \left(\sum_{i=1}^{N} (ax_i + b - y_i)^2 \right) \qquad \text{(A11.3)}$$

Then, the required minimum values may be obtained by differentiating partially the right-hand side of Eq. (A11.3) with respect to a and b and setting the derivatives to zero. Hence,

$$\frac{\partial \sum_i e_i^2}{\partial a} = 2\sum_i \left(ax_i^2 + bx_i - x_i y_i\right) = 0$$

$$\frac{\partial \sum_i e_i^2}{\partial b} = 2\sum_i \left(ax_i + b - y_i\right) = 0$$

(A11.4)

Using $[\dots]$ to represent $\sum_i$ and including a weight w_i for each observation, Eqs. (A11.4) lead to the *normal equations*:

$$a[wx^2] + b[wx] - [wxy] = 0$$

$$a[wx] + b[w] - [wy] = 0$$

(A11.5)

Here, $[w]$ is the sum of the weights; if all weights are set equal to unity, then $[w] = N$, the number of observations. The equations are solved readily to give

$$a = \frac{[w][wxy] - [wx][wy]}{\Delta}$$

(A11.6)

$$b = \frac{[wx^2][wy] - [wx][wxy]}{\Delta}$$

(A11.7)

where

$$\Delta = [w][wx^2] - [wx][wx]$$

(A11.8)

The estimated variances σ^2 in a and b are given by [1]

$$\sigma_a^2 = \frac{[e^2]}{(N-2)}\frac{[w]}{\Delta}$$

$$\sigma_b^2 = \frac{[e^2]}{(N-2)}\frac{[wx^2]}{\Delta}$$

(A11.9)

and the estimated standard deviation (esd) is $\sqrt{\sigma^2}$. It is good practice to plot the derived line. The least-squares procedure will always give the best fit to the observations, even the bad ones.

A11.2.   Correlation Coefficient, r

The Pearson correlation coefficient r is defined as [1]

$$r = \frac{N[wxy] - [wx][wy]}{\{N[wx^2] - [wx]^2\}^{1/2}\{N[wy^2] - [wy]^2\}^{1/2}} \tag{A11.10}$$

A perfect fit corresponds to $r = 1$, but $r > 0.95$ is regarded as very satisfactory.

A11.3.   Propagation of Errors

The number of significant figures in a result is not necessarily similar to the number of figures in the data. It will not be greater, but may be a lesser value than expected. Let $y = p^n$, where $p = 2.0 \pm 0.1$. For $n = 0.1$, y lies between 1.066 and 1.077, whereas for $n = 4$, y lies between 13.0 and 19.4.

Consider a function $y = f(p)$ as exemplified in Fig. A11.1. In the small interval δp, the change δy in y is given by

$$\delta y = \left(\frac{dy}{dp}\right)\delta p \tag{A11.11}$$

Consider next the function $y = f(p_1, p_2)$, where p_1 and p_2 are independent variables. For two small independent changes δp_1 and δp_2, the changes in y are given, by analogy with Eq. (A11.11), as

$$\delta y_{p_1} = \left(\frac{\partial y}{\partial p_1}\right)\delta p_1$$

$$\delta y_{p_2} = \left(\frac{\partial y}{\partial p_2}\right)\delta p_2 \tag{A11.12}$$

Since the two variations in y are uncorrelated, they can be represented along two rectangular axes, as in Fig. A11.2.

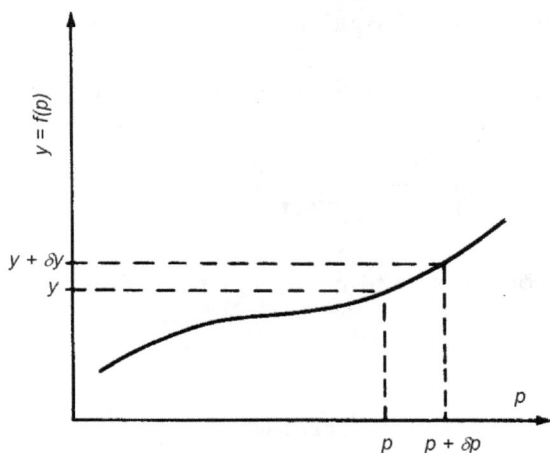

Fig. A11.1.   A function $y = f(p)$; δp and δy are small increments of the variables p and y, respectively.

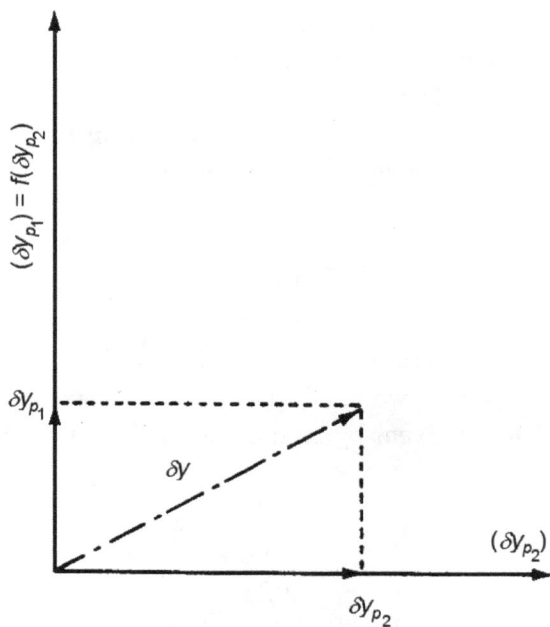

Fig. A11.2.   Representation of the uncorrelated errors δy_{p_1} and δy_{p_2} in p_1 and p_2, respectively; orthogonal components are uncorrelated.

Hence,

$$(\delta y)^2 = (\delta y_{p_1})^2 + (\delta y_{p_2})^2 = \left(\frac{\partial y}{\partial p_1}\right)^2 (\delta p_1)^2 + \left(\frac{\partial y}{\partial p_2}\right)^2 (\delta p_2)^2$$

$$(A11.13)$$

Then, for a generalized function $y = f(p_j)$, $j = 1, 2, 3, \ldots, N$ it follows now that

$$(\delta y)^2 = \sum_{j=1}^{N} \left(\frac{\partial y}{\partial p_j}\right)^2 (\delta p_j)^2 \qquad (A11.14)$$

where $(\delta y)^2$ is identified as the variance σ_y^2 of y so that the esd is σ_y.

Reference

[1] Whittaker ET and Robinson G, *The Calculus of Observations*, Blackie & Son Limited (1924).

Tutorial Solutions

Solutions 1

1.1. (a) $(6\bar{2}3)$; (b) $(\bar{1}03)$; (c) $(612,\bar{1})$.

1.2. (a) $[\bar{1}10]$; (b) $[031]$; (c) $[0\bar{4}8] \equiv [0\bar{1}2]$. *Note:* $[UVW] \equiv [\bar{U}\bar{V}\bar{W}]$ as a zone symbol but not as a direction in a lattice.

1.3. (a) $[\bar{1}\bar{1}2]$; (b) $[\bar{1}10]$, $[031]$; (c) $[0\bar{2}1]$.

1.4. (a) $(\bar{1}10)$; (b) $(2\bar{1}2)$; (c) $(\bar{7}10,\bar{1})$.

1.5. The two planes must satisfy the Weiss zone law, that is, $h - 2k + 3l = 0$. Hence, two possible planes are (210) and (121). Check by cross-multiplication.

1.6. By cross-multiplication, the zone symbol is $(k_1l_2 - k_2l_1)$, $(l_1h_2 - l_2h_1)$, $(h_1k_2 - h_2k_1)$. From the Weiss zone law, $(mh_1 + nh_2)(k_1l_2 - k_2l_1) + (mk_1 + nk_2)(l_1h_2 - l_2h_1) + (ml_1 + nl_2)(h_1k_2 - h_2k_1)$ should equal zero. Expansion of this equation shows that it is equal to zero, so that the two planes are cozonal.

1.7. Begin with pole c, at the intersection of zones $[(010)_-(111)]$ and $[(001)_-(100)]$, that is, $[10\bar{1}]$ and $[010]$ so that by cross-multiplication pole c is 101. Since pole a is given as 210, the symbol for zone a_d_g is $[(210)_-(101)] = [1\bar{2}\bar{1}]$. Pole b lies in zones $[(100)_-(111)]$ and $[1\bar{2}\bar{1}] = [0\bar{1}1]_-[1\bar{2}\bar{1}] = 311$. Pole d lies in zones $[1\bar{2}\bar{1}]$ and $[(001)_-(1\bar{1}0)] = [1\bar{2}\bar{1}]_-[1\,1\,0] = 1\,\bar{1}3$. For pole e, the zones are $[1\bar{2}\bar{1}]$ and $[(010)_-(001)] = 0\bar{1}2$. Pole f lies in zones $[1\bar{2}\bar{1}]$ and $[(111)_-(001)] = [1\bar{2}\bar{1}]_-[1\bar{1}0] = \bar{1}\bar{1}1$. Pole g lies at the intersection of $[1\bar{2}\bar{1}]$ and $[(1\,00)_-(01\,0)] = [1\bar{2}\bar{1}]_-[001] = \bar{2}\bar{1}0$ (evident also from its position on the stereogram).

Summary: a b c d e f g Zone axis

210   311   101 $1\bar{1}3$ $0\bar{1}2$ $\bar{1}\bar{1}1$ $\bar{2}\bar{1}0$ $[1\bar{2}\bar{1}]$

1.8. Equation: $0.13496X + 0.43493Y + 0.09348Z = 1$.

Distances of the points from the calculated plane

Atom	Distance
1	0.6136
2	0.4319
3	0.0330
4	−0.5231
5	−0.7468
6	−0.6595
Rms deviation = 0.5523	

1.9. (a) By cross-multiplication $[(121)\_(231)] = [\bar{1}\,1\,\bar{1}] \equiv [1\bar{1}1]$.

(b) For the (011) plane $(1 \times 0) + (-1 \times 1) + (1 \times 1) = 0$, so that (011) lies in the $[1\bar{1}1]$ zone.

(c) For (121) $p + 2q = -1$, and for (231) $2p + 3q = 1$, so that $q = -1$ and $p = 2$.

1.10.

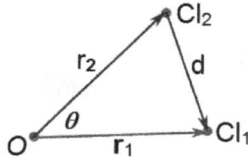

$\mathbf{d} = \mathbf{r}_1 - \mathbf{r}_2 \; \mathbf{d} = \mathbf{r}_1 - \mathbf{r}_2$, so that

$$d = \sqrt{\begin{aligned}(X_1 - X_2)^2 + (Y_1 - Y_2)^2 \\ +(Z_1 - Z_2)^2\end{aligned}}$$

$$= \sqrt{44}\,\text{nm} = 6.633\,\text{nm}$$

$$\cos\theta = \frac{\mathbf{r}_1 \cdot \mathbf{r}_2}{r_1 r_2} = \frac{(\mathbf{i}X_1 + \mathbf{j}Y_1 + \mathbf{k}Z_1) \cdot (\mathbf{i}X_2 + \mathbf{j}Y_2 + \mathbf{k}Z_2)}{\sqrt{(\mathbf{i}X_1 + \mathbf{j}Y_1 + \mathbf{k}Z_1)^2}\,\sqrt{(\mathbf{i}X_2 + \mathbf{j}Y_2 + \mathbf{k}Z_2)}}$$

$$= \frac{8 + 72 + 80}{\sqrt{4 + 36 + 64}\sqrt{16 + 144 + 100}}$$

$$= \frac{160}{10.1980 \times 16.1245} = 0.97301$$

so that $\theta = 13.34°$. [The reader may care to consider the value of d if $\beta(= \widehat{ZX}) = 120°$.]

Solutions 2

2.1. (a) Monoclinic system: m plane normal to y; no other symmetry. (b) Orthorhombic system: m plane normal to x; m plane normal to y; two-fold axis along z. (c) Tetragonal system: Bar-four axis along z; two-fold axes along x and y; m planes normal to $[1\,1\,0]$ and $[1\,\overline{1}\,0]$, i.e. 45° to x and y. (d) Trigonal system: three-fold axis along z; two-fold axes along x, y and u. (e) Hexagonal system: Six-fold axis along z, two-fold axes along x, y and u; two-fold axes normal to x y and u and lying in the x–y plane. (f) Cubic system: two-fold axes along x, y and z; three-fold axes along $[1\,1\,1]$, $[\overline{1}\,1\,1]$, $[1\,\overline{1}\,1]$ and $[\overline{1}\,\overline{1}\,1]$. [*Note*: Do not confuse point groups 23 and 32.]

2.2. 422, $42\overline{2}$, $4\overline{2}2$, $4\overline{2}\overline{2}$, $\overline{4}22$, $\overline{4}2\overline{2}$, $\overline{4}\overline{2}2$, $\overline{4}\overline{2}\overline{2}$ are the permutations. However, a proper rotation combined with an improper rotation leads to another improper rotation. Hence, symbols with one or three negative signs are inadmissible. There remain 422, $4\overline{2}\overline{2}$, $\overline{4}2\overline{2}$ and $\overline{4}\overline{2}2$. Symbols $\overline{4}2\overline{2}$ and $\overline{4}\overline{2}2$ are equivalent under rotation of the x-, y-axes by 45° and correspond to $\overline{4}2m$. Symbol $4\overline{2}\overline{2}$ is written usually as $4mm$. So, the unique groups are 422, $4mm$ and $\overline{4}2m$.

2.3. (a) $\overline{4}3m$; (b) $3m$; (c) E-form $\frac{2}{m}$, Z-form $mm2$; (d) $3m$; (e) $\overline{4}3m$. [A comment on the E/Z notation may be of interest here. The *cis/trans* notation is adequate for simple compounds such as 1,2-dichloroethene. However, with a compound such as bromochloroethene the *cis/trans* notation is not unambiguous, and the E/Z notation was devised. Consider the following structures as exemplars of the E/Z rule:

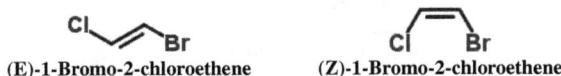

(E)-1-Bromo-2-chloroethene   (Z)-1-Bromo-2-chloroethene

The atom with the higher *priority* is located; in this example, it is bromine, with chlorine at the second level of priority. If the two groups of the higher priorities lie on opposite sides of the double bond then that isomer is the E-form (Ger. *entgegen* = opposite). If the two groups lie on the same side of the double bond, the isomer is the Z-form (Ger. *zusammen* = together). *Cis* and *trans* are not alternatives to E and Z; priorities follow the Cahn–Ingold–Prelog rules. Discussions of the E/Z notation may be found on several websites, for example, https://www.masterorganicchemistry.com/2016/11/03/alkene-nomenclature-cis-and-trans-and-e-and-z/.]

2.4. (a) $\bar{1}$; (b) $\bar{6}$; (c) $4mm$; (d) $\bar{3}m$; (e) mmm; (f) $\frac{2}{m}$.

2.5.

(a)
$$\begin{pmatrix} 1 & 0 & 0 \\ 0 & \bar{1} & 0 \\ 0 & 0 & 1 \end{pmatrix} \begin{pmatrix} 0 & 1 & 0 \\ \bar{1} & 0 & 0 \\ 0 & 0 & \bar{1} \end{pmatrix}$$

$\quad\quad\quad m \perp y \quad\quad \bar{4} \text{ along } z$

(b)
$$\begin{pmatrix} 1 & 0 & 0 \\ 0 & \bar{1} & 0 \\ 0 & 0 & 1 \end{pmatrix} \begin{pmatrix} 0 & 1 & 0 \\ \bar{1} & 0 & 0 \\ 0 & 0 & \bar{1} \end{pmatrix} = \begin{pmatrix} 0 & 1 & 0 \\ 1 & 0 & 0 \\ 0 & 0 & \bar{1} \end{pmatrix}$$

Thus, $m\bar{4} = 2$ along $[1\,1\,0]$.

(c) $\bar{4}m = 2$ along $[1\,\bar{1}\,0]$ which, together with $[1\,1\,0]$, lie in the form $\langle 1\,1\,0\rangle$ in this system.

2.6.

$$\begin{pmatrix} 0 & 0 & \bar{1} \\ 0 & 1 & 0 \\ 1 & 0 & 0 \end{pmatrix} \begin{pmatrix} 1 & 0 & 0 \\ 0 & 0 & \bar{1} \\ 0 & 1 & 0 \end{pmatrix} = \begin{pmatrix} 0 & \bar{1} & 0 \\ 0 & 0 & \bar{1} \\ 1 & 0 & 0 \end{pmatrix}$$

$\quad\quad\quad 4_{[0\,1\,0]} \quad\quad\quad 4_{[1\,0\,0]} \quad\quad\quad 3_{[1\,\bar{1}1]}$

2.7.

(a) $\dfrac{6}{m}mm$ (b) $mm2$ (c) $mm2$ (d) $mm2$ (e) mmm

(f) $mm2$ (g) m (h) $\bar{6}m2$ (i) $mm2$ (j) $mm2$

(k) mmm (l) $mm2$ (m) $\dfrac{6}{m}mm$

2.8.

	F	V	E
(a) Tetrahedron	4	4	6
(b) Cube	6	8	12
(c) Octahedron	8	6	12
(d) Pentagonal dodecahedron	12	20	30
(e) Icosahedron	20	12	30

The relationship is Euler's formula: $V + F = E + 2$.

2.9.

$$32: \quad 2, 1$$
$$\bar{4}2m: \quad \bar{4},\ mm2,\ 222,\ 2,\ m, 1$$
$$m\bar{3}: \quad 23, \bar{3}, 3, mmm, mm2, 222, 2/m, 2, m, \bar{1}, 1$$

2.10. By constructing the stereogram for $6m$, it is clear that a second form of m planes is generated; $6mm$ is the full symbol. In the case of $3m$, no additional form of m planes is generated. The symbol could be written as $3m1$ (or $31m$), but the trivial '1' in the symbol is normally omitted.

2.11. (a) Symmetry operations: **1**, **2**, $\boldsymbol{m} \perp x$, $\boldsymbol{m} \perp y$.
(b) The identity operator **1** commutes with all other operators of the group; **2** commutes with **1**, $\boldsymbol{m} \perp x$, $\boldsymbol{m} \perp y$.

2.12. From any point x, y, z, symmetry **4** along [100] generates four points (point group 4); then, combined with **3** along [111] it becomes twelve points (point group 23). A two-fold rotation along [100] now takes this number to twenty-four (point group 432). Finally, the addition of a centre of symmetry produces the maximum number of 48 points (point group $m\bar{3}m$).

2.13.

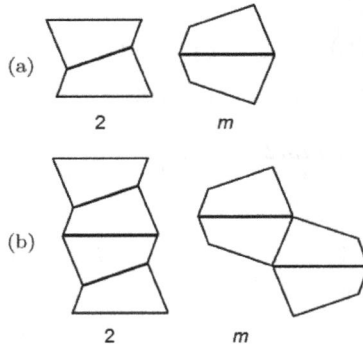

2.14.

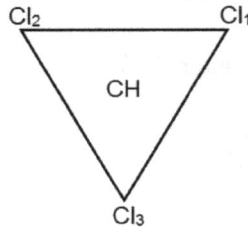

	Operation	Symmetry
	Three-fold rotational symmetry about the centre	$3, 3^{-1}$
	Rotation of $0°$ $(360°)$	1
	Reflection line through Cl_1, bisecting the angle	m
	Reflection line through Cl_2, bisecting the angle	m'
	Reflection line through Cl_3, bisecting the angle	m''

G_{3m}	1	3	3^{-1}	m	m'	m''
1	E	3	3^{-1}	m	m'	m''
3	3	3^{-1}	1	m'	m''	m
3^{-1}	3^{-1}	1	3	m''	m	m'
m	m	m''	m'	1	3^{-1}	3
m'	m'	m	m''	3	1	3^{-1}
m''	m''	m'	m	3^{-1}	3	1

Cayley table for the group G_{3m}

(1) x, y, $z \xrightarrow{2_{[0,\frac{1}{2},0]}}$ (2) $\bar{x}$, $\frac{1}{2} + y$, $\frac{1}{2} - z \xrightarrow{\bar{1}_{[0,0,0]}}$ (3) x, $\frac{1}{2} - y$, $\frac{1}{2} + z$ (4) x, $\frac{1}{2} - y$, $\frac{1}{2} + z$

Solutions 3

3.1. (a) Monoclinic.

(b) $d_{[321]} = [3\,2\,1]G[3\,2\,1]$

$$= \left\{ [321] \begin{pmatrix} 25.00 & 0 & -17.50 \\ 0 & 36.00 & 0 \\ -17.50 & 0 & 49.00 \end{pmatrix} [321] \right\}^{1/2}$$

$$= 17.69 \text{ Å}$$

3.2. (a) Monoclinic C; (b) Orthorhombic C (or A); (c) Tetragonal I; (d) Not a lattice; (e) Triclinic P.

3.3.

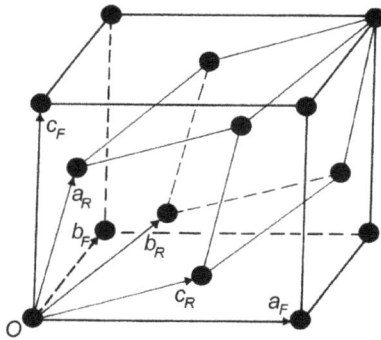

By counting lattice points per unit cell, $V_{\text{rhomb}}/V_{\text{cubic}} = 1/4$.

3.4. The transformation equations are

$$a_R = \frac{2}{3}a_H + \frac{1}{3}b_H + \frac{1}{3}c_H, \quad b_R = -\frac{1}{3}a_H + \frac{1}{3}b_H + \frac{1}{3}c_H$$

$$a_R = -\frac{1}{3}a_H - \frac{2}{3}b_H + \frac{1}{3}c_H$$

In a rhombohedron, $a = b = c$. Thus,

$$a^2 = (2/3, 1/3, 1/3)G(2/3, 1/3, 1/3)^{\mathrm{T}}$$

$$= (2/3, 1/3, 1/3) \begin{pmatrix} 9.95402 & -4.97701 & 0 \\ -4.97701 & 9.5402 & 0 \\ 0 & 0 & 80.37122 \end{pmatrix} (2/3, 1/3, 1/3)^{\mathrm{T}}$$

$$= 12.481, \text{ so that } a_R = 3.4997 \text{ Å}$$

$$\cos\alpha = \frac{1}{a_R^2}\left(\frac{2}{3}a_{\mathrm{H}} + \frac{1}{3}b_{\mathrm{H}} + \frac{1}{3}c_{\mathrm{H}}\right) \cdot \left(-\frac{1}{3}a_{\mathrm{H}} + \frac{1}{3}b_{\mathrm{H}} + \frac{1}{3}c_{\mathrm{H}}\right)$$

$$= \frac{1}{a_R^2}\left(-\frac{1.5}{9}a_{\mathrm{H}}^2 + \frac{c_{\mathrm{H}}^2}{9}\right)$$

$$= \frac{7.27114 \text{ Å}^2}{(12.24818 \text{ Å})^2} = 0.59365$$

Thus, $a_R = 3.500$ Å and $\alpha = 53.58°$.

Now, confirm the results with the program METTENS.

3.5. (a) The given cell I when transformed becomes a C-centred unit cell II. Hence, by counting lattice points per unit cell, $V_{\mathrm{II}}/V_{\mathrm{I}} = 2$. The transformation matrix for the unit cell is

$$\mathbf{S} = \begin{pmatrix} 1 & \bar{1} & 0 \\ 1 & 1 & 0 \\ 0 & 0 & 1 \end{pmatrix}$$

(b)

$$(\mathbf{S}^{\mathrm{T}})^{-1} = \begin{pmatrix} 1/2 & \bar{1/2} & 0 \\ 1/2 & 1/2 & 0 \\ 0 & 0 & 1 \end{pmatrix}$$

so that the transformed coordinates are given by

$$\begin{pmatrix} 1/2 & \bar{1/2} & 0 \\ 1/2 & 1/2 & 0 \\ 0 & 0 & 1 \end{pmatrix} \begin{pmatrix} 0.123 \\ \bar{0.671} \\ 0.314 \end{pmatrix} = \begin{pmatrix} 0.397 \\ \bar{0.274} \\ 0.314 \end{pmatrix}$$

The same result is obtained by using the mnemonic in Section 3.10.

3.6. (a)

$$\begin{pmatrix} 0 & 1 & 0 \\ \bar{1} & 0 & 0 \\ 0 & 0 & \bar{1} \end{pmatrix}$$
$$\bar{4}$$

(b)

$$x, y, z \xrightarrow{\bar{4}} y, \bar{x}, \bar{z} \xrightarrow{\bar{4}} \bar{x}, \bar{y}, z \xrightarrow{\bar{4}} \bar{y}, x, \bar{z}$$

(i)   (ii)   (iii)   (iv)

(c) 2 and 1.

(d) Consider points (i) and (ii) in (b) above. The change in signs of x and y implies perpendicularity between x, y and z; otherwise, cross terms would occur in products such as $\mathbf{a} \cdot \mathbf{b}$. Also, $x\mathbf{a} \cdot y\mathbf{b} = xyab \cos \gamma$ while $(-y)\mathbf{a} \cdot x\mathbf{b} = -xyab \cos \gamma$. Then, indistinguishability requires $xyab \cos \gamma = -xyab \cos \gamma$, which means that $\cos \gamma = 0$, so that $\gamma = 90°$. Similar arguments apply in respect of α and β. Also, the interchange of x and y demonstrates the equality of a and b. In summary, $a = b \neq c$ and $\alpha = \beta = \gamma = 90°$.

3.7. To obtain a C cell, a transformation is $a' = -a/2 + c/2$, $b' = b$, $c' - c$. (a) Then, the Bilbao Server gives $V_F = 314.0156$ Å^3. $a' = 6.1050$ Å, $b' = 6.9850$ Å, $c' = 6.0252$ Å, $\beta' = 142.330°$; $V_C = 157.0078$. (b) $V_C/V_F = 1/2$, as expected.

3.8. $a^* = 0.20017$, $b^* = 0.19335$, $c^* = 0.14433$ Å^{-1}, $\alpha^* = 90°$, $\beta^* = 75.00°$, $\gamma^* = 90°$.
$V = \sqrt{\det G} = \sqrt{3.435 \times 10^4} = 185.34 V = 185.3$ Å^3 and $V^* = 1/V = 5.396 \times 10^{-3}$ Å^{-3}.

3.9. If the cell is rotated about c, no indistinguishable position arises during a rotation of $360°$. Thus, the symmetry at each lattice point is $\bar{1}$, and the cell is a special case of the triclinic system, with $\gamma = 90°$ within the limits of experimental error.

3.10. (a) Matrix

$$\mathbf{S} = \begin{pmatrix} 3 & \bar{1} & 0 \\ 0 & 1 & 0 \\ \bar{2} & 1 & 1 \end{pmatrix}$$

entered into the Server by columns gives
$a' = 18.0278$ Å, $b' = 10.0000$ Å, $c' = 20.6155$ Å, $\alpha' = 60.980°$,
$\beta' = 132.270°$, $\gamma' = 123.690°$.
(b)   (i) $h'k'l' = (10\bar{2})(\mathbf{S}^{\mathrm{T}}) = (30\bar{4})$;   (ii) $[u'v'w']$ $=$
$(\mathbf{S}^{\mathrm{T}})^{-1}[310]^{\mathrm{T}} = [120]$; (iii) $x'y'z' = (\mathbf{S}^{\mathrm{T}})^{-1}[x, y, z]^{\mathrm{T}} = 0.3150$,
$0.4700, -0.5170 = -0.2400, 0.7475, -0.5175$.

3.11. $a^* = b^* = 1/(a \sin \gamma) = 1/(4.990 \text{ Å} \times \sin 120°) = 0.231403$ Å^{-1},
$c^* = 1/c = 1/17.06$ Å $= 0.0586166$ Å^{-1}; $\gamma^* = 60°$.
$d^{*2}(104) = [104]G^*[104]^{\mathrm{T}}$. The reciprocal metric tensor G^*
is constructed in the same manner as G. Using the program
METTENS, $d^{*2}(104) = 0.108522$ Å^{-2}, and $d^*(104)$ is equal
to $d^*(\bar{1}\,14)$ by three-fold symmetry. Also, $\mathbf{d}^*(104) \cdot \mathbf{d}^*(\bar{1}14) =$
$[104]G^*[\bar{1}14]^{\mathrm{T}} = 0.0282008$ Å^{-2}. Hence, the angle $\widehat{[104][\bar{1}\,14]} =$
$\frac{\mathbf{d}^*(104)\cdot\mathbf{d}^*(\bar{1}\,14)}{d^*(104)\cdot d^*(\bar{1}\,14)} = 0.0282008$ Å$^2/0.108522$ Å$^2 = 0.25986$, so that
$\beta^* = 74.94°$.

3.12. From METTENS and the Server: $a = 8.5440$ Å, $b = 5.8310$ Å,
$c = 10.7703$ Å, $\alpha = 142.760°$, $\beta = 69.650°$, $\gamma = 79.590°$,
$V = 180.0000$ Å^3, and the determinant of the transformation
matrix $(\mathbf{S}^{\mathrm{T}}) = \begin{pmatrix} 1 & 1 & 0 \\ 2 & 0 & 1 \\ 0 & \bar{1} & 2 \end{pmatrix}$ is reported as -3.
[*Note*: The results show the power of BILBAO. It provides also
the transformed interaxial angles and reports the determinant
of the transformation matrix as negative, which implies left-
handed axes. If any one axis and its transformation equation
are negated to give, say, $\mathbf{a}' = -\mathbf{a} - 2\mathbf{b}$ to give

$$\begin{pmatrix} \bar{1} & \bar{1} & 0 \\ 2 & 0 & 1 \\ 0 & \bar{1} & 2 \end{pmatrix}$$

then the unit-cell parameters are unaltered but the determinant is now +3.]

3.13. (a) The crystal system is triclinic with one angle of 90° within experimental error. (b) The quantity $d^2_{[35\bar{4}]}$ is given by the metric tensor as

$$(3\,5\,\bar{4})\; G \begin{pmatrix} 3 \\ 5 \\ \bar{4} \end{pmatrix}$$

and using the program METTENS, $d = 72.23$ Å. (c) The angle θ between $[111]$ and $[3\,5\,\bar{4}]$ is given by

$$\cos\theta = \frac{1}{|\mathbf{p}|\,|\mathbf{q}|}[111]G \begin{bmatrix} 3 \\ 5 \\ 4 \end{bmatrix}$$

so that $\theta = 104.8°$.

Solutions 4

4.1. Plane group $c2mm$.

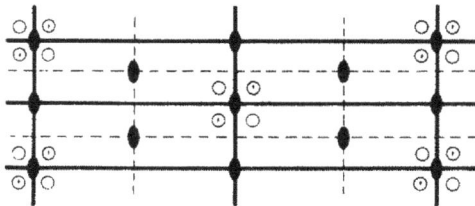

Origin at 2*mm*

$$(0,\,0;\;{}^1\!/_2,\,{}^1\!/_2)+$$

8	f	1	$x, y;\ \bar{x}, y;\ \bar{x}, \bar{y};\ x, \bar{y}$
4	e	m	$0, y;\ 0, \bar{y}$
4	d	m	$x, 0;\ \bar{x}, 0$
4	c	2	${}^1\!/_4, {}^1\!/_4;\ {}^1\!/_4, {}^3\!/_4$
2	b	$2mm$	$0, {}^1\!/_2$
2	a	$2mm$	$0, 0$

4.2.

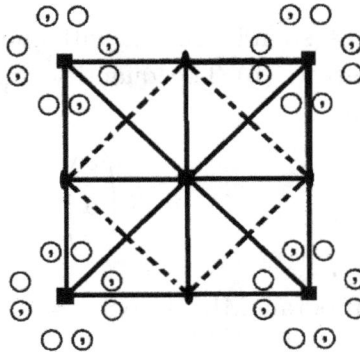

Origin on 4*mm*

8	g	1	$x, y;\ y, x;\ \bar{y}, x;\ \bar{x}, y;$ $\bar{x}, \bar{y};\ \bar{y}, \bar{x};\ y, \bar{x};\ x, \bar{y}$
4	f	m	$x, x;\ \bar{x}, x;\ \bar{x}, \bar{x};\ x, \bar{x}$
4	e	m	$x, 1/2;\ \bar{x}, 1/2;\ 1/2\ x;\ 1/2, \bar{x}$
4	d	m	$x, 0;\ \bar{x}, 0;\ 0, x;\ 0, \bar{x}$
2	c	mm2	$1/2, 0;\ 0, 1/2$
1	b	4mm	$1/2, 1/2$
1	a	4mm	$0, 0$

4.3. (a) ρ_o = mass/volume = $\frac{Z_c M_r u}{V}$, where Z_c is the number of molecules in the unit cell and the other symbols have their usual values. Thus, 1190 kg m^{-3} = $\frac{Z_c \times 154.208 \times 1.6605 \times 10^{-27}\ \text{kg}}{(8.124\text{Å} \times 5.635\text{Å} \times 9.513\text{Å} \times 0.99604) \times (10^{-30}\ \text{m}^3\text{Å}^{-3})}$, so that $Z_c = 2.016$, or 2 to the nearest integer. By referring to Fig. 4.14, it is evident that the molecules occupy a set of special equivalent positions of $\bar{1}$ symmetry, such as 0, 0, 0 and 0, 1/2, 1/2. It follows that the molecule is centrosymmetric and, therefore, planar. (The expected conjugation throughout the molecule is supported by the fact that the C–C′ bond length is 1.49 Å, whereas the average C–C distance in the ring is 1.40 Å.)

(b) Using the above equation, but with $Z = 2$, gives $D_c = $ 1181 kg m^{-3}.

(c)

Vector coordinates	Multiplicity	Vector coordinates	Multiplicity
$2x,\ 2y\ 2z$	1	$2x,\ \tfrac{1}{2},\ \tfrac{1}{2}+2z$	2
$2\bar{x},\ 2\bar{y},\ 2\bar{z}$	1	$2\bar{x},\ \tfrac{1}{2},\ \tfrac{1}{2}-2z$	2
$2x,\ 2\bar{y},\ 2z$	1	$0,\ \tfrac{1}{2}+2y,\ \tfrac{1}{2}$	2
$2\bar{x},\ 2y,\ 2\bar{z}$	1	$0,\ \tfrac{1}{2}-2y,\ \tfrac{1}{2}$	2

4.4. (a) The three t-generators $t_{[100]}$, $t_{[010]}$ and $t_{[001]}$ extend a given unit cell to three-dimensional space. In space group $P2_1/c$, the two-fold generator here has the action

$$x, y, z \xrightarrow{\ 2_{\,[0,\,y,\,1/4]}\ } \bar{x}, y, \bar{z} + (0, \tfrac{1}{2}, \tfrac{1}{2}).$$

The $\bar{1}$ generator has the action

$$x, y, z \xrightarrow{\ \bar{1}_{\,[0,\,0,\,0]}\ } \bar{x}, \bar{y}, \bar{z}.$$

The totality of these results gives the coordinates $\pm(x, y, z; \bar{x}, \tfrac{1}{2}+y, \tfrac{1}{2}-z)$ for $P2_1/c$, thus revealing the c-glide planes at $(x, \pm\tfrac{1}{4}, z)$.

(b) In order to obtain $C2/c$, the additional generator needed is $t_{[1/2,\,1/2,\,0]}$.

(c) Applying this to, say, $\bar{x}, \tfrac{1}{2}+y, \tfrac{1}{2}-z$ and completing the unit-cell diagram shows an n-glide at $y = 0$. [The standard orientation of $C2/c$ places the c-glide at $y = 0$; then, the n-glide is at $y = \tfrac{1}{4}$].

4.5. (a) Adopting the argument as before:

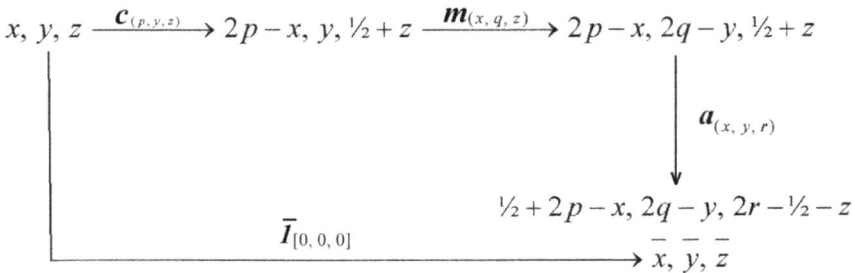

$$x, y, z \xrightarrow{\ C_{(p,\,y,\,z)}\ } 2p - x, y, \tfrac{1}{2} + z \xrightarrow{\ m_{(x,\,q,\,z)}\ } 2p - x, 2q - y, \tfrac{1}{2} + z$$

with a downward arrow $a_{(x,\,y,\,r)}$ to

$$\tfrac{1}{2} + 2p - x, 2q - y, 2r - \tfrac{1}{2} - z$$

and $\bar{1}_{[0,\,0,\,0]}$ leading to $\bar{x}, \bar{y}, \bar{z}$.

It follows now that $p = r = \tfrac{1}{4}$, $q = 0$.

From the above scheme, the coordinates of the general equivalent positions are as follows:

$$x,\ y,\ z;\ x,\ \bar{y},\ z;\ 1/2+x,\ y,\ 1/2-z;\ 1/2-x,\ y,\ 1/2+z;$$

$$\bar{x},\ \bar{y},\ \bar{z};\ \bar{x},\ y,\ \bar{z};\ 1/2-x,\ \bar{y},\ 1/2+z;\ 1/2+x,\ \bar{y},\ 1/2-z$$

(b) The coordinates show: 2_1 at $[x.0,\ z]$, 2 at $[0,\ y,\ 0]$ and 2_1 at $[1/4,\ 0,\ z]$, so that the full symbol is $P\frac{2_1}{c}\frac{2}{m}\frac{2_1}{a}$.

(c) Following the arguments in Section 4.9 and Self-assessment 4.9, *Pcma* (**bca**) transforms to *Pbam* (**abc**):

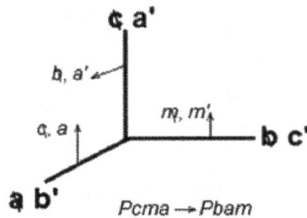

Pcma → Pbam

4.6. Special positions are sites of point-group symmetry. In space group $Pn\bar{3}$, their sites and typical coordinates are as follows:
2, $x,\ 3/4,\ 1/4$; 2, $x,\ 1/4,\ 1/4$; 3, $x,\ x,\ x$; 222, $1/4, 3/4, 3/4$;
$\bar{3}, 1/2, 1/2, 1/2$ $\bar{3}, 0, 0, 0$ 23, $1/4, 1/4, 1/4$.

Number of sites	Site symmetry	Coordinates of special equivalent positions
12	2	$\pm\{x, 3/4,\ 1/4;\ 1/4, x, 3/4; 3/4, 1/4, x;$ $1/2+x, 1/4, 3/4; 3/4, 1/2+x, 1/4;$ $1/4, 3/4, 1/2+x\}$
12	2	$\pm\{x, 1/4, 1/4; 1/4, x, 1/4; 1/4, 1/4, x;$ $1/2+x, 3/4, 3/4; 3/4, 1/2+x, 3/4;$ $3/4, 3/4, 1/2+x\}$
8	3	$\pm\{x, x, x; x, 1/2-x, 1/2-x; 1/2-x, x,$ $1/2-x; 1/2-x, 1/2-x, x\}$
6	222	$\pm\{1/4, 3/4, 3/4; 3/4, 1/4, 3/4; 3/4, 3/4, 1/4\}$
4	$\bar{3}$	$1/2, 1/2, 1/2; 1/2, 0, 0; 0, 1/2, 0; 0, 0, 1/2$
4	$\bar{3}$	$0, 0, 0; 0, 1/2, 1/2; 1/2, 0, 1/2; 1/2, 1/2, 0$
2	23	$1/4, 1/4, 1/4; 3/4, 3/4, 3/4$

4.7. The coordinates of the general equivalent positions in space group $P432$ are listed below, together with their numbers on the stereogram:

(1) x, y, z (2) $\bar{x}, \bar{y}, z$ (3) $\bar{x}, y, \bar{z}$ (4) $x, \bar{y}, \bar{z}$

(5) z, x, y (6) $z, \bar{x}, \bar{y}$ (7) $\bar{z}, \bar{x}, y$ (8) $\bar{z}, x, \bar{y}$

(9) y, z, x (10) $\bar{y}, z, \bar{x}$ (11) $y, \bar{z}, \bar{x}$ (12) $\bar{y}, \bar{z}, x$

(13) $y, x, \bar{z}$ (14) $\bar{y}, \bar{x}, \bar{z}$ (15) $y, \bar{x}, z$ (16) $\bar{y}, x, z$

(17) $x, z, \bar{y}$ (18) $\bar{x}, z, y$ (19) $\bar{x}, \bar{z}, \bar{y}$ (20) $x, \bar{z}, y$

(21) $z, y, \bar{x}$ (22) $z, \bar{y}, z$ (23) $\bar{z}, y, x$ (24) $\bar{z}, \bar{y}, \bar{x}$

Stereogram of point group 432: the numbers refer to the coordinates listed above

4.8. With $Z_c = 8$, the atoms must occupy special equivalent positions in the space group. From the International Tables, data on the appropriate special equivalent positions of the space group $Fd\bar{3}m$, origin on $\bar{4}3m$, are as follows:

$$(0, 0, 0; \ 0, {}^1/_2, {}^1/_2; \ {}^1/_2, 0, {}^1/_2; \ {}^1/_2, {}^1/_2, 0)+$$

8	a	$\bar{4}3m$	$0, 0, 0; \ {}^1/_4, {}^1/_4, {}^1/_4.$
16	d	$\bar{3}m$	${}^5/_8, {}^5/_8, {}^5/_8; \ {}^5/_8, {}^7/_8, {}^7/_8; \ {}^7/_8, {}^5/_8, {}^7/_8; \ {}^7/_8, {}^7/_8, {}^5/_8.$

$$x, x, x; \ {}^1/_4 - x, {}^1/_4 - x, {}^1/_4 - x;$$
$$x, \bar{x}, \bar{x}; \ {}^1/_4 - x, {}^1/_4 + x, {}^1/_4 + x;$$

32	e	$3m$	$\bar{x}, x, \bar{x}; {}^1/_4 + x, {}^1/_4 - x, {}^1/_4 + x;$

$$\bar{x}, \bar{x}, x; \ {}^1/_4 + x, {}^1/_4 + x, {}^1/_4 - x$$

A probable allocation is 8 Mg in a (or a similar set in b), 16 Al in d (or a similar set in c) and 32 O in e. Thus, the structure is determined by a single parameter, x. [In fact, it is almost a close packed array of oxygen atoms. with the cations in the interstices. For a truly close-packed structure, the fractional value of x is 0.375, By experiment for this example, $x = 0.387$.]

4.9. (a) The product $c\ m$ is given by

$$
\begin{pmatrix} 1 & 0 & 0 \\ 0 & \bar{1} & 0 \\ 0 & 0 & 1 \end{pmatrix}
\begin{pmatrix} \bar{1} & 0 & 0 \\ 0 & 1 & 0 \\ 0 & 0 & 1 \end{pmatrix}
+ \begin{pmatrix} 2p \\ 0 \\ 0 \end{pmatrix}
+ \begin{pmatrix} 0 \\ 2q \\ \tfrac{1}{2} \end{pmatrix}
$$

$$
= \begin{pmatrix} \bar{1} & 0 & 0 \\ 0 & \bar{1} & 0 \\ 0 & 0 & 1 \end{pmatrix}
+ \begin{pmatrix} 2p \\ 2q \\ \tfrac{1}{2} \end{pmatrix}
$$

Then, $b\ c\ m$ becomes

$$
\begin{pmatrix} 1 & 0 & 0 \\ 0 & 1 & 0 \\ 0 & 0 & \bar{1} \end{pmatrix}
\begin{pmatrix} \bar{1} & 0 & 0 \\ 0 & \bar{1} & 0 \\ 0 & 0 & 1 \end{pmatrix}
+ \begin{pmatrix} 2p \\ 2q \\ \tfrac{1}{2} \end{pmatrix}
+ \begin{pmatrix} 0 \\ \tfrac{1}{2} \\ 2r \end{pmatrix}
$$

$$
= \begin{pmatrix} \bar{1} & 0 & 0 \\ 0 & \bar{1} & 0 \\ 0 & 0 & \bar{1} \end{pmatrix}
+ \begin{pmatrix} 2p \\ 2q + \tfrac{1}{2} \\ 2r + \tfrac{1}{2} \end{pmatrix}
$$

But this is equivalent to the identity operator

$$
\begin{pmatrix} \bar{1} & 0 & 0 \\ 0 & \bar{1} & 0 \\ 0 & 0 & \bar{1} \end{pmatrix}
+ \begin{pmatrix} 0 \\ 0 \\ 0 \end{pmatrix}
$$

since $\bar{1}$ is at the origin. Hence, $p = 0$ and $q = r = \tfrac{1}{4}$. (b) The full symbol is $P\frac{2}{m}\frac{2_1}{c}\frac{2_1}{b}$. (c) In the **abc** setting, using the transformation discussed in Section 4.9, the symbol becomes *Pbam*, which is $P\frac{2_1}{b}\frac{2_1}{a}\frac{2}{m}$ in full.

4.10. (a) x, y, $z \to \bar{x}$, $^1\!/_2 + y$, $^1\!/_2 - z : 2_1$ screw axis along $[0, y, ^1\!/_4]$;

(b) x, y, $z \to ^1\!/_2 - y$, x, $^3\!/_4 + z : 4_3$ screw axis along $[^1\!/_4, ^1\!/_4, z]$;

(c) x, y, $z \to y$, x, $^1\!/_2 + z : c$-glide plane $\parallel (x, x, z)$.

4.11. (a)

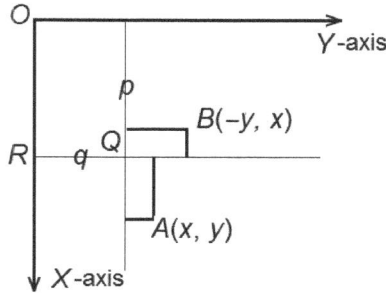

In the diagram, $p = OR$ and $q = RQ$. The X-coordinate of B is $p - y + q$ and the Y-coordinate of B is $q + X - p$. Thus, B is the point $(p + q) - y$, $(q - p) + x$.

(b) x, y, $z \xrightarrow{\,4_3[1/4,3/4,z]\,} \bar{y}$, $^1\!/_2 + x$, $^3\!/_4 + z$.

4.12. (a) Reference to Appendix A4 leads to the results

$$
6_3 = \begin{pmatrix} 1 & \bar{1} & 0 \\ 1 & 0 & 0 \\ 0 & 0 & 1 \end{pmatrix} + \begin{pmatrix} 0 \\ 0 \\ ^1\!/_2 \end{pmatrix}
$$

(b) The c-glide plane orientation is $(2x, x, z)$, so that its matrix is

$$
\begin{pmatrix} 1 & 0 & 0 \\ 1 & \bar{1} & 0 \\ 0 & 0 & 1 \end{pmatrix} + \begin{pmatrix} 0 \\ 0 \\ ^1\!/_2 \end{pmatrix}
$$

(c)

$$
c\,6_3 = \begin{pmatrix} 1 & 0 & 0 \\ 1 & \bar{1} & 0 \\ 0 & 0 & 1 \end{pmatrix} \begin{pmatrix} 1 & \bar{1} & 0 \\ 1 & 0 & 0 \\ 0 & 0 & 1 \end{pmatrix} + \begin{pmatrix} 0 \\ 0 \\ ^1\!/_2 \end{pmatrix} + \begin{pmatrix} 0 \\ 0 \\ ^1\!/_2 \end{pmatrix}
$$

$$
= \begin{pmatrix} 1 & \bar{1} & 0 \\ 0 & \bar{1} & 0 \\ 0 & 0 & 1 \end{pmatrix} + \begin{pmatrix} 0 \\ 0 \\ 0 \end{pmatrix}
$$

which is a mirror plane with the orientation $(x, 0, z)$. (d) The space group is $P6_3cm$.

Solutions 5

5.1. $\underline{k+l = 2n}$: $A_{hkl} = 4\cos 2\pi(hx + lz)\cos 2\pi ky; \ B_{hkl} = 0.$
$\underline{k + l = 2n + 1}$: $A_{hkl} = -4\sin 2\pi(hx + lz)\sin 2\pi ky; \ B_{hkl} = 0.$
By permuting the $\pm$ signs on $h\,k$ and l,

$$|\mathbf{F}(hkl)| = |\mathbf{F}(\bar{h}\,\bar{k}\,\bar{l})| = |\mathbf{F}(h\bar{k}l)| = |\mathbf{F}(\bar{h}k\bar{l})| \neq |\mathbf{F}(\bar{h}kl)|;$$

$$|\mathbf{F}(\bar{h}kl)| = |\mathbf{F}(h\bar{k}\,\bar{l})| = |\mathbf{F}(hk\bar{l})| = |\mathbf{F}(\bar{h}\,\bar{k}l)|$$

which may be given concisely as

$$|\mathbf{F}(hkl)| = |\mathbf{F}(\bar{h}\,\bar{k}\,\bar{l})|$$

and
$\underline{k + l = 2n}$:

$$|\mathbf{F}(hkl)| = |\mathbf{F}(h\bar{k}l)| \neq |\mathbf{F}(\bar{h}kl)|; |\mathbf{F}(\bar{h}kl)| = |\mathbf{F}(hk\bar{l})|$$

on account of Friedel's law

and
$\underline{k + l = 2n + 1}$:

$$|\mathbf{F}(hkl)| = -|\mathbf{F}(h\bar{k}l)| \neq |\mathbf{F}(\bar{h}kl)|; |\mathbf{F}(\bar{h}kl)| = -|\mathbf{F}(hk\bar{l})|$$

5.2.

$$A(hkl) = \cos 2\pi(hx + ky + lz) + \cos 2\pi(-hx - ky + lz)$$
$$+ \cos 2\pi(-hx + ky + lz + h/2)$$
$$+ \cos 2\pi(hx - ky + lz + h/2)$$

Combining the first and third and the second and fourth terms (Appendix A7):

$$A(hkl)/2 = \cos 2\pi(ky + lz + h/4)\cos 2\pi(hx - h/4)$$
$$+ \cos 2\pi(-ky + lz + h/4)\cos 2\pi(hx + h/4)$$

In order to obtain a common factor, $h/2$ in the fourth term of the first equation for $A(hkl)$ is changed to $-h/2$. This change is equivalent to changing the fourth general equivalent position to $-\frac{1}{2} + x, \bar{y}, z$, which implies a change of one repeat distance in the negative a direction, a crystallographically valid change and generally useful tactic. Thus, remembering that $\cos -\theta = \cos \theta$,

$$A(hkl)/2 = \cos 2\pi(ky + lz + h/4) \cos 2\pi(hx - h/4)$$
$$+ \cos 2\pi(-ky + lz - h/4) \cos 2\pi(hx - h/4)$$

which simplifies to

$$A(hkl)/2 = \cos 2\pi(hx - h/4)[\cos 2\pi(ky + lz + h/4)$$
$$+ \cos 2\pi(-ky + lz - h/4)]$$

Combining again gives

$$A(hkl) = 4 \cos 2\pi(hx - h/4) \cos 2\pi(ky + h/4) \cos 2\pi lz$$

By a similar analysis, the B term is found to be

$$B(hkl) = 4 \cos 2\pi(hx - h/4) \cos 2\pi(ky + h/4) \sin 2\pi lz$$
$$(B_{hkl} = 0 \text{ if } l = 0)$$

It is evident that these results should be considered now for h even and odd.

$\underline{h=2n}$

$$A(hkl) = 4 \cos 2\pi hx \, \cos 2\pi ky \, \cos 2\pi lz$$
$$B(hkl) = 4 \cos 2\pi hx \, \cos 2\pi ky \, \sin 2\pi lz$$

$\underline{h = 2n + 1}$

$$A(hkl) = -4 \sin 2\pi hx \, \sin 2\pi ky \, \cos 2\pi lz$$
$$B(hkl) = -4 \sin 2\pi hx \, \sin 2\pi ky \, \sin 2\pi lz$$

Hence, $A = B = 0$ if $k = 0$, that is, systematic absences occur only in the $h0l$ data for $h = 2n + 1$, or the only *limiting condition* is $h0l$: $h = 2n$.

5.3.

$$A' = \sum 10 \cos 15° + 20 \cos 150° + 40 \cos(-110°),$$

$$B' = \sum 10 \sin 15° + 20 \sin 150° + 40 \sin(-110°)$$

$$|\mathbf{F}| = \sqrt{A'^2 + B'^2} = 32.87; \quad \phi = \tan^{-1}(B'/A') = 49.51°$$

5.4. (a) In A-centring, the atoms occur in $N/2$ pairs, x, y, z and x, $1/2 + y$, $1/2 + z$. The result may be achieved readily with the Eq. (5.29), omitting the atomic scattering factor:

$$
\begin{aligned}
\mathbf{F}(hkl) = \sum_{N/2} & \Big\{ \exp 2\pi i(hx + ky + lz) \\
& + \exp 2\pi i \left(hx + ky + lz + \frac{k+l}{2} \right) \Big\} \\
= \sum_{N/2} & \exp 2\pi i(hx + ky + lz) \left(1 + \exp 2\pi i \frac{k+l}{2} \right)
\end{aligned}
$$

where the $N/2$ atoms are those not related by the A-centring. (b) The term $1 + \exp 2\pi i \frac{k+l}{2} = 1 + \cos \pi(k+l)$, since $(k+l)$ is integral. Hence, the resultant is zero if $k+l$ is odd, but 2 (the G-factor) if $k+l$ is even.

5.5.

Range, n	1	2	3	4	5	6
$\langle (\sin^2 \theta_n)/\lambda^2 \rangle$	0.091	0.149	0.192	0.256	0.297	0.362
$\ln\{\langle \|\mathbf{F}_{\mathrm{obs},\,\theta_n}\|^2 \rangle / \sum_j f_j^2\}$	0.1947	0.4650	0.6596	0.9944	1.1743	1.3902

Least-squares fit: Slope $= 4.523$ Å^2, Intercept $= -0.2025$; Pearson's $r = 0.998$. Hence, $B = 2.26$ Å^2 and $K = 0.903$.
In the following, the Wilson plot is shown; not all plots are such a good fit to the theory.

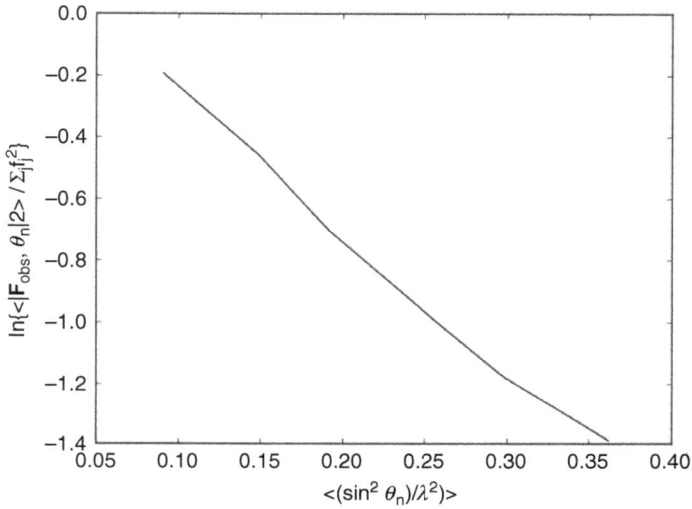

Wilson plot: Slope = 4.523 Å2; Intercept = –0.2025

5.6. The coordinates imply a centrosymmetric structure with $\bar{1}$ at the origin. Hence, $|\mathbf{F}(hk0)| = 2f_U\{\cos 2\pi(ky) + \cos 2\pi[(h + k)/2 + ky]\} = 4f_U \cos 2\pi[ky + (h+k)/4] \cos \pi[(h+k)/2]$. Since $(h + k)$ is an even integer for the given data, $|\mathbf{F}(hk0)| = 4f_U \cos 2\pi(ky)$. Thus, the following results emerge:

$y = 0.05$		110	020	$y = 0.10$	110	020
$|\mathbf{F}_{obs}|$		270	90		270	90
$|\mathbf{F}_{calc}| = 4f_U \cos 2\pi(ky)$		304.3	226.5		258.9	86.5
$y = 0.15$	110	020				
	270	90				
	188.1	86.5				

Thus, the best value for y is 0.10.

5.7. For a cubic crystal, $\frac{\sin\theta}{\lambda} = \frac{\sqrt{h^2+k^2+l^2}}{2a}$:

NaCl, $\sin\theta(111) = 0.1539$ Å^{-1}, $\sin\theta(222) = 0.3078$ Å^{-1}.
KCl, $\sin(111) = 0.1379$ Å^{-1}, $\sin\theta(222) = 0.2759$ Å^{-1}.

The amplitudes of the structure factors for these crystals are given by $|F(111)| = 4|(f_{Na^+, K^+} - f_{Cl^-})|$ and $|F(222)| = 4|(f_{Na^+, K^+} + f_{Cl^-})|$. Then, the following results emerge:

	111		222	
hkl	NaCl	KCl	NaCl	KCl
$(\sin\theta)/\lambda$	0.1539	0.1379	0.3078	0.1379
f_+	8.979	15.652	6.777	11.576
f_-	13.593	14.207	9.387	9.997
$|\mathbf{F}_{calc}|$	18.46	5.78	64.66	86.29
I_{calc}	341	33	4181	7446

Since $I(hkl) \propto |\mathbf{F}(hkl)|^2$, reflection 111 and, indeed, all reflections with $(h+k+l)$ equal to an odd integer, are very much weaker for KCl than for NaCl because they are proportional to $(f_+ - f_-)$, and K^+ and Cl^- are isoelectronic (18 electrons each).

5.8. (a) $(\sin\theta)/\lambda = 0.61676/1.5418 = 0.4000$, so that f_C (at rest) $= 1.948$. At ambient temperature, the Debye–Waller factor is $\exp = [-7.0 \text{ Å}^2 \times (0.4000 \text{ Å})^2] = 0.32628$, so that f_C (ambient) $= 0.6356$. The fractional reduction is the exponential factor of 0.3263, or 32.6%. (b) Since $B = 8\pi^2\langle u^2 \rangle$, the rms u-value is $\sqrt{7.0 \text{ Å}^2/8\pi^2}$, or 0.30 Å.

5.9. The distinguishing features of space groups $P2$, Pm and $P\frac{2}{m}$ are summarized in their distributions below:

	$P2$	Pm	$P\frac{2}{m}$
hkl	A	A	C
h0l	C	A	C
0k0	A	C	C

Thus, in principle, it should be possible to distinguish between these three space groups by examining the *hkl* and *h0l* data. [Note that when averaging, the *hkl*, *h0l* and *0k0* data should not be mixed, which means that *hkl* should not include *h0l* or

$0k0$. In practice, although there may be too few $0k0$ reflections for a significant deduction to be made from them, the other two classes of reflections should provide the necessary result.]

5.10. (a) Substitute for $|\mathbf{F}|$ in Eq. (A8.12) by $|\mathbf{E}|$ from Eq. (5.74), noting that $\mathrm{d}|\mathbf{F}| = \sum(|\mathbf{E}|/|\mathbf{F}|)\,\mathrm{d}|\mathbf{E}| = \sum[|\mathbf{E}|/(\sum^{1/2}|\mathbf{E}|)]\mathrm{d}|\mathbf{E}| = \sum^{1/2}\mathrm{d}|\mathbf{E}|$. Then Eq. (A8.12) becomes $P_{\mathrm{c}}|\mathbf{E}|\mathrm{d}|\mathbf{E}| = (2/\pi)^{1/2}\exp-|\mathbf{E}|^2/2)\mathrm{d}|\mathbf{E}|$ or $P_{\mathrm{c}}|\mathbf{E}| = (2/\pi)^{1/2}\exp-|\mathbf{E}|^2/2$.
(b) From Eq. (A8.1) recalling that the distribution for $P_{\mathrm{c}}|\mathbf{E}|$ is normalized, it follows that

$$\langle|\mathbf{E}|\rangle = (2/\pi)^{1/2}\int_0^\infty |\mathbf{E}| = \exp-(|\mathbf{E}|^2/2)\mathrm{d}|\mathbf{E}|.\ \text{Let } t = |\mathbf{E}|^2/2,$$

so that $|\mathbf{E}| = (2t)^{1/2}$ and $\mathrm{d}|\mathbf{E}| = (2t)^{-1/2}\mathrm{d}t$. Then,

$$\langle|\mathbf{E}|\rangle = 2^{-1/2}2^{1/2}(2/\pi)^{1/2}\int_0^\infty t^0\exp(-t)\mathrm{d}t = (2/\pi)^{1/2}$$

$$\times \int_0^\infty t^{1-1}\exp(-t)\mathrm{d}t = (2/\pi)^{1/2}\Gamma(1)$$

where Γ is the gamma function (Appendix A9), and $\Gamma(1) = 1$, so that $\langle|\mathbf{E}|\rangle = (2/\pi)^{1/2} = 0.798$.

Solutions 6

6.1. x_{U}: From 200, $|\mathbf{F}|_{\mathrm{U}} \approx \cos 2\pi(2x_{\mathrm{U}} - 1/2)$. For zero $|\mathbf{F}|$, $2\pi(2x_{\mathrm{U}} - 1/2) \approx (2n+1)\pi/2$; taking $n = 1$, $x_{\mathrm{U}} \approx 5/8$, and by symmetry $1/8$, $3/8$ and $7/8$ also.
y_{U}: Choose $x_{\mathrm{U}} = 1/8$; then for 111, For $I(111)$ to be of large magnitude, $2\pi(y_{\mathrm{U}} + 3/4) \approx 0$ or $n\pi$, so that $y_{\mathrm{U}} \approx 1/4(\equiv -3/4)$. From reflection 231, $y_{\mathrm{U}} \approx 1/6$, and from reflection 040, $y_{\mathrm{U}} \approx 3/16$. The mean value for y_{U} is *ca.* 0.2.

6.2.

$$\int_{-\pi}^{\pi}\cos 2\pi mx\,\cos 2\pi nx\,\mathrm{d}x = 1/2\int_{-\pi}^{\pi}\cos 2\pi(m+n)x\,\mathrm{d}x$$

$$+ 1/2\int_{-\pi}^{\pi}\cos 2\pi(m-n)x\,\mathrm{d}x$$

$$= \frac{1/2}{2\pi(m+n)} \sin 2\pi(m+n)x\big|_{-\pi}^{\pi}$$

$$+ \frac{1/2}{2\pi(m-n)} \sin 2\pi(m-n)x\big|_{-\pi}^{\pi}$$

$$= 0 \text{ for } m \neq n, \text{ since } m \text{ and } n$$

$$\text{are integers}$$

For $m = n$,

$$\int_{-\pi}^{\pi} \cos 2\pi mx \cos 2\pi mx \, dx = \int_{-\pi}^{\pi} \cos^2 2\pi mx \, dx$$

$$= \tfrac{1}{2} \int_{-\pi}^{\pi} (1 + \cos 4\pi mx) \, dx = \frac{x}{2}\Big|_{-\pi}^{\pi} + \tfrac{1}{2}\frac{\sin 4\pi mx}{4\pi m}\Big|_{-\pi}^{\pi} = \pi$$

6.3. (a) Asymmetric unit of $P2_1/m$:

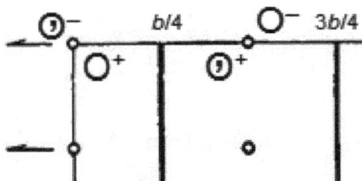

General equivalent positions: $\pm(x,\ y,\ z;\ x,\ {}^{1}\!/_{2} - y,\ z)$.

(b) For $Z = 2$, the molecules occupy special equivalent positions of symmetry $\bar{1}$ or m. Since the molecule cannot have a centre of symmetry, it is located across the m planes $(x,\ \pm^{1}\!/_{4},\ z)$. The probable arrangement is B, N, C, one F and one H atoms lying on one and the same m plane; with the remaining two F and four H atoms arranged symmetrically across that plane.

6.4. (a) Patterson vectors. Single weight: $2x,\ 2y,\ 2z;\ 2x,\ 2\bar{y},\ 2z$. Double weight: $\pm({}^{1}\!/_{2} - 2x,\ {}^{1}\!/_{2},\ {}^{1}\!/_{2} - 2z;\ {}^{1}\!/_{2},\ {}^{1}\!/_{2} - 2y,\ {}^{1}\!/_{2})$. (b) From measurement on the Harker section, ${}^{1}\!/_{2} - 2x \approx 0.14$, so that $x \approx 0.18$. Also, ${}^{1}\!/_{2} - 2z \approx 0.10$, so that $z \approx 0.20$. On the section at $v = 0.10$, ${}^{1}\!/_{2} - 2y = 0.06$, so that $y \approx 0.22$. Thus, the heavy atom coordinates in the unit cell are *ca*: $\pm(0.18,\ 0.20,\ 0.22;\ 0.68,\ 0.30,\ 0.72)$.

6.5. (a) The coordinates for Na^+ are $(0, 0, 0)+F$ and for $H^-(D^-)(0, 0, 1/2)+F$.

$$|\mathbf{F}(hkl)| = 4f_{Na^+}\{\exp(i\pi0) + \exp[i\pi(h + k)]$$
$$+ \exp[i\pi(k + l)] + \exp[i\pi(l + h)]\}$$
$$+ 4f_{H^-/D^-}\{\exp(i\pi h) + \exp(i\pi k)$$
$$+ \exp(i\pi l) + \exp[i\pi(h + k + l)]\}$$

(b) For h, k and l all even: $|\mathbf{F}(hkl)| = 4(f_{Na^+} + f_{H^-/D^-})$; for h, k and l all odd: $|\mathbf{F}(hkl)| = 4(f_{Na^+} - f_{H^-/D^-})$.

	$|\mathbf{F}(111)|$		$|\mathbf{F}(220)|$	
hkl	NaH	NaD	NaH	NaD
X-rays	30.7	30.7	27.7	27.7
Neutrons	2.88	−1.30	−0.096	4.08

6.6. (a) As discussed earlier, the origin in this space group is along $[0, y, 0]$ and so is undetermined in y. Thus, y may be chosen for the first atom to be placed, iodine in this example, and $y = 1/4$ is conventional, and convenient. From the Harker section, $u \approx 0.845$ and $w \approx 0.290$, so that the iodine coordinates in the unit cell are *ca.* $(0.423, 1/4, 0.145; 0.577, 3/4, 0.845)$.
(b) The shortest I–I distance d_{I-I} is given by $d_{I-I}^2 = \{[(0.845 \times 7.26 \text{ Å})^2 + (1/2 \times 11.55 \text{ Å})^2 + (0.290 \times 19.22 \text{ Å})^2 + [2 \times (0.845/2) \times (0.290/2) \times \cos 94.04°]\}$, so that $d_{I-I} = 10.1$ Å.
(c) In this space group, $\mathbf{F}(h0l) = 2f_I \cos 2\pi(hx_I + lz_I)$.

h0l	001	0014	300	106
$(\sin\theta/\lambda)$	0.026	0.364	0.207	0.175
$2f_I$	105	67	84	88
$|\mathbf{F}_{obs}|$	40	37	35	33
$|\mathbf{F}_{calc_I}|$	65	67	−8	−20

For 300, $|\mathbf{F}_{\text{calc}_\text{I}}|$ is small compared to its $|\mathbf{F}_{\text{obs}}|$, and the magnitude from the rest of the structure could reverse the sign given by $|\mathbf{F}_{\text{calc}_\text{I}}|$; thus, the signs of 001, 0014 and 106 are most probably $+$, $+$ and $-$.

6.7. The equations for the anomalous scattering species in terms of the familiar $A'(002)$ and $B'(002)$ terms are as follows:

$$A'(002) = 90.0 \cos \phi + (f_0 + \Delta f') \cos 2\pi 2z$$
$$+ \Delta f'' \cos(2\pi 2z + \pi/2)$$
$$B'(002) = 90.0 \sin \phi + (f_0 + \Delta f') \sin 2\pi 2z$$
$$+ \Delta f'' \sin(2\pi 2z + \pi/2)$$

Introducing the given data for reflection 002:

$$A'(002) = 90.0 \cos 30° + 52.2 \cos 144° + 9.0 \cos 234° = 30.422$$
$$B'(002) = 90.0 \sin 30° + 52.2 \sin 144° + 9.0 \sin 234° = 68.401$$

Then,

$$|\mathbf{F}(002)| = \sqrt{A'(002)^2 + B'(002)^2} = 74.9;$$
$$\phi(002) = \tan^{-1} \frac{B'(002)}{A'(002)} = 66.0°$$

Introducing the given data for reflection $00\bar{2}$:

$$A'(00\bar{2}) = 90.0 \cos 30° + 52.2 \cos 144° + 9.0 \cos 234° = 41.002$$
$$B'(00\bar{2}) = 90.0 \sin(-30°) + 52.2 \sin(-144°) + 9.0 \sin(-54°)$$
$$= -83.0$$

Then,

$$|\mathbf{F}(00\bar{2})| = \sqrt{A'(00\bar{2})^2 + B'(00\bar{2})^2} = 92.5;$$
$$\phi(00\bar{2}) = \tan^{-1} \frac{B'(00\bar{2})}{A'(00\bar{2})} = -63.7°$$

6.8. A $\sum_2$ listing is prepared as follows:

h	k	h − k	|E(h)||E(k)||E(h − k)|
038	081	011,9	10.2
	059	0817	7.2
	081	011,7	6.0
0310	081	011,9	9.2
	059	081	7.9
011	024	035	5.0
	026	035	0.5
021	038	059	0.4
	0310	059	0.4
0018	081	0817	9.5
024	035	059	9.6
0018	081	0817	9.5

An origin is specified by the reflections 081 + and 011,9 +. Sign determination follows:

h	k	h − k	Result
011,9 (+)			(Origin fixing)
081 (+)			(Origin fixing)
011,9 (+)	081 (+)	038	$s(038) = +$
011,9 (+)	08$\bar{1}$ (+)	0310	$s(0310) = +$
038 (+)	0$\bar{8}$1 (+)	011,7	$s(011,7) = +$
0310 (+)	081 (+)	0$\bar{5}$9	$s(059) = -$
059 (−)	0$\bar{3}\bar{8}$ (+)	0817	$s(0817) = -$
038 (+)	059 (−)	0$\bar{2}\bar{1}$	$s(021) = -$

(Continued)

(*Continued*)

h	k	h − k	Result
0310 (+)	059 (−)	0$\bar{2}$1	$s(021) = -$
0817(−)	08$\bar{1}$ (+)	0018	$s(0018) = -$
Introduce $s(035) = a$			
059 (−)	035 (a)	024	$s(024) = -a$
035 (a)	024 (−a)	011	$s(011) = -$
035 (a)	0$\bar{1}$1 (+)	026	$s(026) = a$

Sign propagation with the given data set halted at this stage. The two indications for 021 and one indication for 026 have low probabilities (see $\sum_2$ listing) and are unreliable.

6.9. (a) Patterson S–S vectors distant 0 to $\pm^{1}/_{2}$ from the origin along both a and b; P = single weight peak and $2P$ = double weight peak:

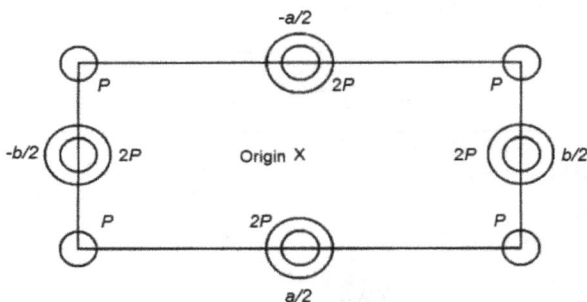

S(1)–S(2) Patterson vector positions

The double-weight peaks occur at the midpoints of the lines that join the single-weight peaks. Four S–S vectors are superimposed at the origin x, making 4^2 vectors in all.

(b) $\mathbf{d}_{S-S} = (x_2 - x_1)\mathbf{a} + (y_2 - y_1)\mathbf{b} + (z_2 - z_1)\mathbf{c}$. Using the dot product $\mathbf{d} \cdot \mathbf{d}$, d is evaluated as $d_{S-S}^2 = (x_2 - x_1)^2 a^2 + (y_2 - y_1)^2 b^2 + (z_2 - z_1)^2 c^2 + 2(z_2 - z_1)(x_2 - x_1)ca \cos \beta$. Evaluating with the given data: $d_{S-S} = 2.163$ Å.

6.10. (a) From Eq. (6.51), $\phi(771)$ is *ca.* $(-37 - 3 - 54 + 38 + 13)/6 = -7.2°$.

(b) From Eq. (6.54), $\tan \phi_{771} = \frac{\sum_{\mathbf{h}} |\mathbf{E(k)}||\mathbf{E(h-k)}| \sin[\phi(\mathbf{k})+\phi(\mathbf{h-k})]}{\sum_{\mathbf{h}} |\mathbf{E(k)}||\mathbf{E(h-k)}| \cos[\phi(\mathbf{k})+\phi(\mathbf{h-k})]}$. Evaluating this expression for the six data gives $\phi(771) = -5.7°$, and this result is preferred.

6.11. A $\sum_2$ listing is given below, together with the probability that the $|\mathbf{E(h)}||\mathbf{E(k)}||\mathbf{E(h - k)}|$ product is positive in sign.

TPR	h	k	h − k	$|\mathbf{E(h)}||\mathbf{E(k)}||\mathbf{E(h-k)}|$	$P_{+(hk0)}/\%$
1	800	670	$2\bar{7}0$	10.095	98.7
2		340	$5\bar{4}0$	7.672	96.5
3		411,0	$4\bar{1}\bar{1},0$	4.860	89.1
4		040	$8\bar{4}0$	4.176	85.9
5	300	040	$3\bar{4}0$	3.497	81.9
6		840	$\bar{5}\bar{4}0$	6.014	93.1
7		570	$\bar{2}\bar{7}0$	10.046	98.7
8	730	$0\bar{4}0$	770	3.069	79.0
9		$5\bar{4}0$	270	6.924	95.2
10	700	570	$2\bar{7}0$	12.974	99.6
11	340	$7\bar{7}0$	$\bar{4}11,0$	4.051	85.2

No more TPRs are available with the given $|\mathbf{E}|$data. The reflections 270 and 540 occur in several TPRs and form a satisfactory origin set (e o and o e). Signs/symbols are allocated as needed:

$hk0$	TPR	s_{hk0}	Comments
270		$+$	Origin fixing
540		$+$	Origin fixing
730	9	$+$	
800	3	$-$	$s_{411,0} = -s_{411,0}$
670	1	$+$	
340	2	$-$	No further signs; let $s_{040} = a$
040		a	First letter symbol
300	5	$-a$	
840	4	$-a$	
570	7	$-a$	
700	10	a	
840	4	$-a$	
770	8	a	
411,0	11	a	

The sign of symbol a may evolve in a more extended set of TPRs, or $|\mathbf{E}|$-maps with $+a$ and $-a$ would be needed in the structure analysis.

6.12. The determination of ϕ_P depends on the equation $|\mathbf{F}_P| \exp i\phi_P = |\mathbf{F}_P| \exp i\phi_{PH} - \mathbf{F}_H$. (a) Draw a circle centre O of radius $|\mathbf{F}_P|$. From O, construct the *vector*$-\mathbf{F}_H$, and from the end of this vector as centre, draw a circle of radius $|\mathbf{F}_{PH}|$. The angles formed by OX and OY with the real axis are then possible values for the phase angle ϕ_P, *ca.* 84° and −51°.

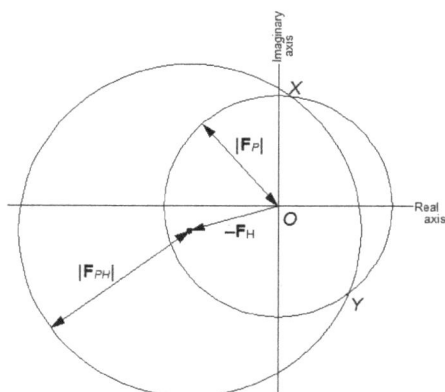

Circles for |F$_{PH}$|, |F$_P$| and F$_H$

(b) On the same diagram, construct the vector $\mathbf{F}_K$ relating to the derivative PK. Now, the evidence is that the phase of $|\mathbf{F}_P|$ is given by the angle between OX and the real axis, *ca.* 84°. However, it should be noted that in practice the evidence is rarely as clear cut as in this example; usually, there exists a circle of error with most of such determinations.

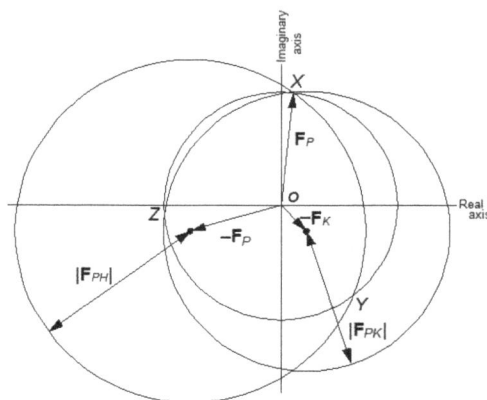

Circles for |F$_{PK}$|, |F$_P$| and F$_K$

6.13. The flat field around the site of N_2 indicates that it is –N= or =N– rather than >NH. Six fairly round peaks correspond to hydrogen atoms attached to carbon atoms. The density density at the site allocated to N_1 indicates an atom of slightly higher atomic number than nitrogen, probably –O– rather than >NH. The density at the carbon atom marked C* indicates an atom of slightly higher atomic number than carbon, and with an elongation representing a hydrogen atom, that is, >NH rather than >CH. A possible structure based on these changes is as follows:

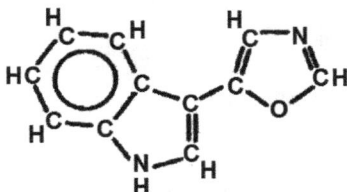

Revised structure, now $C_{11}H_8N_2O$ C11H8N2O

This structure, 5-(3-indolyl)oxazole, including the hydrogen atoms in riding mode contributing to $|\mathbf{F}|$, was refined to $R = 5.5\%$, which is a satisfactory value.

Solutions 7

7.1. (a)

$$\ln\left(\sum_j f_j^2(hkl)/\langle|\mathbf{F}_{\text{obs},j}(hkl)|^2\rangle\right) = y \text{ and } \langle[\sin^2\theta_j(hkl)]/\lambda^2\rangle = x$$

y	x
3.994	0.100
5.605	0.200
6.497	0.300
7.952	0.400
9.389	0.500

Using program LSLI, the line equation is $y = 13.157x + 2.7463$, with Pearson's $r = 0.997$. Hence, $B = 13.157$ Å^2/2) $= 6.58$ Å^2 and $K = \ln^{-1}(2.7463/2) = 3.95$.

(b)

$$\sqrt{\langle U^2\rangle} = \sqrt{\frac{13.157 \text{ Å}^2}{2 \times 8\pi^2}} = 0.29 \text{ Å}$$

7.2. A sketch of the atomic positions indicates that the shortest Cl- - -Cl distance is between $1/4$, y, z and $3/4$, $\bar{y}$, z. Then, the required distance d is given by $d^2 = \frac{a^2}{4} + 4y^2b^2 = \frac{(7.210\text{ Å})^2}{4} +$ $[4 \times 0.140^2 \times (10.43\text{ Å})^2]$, so that $d = 4.6395$ Å. From Section 7.4.4, specifically Eq. (7.25),

$$[2d\sigma(d)]^2 = \left[\frac{2a\sigma(a)}{4}\right]^2 + [8b\sigma(b)y^2]^2 + [8b^2y\sigma(y)]^2$$

$$= \left(\frac{2 \times 7.210\text{ Å} \times 0.004\text{ Å}}{4}\right)^2$$

$$+ (8 \times 10.43\text{ Å} \times 0.01\text{ Å} \times 0.140^2)^2$$

$$+ [8 \times (10.43\text{ Å})^2 \times 0.140 \times 0.002]^2$$

which evaluates to $5.98544 \times 10^{-2}\text{Å}^4$, so that $\sigma(d) = \frac{\sqrt{5.98544 \times 10^{-2}\text{ Å}^4}}{2 \times 4.6395\text{ Å}} = 0.02637$ Å. Note that if the third term alone is used, the value for $\sigma(d)$ is 0.2626 Å indicating that the error in a bond length arises mainly from the error in the atomic coordinates. Summary: $d = 4.640(3)$ Å.

7.3. Benzene is planar. By making the plane the X_Y field, all Z coordinates are zero. If the Cartesian axes X and Y are set as shown in the diagram below, with the origin O at the centre of the ring, then simple geometric arguments show that the coordinates of the carbon atoms are as follows:

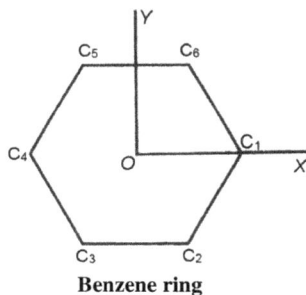

Benzene ring

Atom	X	Y	Z
1	1.400	0.000	0.000
2	0.700	−1.212	0.000
3	−0.700	−1.212	0.000
4	−1.400	0.000	0.000
5	−0.700	1.212	0.000
6	0.700	1.212	0.000

7.4. (a) $B_j = \frac{8\pi^2}{3}\mathrm{trace}(\mathbf{U}_j) = \frac{8\pi^2}{3}\sum_{i=1}^{3} U_{ii} = \frac{8\pi^2}{3} \times 0.090 \text{ Å}^2 = 2.3687 \text{ Å}^2$. (b) $(\sin^2\theta)/\lambda^2 = 0.10517 \text{ Å}^{-2}$, so that

$$T_j = \exp[-B_j(\sin^2\theta)/\lambda^2] = \exp(-2.6387 \text{ Å}^2 \times 0.10517 \text{ Å}^{-2})$$

$$= 0.75767. \text{ Thus, } B_j = 2.37 \text{ Å}^2, \ T_j = 0.758.$$

7.5. (a) Using the program MOLGOM with the constraint $d < 2.0$ Å the following values were obtained; the fourth figure is not significant.

Atom pair		d/Å
1	3	1.8974
2	3	1.3651
2	18	1.3666
3	4	1.4037
4	5	1.3820
5	6	1.3935

(*Continued*)

(*Continued*)

Atom pair		$d/\text{Å}$
6	7	1.3813
6	18	1.4191
7	8	1.3582
8	9	1.3429
8	16	1.4844
9	10	1.4468
10	11	1.3899
10	15	1.4308
11	12	1.3813
12	13	1.3885
13	14	1.3901
14	15	1.3958
15	16	1.4537
16	17	1.3202
17	18	1.4617

(b) Manipulating the bond lengths, given that the compound is tetracyclic with atom 1 = Br and atom 3 = O, the following molecular formula evolves:

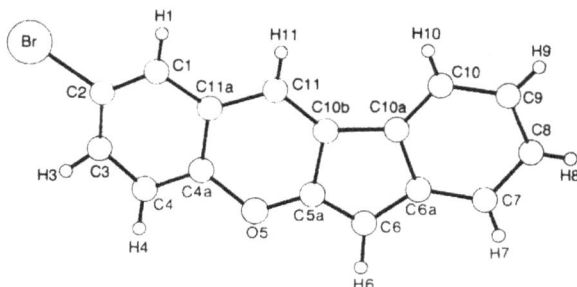

The molecule $C_{16}H_9OBr$; circles in order of decreasing size are $Br > O > C > H$

7.6. (a) The plane group for each of the principal projections of space group $P2_12_12_1$ is $p2gg$.

(b) In this plane group, the vector peaks have the following coordinates:

$$\pm(2x, 2y), \pm(2x, 2\overline{y}), \pm(\tfrac{1}{2}, \tfrac{1}{2} \pm 2y), \pm(\tfrac{1}{2} \pm 2x, \tfrac{1}{2})$$

For single-weight peaks, the heights are expected in the order Ni–Ni > Ni–S > S–S, but irregularities may occur if peaks overlap. The origin peak is negligible because the map is an origin-removed Patterson synthesis. An assignment of peaks is given below; a little more than a single asymmetric unit is shown. The correctness, or otherwise, of this assignment will be determined by completing the structure analysis in a subsequent exercise.

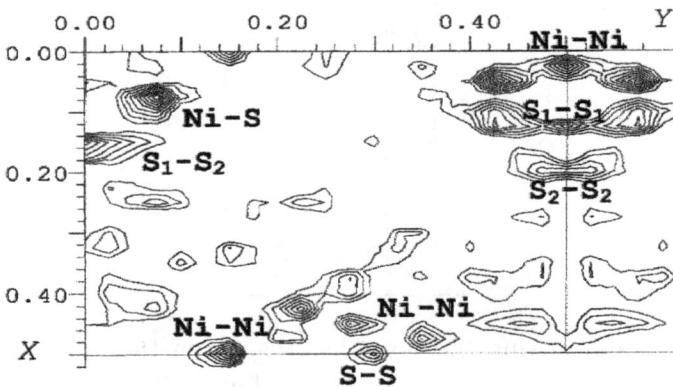

Atom	Peak type	Coordinates	
Ni	$2x, 2y$	$x = 0.235$	$y = 0.174$
	$\tfrac{1}{2}, \tfrac{1}{2}{-}2y$		$y = 0.176$
	$\tfrac{1}{2} - 2x, \tfrac{1}{2}$	$x = 0.237$	
S(1)	$\tfrac{1}{2}, \tfrac{1}{2} - 2y$		$y = 0.101$
	$\tfrac{1}{2} + 2x, \tfrac{1}{2}$	$x = 0.312$	

(*Continued*)

(*Continued*)

Atom	Peak type	Coordinates	
S(2)	$\frac{1}{2}$, $\frac{1}{2} - 2y$		$y = 0.101$
	$\frac{1}{2} - 2x$, $\frac{1}{2}$	$x = 0.150$	
S(1), S(2)	S(1) − S(2)	$x = 0.157$	$y = 0.000$
S(1), Ni	S(1) − Ni	$x = 0.075$	$y = 0.070$

The following self-consistent assignment forms a first trial:

Ni	S(1)	S(2)
0.236, 0.175	0.312, 0.101	0.154, 0.101

Alternatively, phases could be developed on the nickel atom alone, and an electron density calculation would be expected then to reveal the sulphur atoms with more certainty.

7.7. For C_4HO_4K, M_r is 152.15 and $V = abc\sin\beta = 609.61$ Å^3. Then $Z_c = \frac{1.855 \text{ gcm}^{-3} \times 609.61 \times 10^{-24} \text{ cm}^3}{152.15 \times 1.66054 \times 10^{-24} \text{ g}} = 4.48$. The nearest integer is 4, but the discrepancy is large and needs explaining. Assuming $Z_c = 4$, then the calculated M_r is given by $\frac{1.855 \text{ g cm}^{-3} \times 609.61 \times 10^{-24} \text{ cm}^3}{4 \times 1.66054 \times 10^{-24} \text{ g}} = 170.25$. $M_r(C_4HO_4K) = \frac{1.85 \text{ gcm}^{-3} \times 609.613 \times 10^{-24} \text{ cm}^3}{4 \times 1.66054 \times 10^{-24} \text{ g}} = 169.79$. The difference between the two calculations for M_r is 18.1, which is very close to the relative molecular mass of H_2O, thus leading to the formula $C_4HO_4K \cdot H_2O$. This result was confirmed by a crystal structure analysis.

7.8. Continuing the analysis of the nickel compound featured in Problem 7.6, there are several ways in which the analysis could proceed depending upon what has been deduced:

1. Phase on the nickel atom followed by successive Fourier synthesis (SFS);
2. Phase on the nickel and sulphur atoms followed by SFS;
3. Refine the position of the nickel coordinates by least squares followed by SFS;

4. Refine the position of the nickel and sulphur coordinates by least squares followed by SFS.

In this answer, route (4) has been chosen in order to first confirm the sulphur atom positions.

The initial R-factor of 40% was reduced by least-squares refinement to 35% and the subsequent electron density map (one asymmetric unit) is shown below.

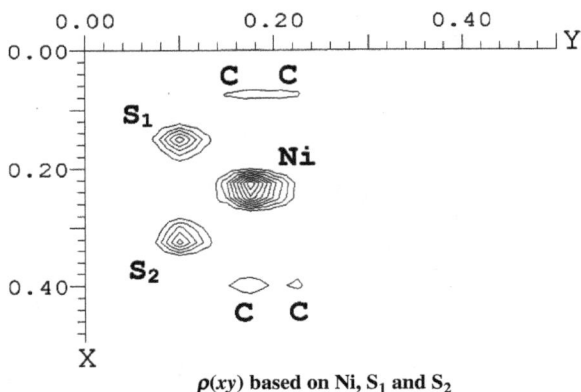

$\rho(xy)$ based on Ni, S_1 and S_2

An allocation of atomic positions was made as shown, and by the successive calculations of structure factors and electron density maps, the molecule was revealed. The atoms of the benzene ring were not well resolved, and it was necessary to print the map and plot the level 5 contour by hand on the electron density field figures (the program plots contours from -10 units in steps of $+10$). The difference in peak heights and a study of B-values of the atoms (initially taken as carbon) permitted a distinction between carbon and nitrogen atoms to be made. Further cycles of least squares led to a final R-factor of *ca.* 9.5%. It is useful with the approximate diagonal least squares procedure to decrease the amount of shift for the last few cycles to, say 30%, instead of using the built-in 60%.

A portion of the electron density field figures around the nickel (9.4/40, 7.1/40) and sulphur (6.2/40, 4.0/40; 12.7/40, 4.1/40) atoms is in the following table.

		0	1	2	3	4	5	6	7	8	9 → Y in 40^{ths}
0		−9	−8	3	10	10	12	16	0	−2	−9
1		−2	−6	4	3	0	6	10	−2	−6	−6
2		−5	−5	−1	−2	−4	−2	0	−3	−5	6
3		−9	−1	−8	−8	−7	−2	13	15	11	18
4		−8	−5	−7	−7	−8	−3	3	0	1	10
5		−6	−4	−3	0	15	−2	−8	−3	−6	6
6		0	−5	−11	21	72	24	−7	−6	−7	15
7		−2	−6	−4	4	26	4	−2	−8	−5	2
8		−7	−11	−4	−10	−3	−9	−5	7	−5	−7
9		−6	−4	−10	−9	1	−11	24	100	41	−5
10		−6	−6	−5	−2	−4	−6	17	79	34	−7
11		−9	−11	−5	−5	−6	−4	−13	0	−7	−7
12		−4	−4	−11	0	39	12	−10	5	−4	7
13		−1	−5	−7	13	67	27	−10	−6	−5	11
14	↓ X in	−6	−7	−2	−6	4	−8	−2	−3	1	0
15	40^{ths}	−11	−7	−9	−8	−2	−9	6	4	0	15

The map below shows $\rho(xy)$ for a complete unit cell using the 'print' function in the XRAY program. Contouring the field figures by hand at level 5 revealed the remainder of the benzene ring.

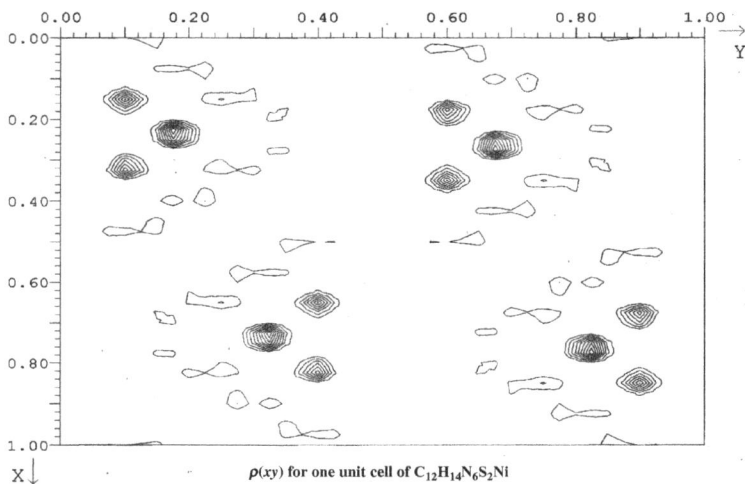

$\rho(xy)$ for one unit cell of $C_{12}H_{14}N_6S_2Ni$

The final coordinates and B-values from the refinement are listed below together with a drawing of the molecule of the nickel complex; POP is the population parameter and is different from unity only for atoms in special equivalent positions in a unit cell or for disordered species.

Atom, j	Type	x	y	Pop	B_j
1	Ni	0.23494	0.17797	1.000	3.05
2	S1	0.15391	0.09899	1.000	2.15
3	S2	0.31759	0.10345	1.000	2.01
4	N1	0.14710	0.23365	1.000	1.68
5	N2	0.32203	0.24614	1.000	2.97
6	N3	0.39233	0.23120	1.000	1.94
7	N4	0.46375	0.14686	1.000	2.82
8	N5	0.07786	0.22043	1.000	3.00
9	N6	0.00397	0.12936	1.000	3.30
10	C1	0.15286	0.38309	1.000	1.33
11	C2	0.18844	0.43700	1.000	2.24
12	C3	0.27425	0.44200	1.000	3.07
13	C4	0.30712	0.40169	1.000	2.46
14	C5	0.27813	0.33483	1.000	2.31
15	C6	0.18834	0.33936	1.000	2.09
16	C7	0.32995	0.29901	1.000	1.84
17	C8	0.39495	0.16385	1.000	2.00
18	C9	0.47464	0.08117	1.000	1.64
19	C10	0.14075	0.28965	1.000	1.26
20	C11	0.07067	0.15736	1.000	4.81
21	C12	0.00112	−0.07327	1.000	2.99

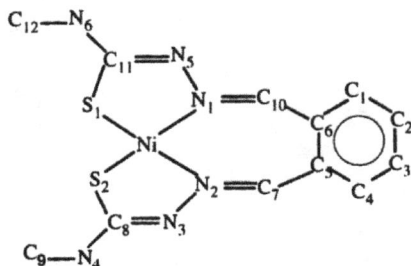

The nickel *o*-phenanthroline complex$C_{12}H_{14}N_6S_2Ni$

7.9. (a) Plane through atoms C_1–C_{17}: $0.42896X + 0.00786Y + 0.01953Z$

Deviations	
Atom	Distance/Å
1	−0.0849
2	0.1298
3	−0.3307
4	−0.6579
5	−0.3146
6	−0.5192
7	−0.6292
8	0.1810
9	−0.0928
10	0.3583
11	0.4831
12	0.1589
13	0.5699
14	−0.1889
15	−0.0025
16	−0.1190
17	0.1346

(b) Calculation of Inter-atomic Angles

Atoms			Inter-atomic angle
1	2	3	113.66
2	3	4	117.36
3	4	5	123.48
4	5	10	121.38
5	10	1	108.19
10	1	2	113.84
5	6	7	119.83

(*Continued*)

(Continued)

	Atoms		Inter-atomic angle
6	7	8	114.25
7	8	9	109.63
8	9	10	111.43
9	10	5	109.95
10	5	6	118.42
9	8	14	109.76
8	14	13	112.23
14	13	12	108.69
13	12	11	109.74
12	11	9	114.19
11	9	8	113.12
14	15	16	102.82
15	16	17	105.86
16	17	13	108.51
17	13	14	100.37
13	14	15	103.94

(c) The results from PLANE indicate a possible lack of planarity in the regions of C_4, C_6, C_7 and C_{13}. The following torsion angles reveal this lack in the ring junction near these atoms.

	Ato	ms		Torsion angle
3	4	5	6	171.7
5	4	3	2	14.0
5	6	7	8	31.4
12	13	14	15	165.2
16	17	13	12	−147.5
11	12	13	17	171.3

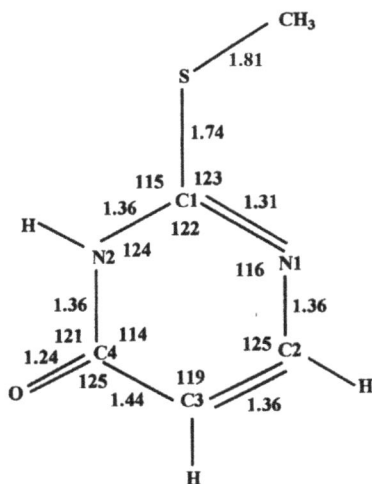

Structure of 2-*S*-methylthiouracil

Bond lengths and bond angles from the crystal structure analysis of 2-*S*-methylthiouracil, using the data sets *SMTX* and *SMTY* (see Chapter 8, page 368).

Index

(The letter 'T' before a page number indicates a tabular presentation)

485